The Geography of Life

THE GEOGRAPHY OF LIFE *Wilfred T. Neill*

COLUMBIA UNIVERSITY PRESS

NEW YORK AND LONDON   1969

To My Son,

Trammell Neill

INTRODUCTION

When the naturalist Alexander Von Humboldt returned in 1804 from a South American expedition he wrote an account of the trip, and it ran to thirty over-sized volumes, including 1,400 illustrations. I suspect that when the Baron finally laid his pen aside, it was with a sigh—a sigh not of relief but of regret that so much had to be left unsaid. For the scientist enjoys writing about his work, especially if it is the sort of work that takes him to remote places, where the animals and plants are strange and the people have unfamiliar ways.

In the present book I have tried chiefly to portray the broad outlines of animal and plant distribution in all parts of the world; to review the geography of life on land, in fresh water, and in the ocean; to mention at least a few generally unfamiliar organisms, although concentrating on the better-known ones. I have tried also to describe some of the environmental factors—rain and snow, ice, high temperatures and low, the soils, mountainous uplifts, great ocean currents—that influence the spread of living things, and to discuss the forces of nature that have operated throughout the ages to push animals and plants around the globe. Finally I have drawn attention to some remarkable people who explored the physical world, and the world of ideas, to develop the science of biogeography.

Still I would gladly have penned a few extra chapters, in a more personal vein. I should like especially to convey something of the biogeographer's enthusiasm for out-of-the-way places; to write at length about certain localities that have impressed me above most others. I recall, for example, a trip along the road that skirts Goenoeng Agoeng on the eastern tip of Bali in Indonesia. Agoeng was a well-mannered volcano in those days, and from its base I could look across the Strait of Lombok. The naturalist Alfred Russel Wallace had explored the same region 80-odd years before, and concluded that this otherwise unremarkable stretch of water separated two completely different worlds, as far as animal life was concerned. The differences in animal life between England and Japan, countries half the globe apart, were not as profound as the differences from one side to the other of this 20-mile strait—so thought Wallace. Today we would not describe the situation in just this way, but it was exciting to follow Wallace's trail as well as his train of thought.

And I recall an occasion, nearly 7,000 miles from Bali and in quite another clime, when someone said a polar bear had drifted up on Little Diomede, an island off Alaska's Seward Peninsula. We saw no bear, just ducks and gulls, but the island was fascinating anyway. For one thing, you could stand in an American Monday and look at a Russian Tuesday; for the International Date Line passes between Little Diomede and Russia's Big Diomede just two miles away. But more significantly, the Diomedes lie like steppingstones between Siberia and Alaska, and are remnants of a great land bridge between the Old World and the New, a bridge across which elephants, rhinos, camels, and bison passed in a bygone age when the far-northern climate was milder than at present.

I also remember the strange sense of familiarity that came over me the first time I paddled a boat out into the labyrinthine waterways of Waigani, a swamp in southeastern Papua. The water-lily blossoms were purple rather than white, and a distant ripple was made by a crocodile rather than an alligator; but otherwise I could have been back in Georgia, somewhere east of Trail Ridge in Okefinokee. At the water's edge there were reeds and coarse sedges and cattails; the boat slipped across water-lily pads and what I took to be the floating leaves of water pennywort; the paddle came up tangled with what was surely coontail, pondweed, and parrot's-feather. The plants of the higher ground beyond the swamp—mostly *kunai* grass with a scattering of curly-barks and Eucalyptus trees —struck no familiar note; why should the aquatic plants have done so?

While yielding to the fascination of the faraway, I would not omit mention of the close-at-home. Every state of the Union has localities to intrigue the biogeographer, and often the casual visitor as well. Take, for instance, Mount Washington in the White Mountains of New Hampshire. Here you can, in a sense, visit the "Arctic" not by going north but by going up; for between 4,000 and 5,000 feet on the mountain the trees become stunted and then vanish, and above the timberline is a counterpart of Arctic tundra and stony fell-field. And the boulders atop the peak are not all of native stone; many were carried to their present location by a vast glacier that once swept down out of the north, reaching southward even to Ohio and Illinois. In the northern United States, many aspects of plant and animal distribution still reflect the coming and going of the ice sheets.

Some 800 miles south of the White Mountains, the Smokies command attention. I particularly like the last 50 yards of climb to the top of Tennessee's Mount Le Conte. You leave a dim, dripping forest of spruce and balsam fir, where little wet salamanders hide under cold stones, and you emerge suddenly into bright sun and a scattering of sun-warmed rocks, where the thin air is filled with the pleasant humming of bees around the blossoming heath shrubs. It is interesting to reflect that ecologists—students of the environment and its impact on living things—are not sure why an occasional peak is capped with a "heath bald."

Another 700 miles south and the land becomes low and flat, the climate almost tropical. The once-famous Everglades of Florida have nearly vanished, converted to farms; but a scrap of gladeland—a mere 7 per cent of the original expanse—persists in fairly natural condition as the Everglades National Park. Here, without leaving the States, you can see plant life that is typically West Indian: the tall, straight mahogany; the gumbo-limbo, its trunk supported by buttress roots; the strangler fig, perhaps enfolding a palm in deadly embrace; Jamaica dogwood, whose crushed leaves yield a fish poison; a wild pepper that grows orchid-like on tree branches, along with air plants and true orchids; the manchineel with its blistering sap; and dozens more. And yet, as though to confound the biogeographer, the animal life of the same region includes extremely few West Indian species.

But the geography of life is a topic of such kaleidoscopic diversity that there is scarcely room for personal narrative. In any event, the greatest adventures of science are those of the mind, and are independent of locality. I can think of no spot less exotic than Long Island Sound, bordered as it is by great urban centers such as New Haven, Bridgeport, and New York City; and no living organism less impressive to the eye than the tiny, wriggling thing that was dredged from the bottom of the sound in 1954. But this little animal, overlooked for so long, turned out to be both a "missing link" and "living fossil"; something very like it must have been ancestral to crabs, lobsters, shrimp, and all the other crustaceans.

Even fossil crustaceans 375 million years old are not as primitive, structurally, as this little bottom-dweller, which is often disturbed by the passage overhead of ferries, tankers, and tugs. The intellect is staggered by a drab organism from a prosaic environment.

In fact, even the common, widespread organisms occasionally surprise us with their potentialities. Along the beach almost anywhere in the world there live small animals called sand hoppers, or beach fleas. Whether on Jones Beach or Copacabana, on Bondi or the Cap d'Antibes, in the proper season you can kick over a bit of tidal wrack and put these crustaceans to hopping. In Florida we use the biggest ones for fish bait, but otherwise these animals seem un-inspiring. Yet it has lately been announced that sand hoppers, and their diet of primitive plant life, have survived in a laboratory duplication of the moon's environment, and so might be raised as food for lunar explorers. This possibility evidently has received considerable attention, for we are assured that minced sand hopper tastes very much like shrimp.

The word "environment" must enter frequently into any account of animal and plant distribution, for every living thing has imperative needs that can be met with in some parts of the world but not in others. Thus the geography of life is best understood when we have some idea of the natural landscape. We can speak of organisms that are confined to the Arctic, or the Alps, or the Coast Ranges; but the account is more meaningful if we can visualize the Arctic tundra stretching treeless to the horizon, the snow-capped Alpine peaks in jagged array, the ferny dimness beneath towering redwoods. In lieu of much detailed geographic description, the present book offers many illustrations; these portray not only the wildlife but also the look of the land in various parts of the world.

A few of the illustrations show people doing something with, or to, the local animals and plants; and the landscapes may include at least a trace of man's activities. In an account of biogeography it would be unrealistic to ignore man as a factor influencing distribution. Almost everywhere on land, as well as in the fresh water and the shallows of the sea, man is altering the environment in a direction that suits his needs of the moment, and spreading a few organisms while decimating many. The resources of the world are going more and more to support one species, man; and it seems likely that the only other species to survive, at least on land and in the fresh water, will be the ones that man likes to eat, pet, kill for sport, or use for decoration. Bits of fairly unspoiled landscape will persist, if at all, only because man wants such places in which to hunt, fish, swim, hike, honeymoon, paint, or relax. Feeling this way about the future course of events, I have also slipped into the bibliography a goodly number of books that describe not just the land and wildlife but also their interactions with man.

The illustrations and references need no further commentary, but I should explain certain features of the text. Newspapers and magazines tell us that parents are baffled by the "new math" their children learn in school; but what about the new geography? Perhaps you remember a geography book with countries called Estonia, Persia, Siam, French Indochina, Dutch Guiana; a book with just one Germany, and no mention of Laos, South Vietnam, Pakistan (East and West), Ghana, Tanzania, Botswanaland, or West Irian? So many countries have changed name, outline, and ownership. The biogeographer, as such, is not concerned with political boundaries but only with natural ones. In this connection I have retained a few useful geographic names, without reference to new political terminology. Obvious examples are the islands of Borneo and New Guinea, and the peninsula of Indochina.

The phrase "for example" must also appear frequently in a survey of distribution. Scientists have discovered and named about 1,700,000 species of living things, and obviously only a small percentage of these can receive individual mention. Among the flowering plants that I have singled out for discussion, a majority are grown or somehow utilized by man. It is not meant to imply that horticultural and commercial plants are of exceptional importance in biogeography, although it is entertaining to trace the diverse origins of the flowers in our gardens, the fruits and vegetables on our tables. But of plants that have not reached the garden encyclopedia, most have unfamiliar names, and there is not much point in listing them unless they are so spectacular as to warrant a bit of description as well.

The situation is different with animals, at least with the ones that especially concern us: the mammals and birds, reptiles, amphibians, and fishes. Most of these have common names which, if not really vernacular, are often seen in popular handbooks and guides. Thus the animals singled out for mention are, in general, simply the ones that illustrate a point most sharply.

As is well known, scientists have tried to give every species of plant and animal a Latinized name. It is astonishing to see how these names have crept into common use, although they were not particularly intended to do so. In the tropics of southern Mexico and Guatemala, the cool, cloud-moistened summits of many high peaks are capped with forests of sweet gum, the same tree that grows widely in the eastern United States. This distribution is remarkable enough, but it is even more surprising to hear the local residents call the tree "diquidumbe" or "liquidumbe," obviously from the tree's technical name of Liquidambar. And around Georgia's Okefinokee Swamp the crepe myrtle bush graces many a dooryard; it is locally called "logger-streamer," but who brought the scientific name of Lagerstroemia to the big swamp country, I could not guess. A great many of the Latinized names are used also in our everyday speech, either unchanged (such as alligator, aloe, aster, fuchsia, geranium, iris, and petunia) or slightly modified in spelling (such as coffee, crocodile, heliotrope, hyacinth, jasmine, nicotine, orchid, rose, and tea). Still, in the present book I have tended to subordinate the scientific names to common ones, inserting the former just here and there as needed to forestall misunderstanding. To me, "balsam" is a kind of fir tree, but to the author of my seed catalogue it is a garden flower, while to you it may be a kind of poplar; and in a case like this the scientific name, neatly parenthesized, will make clear just what plant is really meant.

As science advances, its terminology changes. Even the Latin names of living things are not exempt from replacement. If you pursue some aspect of biogeography at considerable length, in the books whose titles are appended, you might find the sequoia tree, the green leek, or the saltwater crocodile turning up under some Latin name other than the one I have used, or implied, for it. This should not bother anyone. Tennyson said it best:

What is it? A learned man | Could give it a clumsy name. |
Let him name it who can, | The beauty would be the same.

Wilfred T. Neill

New Port Richey, Florida
May, 1968

CONTENTS

The Geography of Life

1 *"They told of prodigies, as one who has returned from far countries,*

the force of whirlwinds, and unheard-of birds, monsters of the deep,

uncertain combinations of men and beasts—things seen, or believed

*through fear."** **THE SCOPE OF**

BIOGEOGRAPHY

As soon as man learned to make long sea voyages, he found that animals and plants were not distributed uniformly over the earth. About 520 B.C. the Carthaginian navigator Hanno, with a fleet of 60 ships, followed the African west coast southward from the Straits of Gibraltar, as far as what is now Liberia. On an island—probably Sherbro Island off the coast of Sierra Leone—he caught some chimpanzees, which he took to be wild, hairy people; and finding them uncooperative, he brought their skins back to Carthage as an offering to Juno. In the next few centuries, various animals were imported from Africa and Asia by the Greeks and the Romans. The Roman emperors urged each provincial governor to trap the beasts of his own area and send them to Rome for gladiatorial spectacles; and by A.D. 80, when Titus celebrated the opening of the Flavian Amphitheater—later the Colosseum—with a display of 5,000 wild animals, the Roman citizen could see crocodiles, leopards, hippopotamuses, rhinoceroses, elephants, giraffes, and ostriches from Africa, lions from Asia Minor, even tigers from India.

Today almost everyone knows that each kind of animal or plant is likely to

* TACITUS, Roman historian of the first century.

1

be confined to some particular region. Thus the northern tourist, vacationing in Florida, will remark on the unfamiliar cabbage palms, the gloomy live oaks with a drapery of Spanish moss, a cypress growing in water and ringed about by curious "knees," a magnolia with great waxy blossoms, little green lizards on the window screen, an alligator basking in a roadside ditch, a flight of white ibis over a sawgrass marsh. These plants and animals were certainly not to be seen back home in Detroit or Cleveland or New York. The easterner, traveling westward across the United States, notes a change in the highway signs: somewhere around mid-continent, the familiar admonition, "Prevent Forest Fires," suddenly becomes "Prevent Grass Fires," as the forests of the east give way to tall-grass prairie.

If the gardening enthusiast moves from one state to another, he may find that many favorite plants refuse to grow about the new homestead. For example, the New Yorker who retires to peninsular Florida will probably see no more lilacs, tulips, hyacinths, or crocuses, but instead an abundance of periwinkles, hibiscus, poinsettias, allamandas, flame vines, sagos, and palms. The outdoorsman knows that, even within a single area, a given kind of plant or animal is usually to be found only in certain environments. The angler, fly-fishing for brook trout, will drop his lure into the cold waters of a rocky upland stream, not into some tepid pond of the nearby lowlands. The hunter, whether gunning for big-game trophies or merely potting woodchucks in a farm lot, soon learns to recognize the characteristic habitat of his quarry.

Although regional differences in wildlife have been known since Classical times, the reasons for these differences have never been widely understood. We might guess that adaptation to a special environment must have something to do with distribution. Thus we could surmise that the thick white fur of the polar bear, so useful in the cold and snowy north, would prove intolerably hot and overly conspicuous in a tropical jungle. But adaptation to environment will not explain all the peculiarities of animal and plant distribution.

Take, for example, the surprising similarities between certain areas on opposite sides of the globe. Why should many animals and plants of the eastern United States find their nearest relatives in faraway China? We have mentioned the alligator, basking in a Florida ditch. This big reptile is found (where not recently killed off) over our southeastern lowlands, from Albemarle Sound in North Carolina to the Rio Grande in Texas, and up the Mississippi River

A ranch near Galisteo, New Mexico. In the United States, western landscapes often contrast sharply with eastern ones. Such unlike areas will differ in plant and animal life. Man's activities modify the landscape almost everywhere. (New Mexico State Tourist Bureau)

drainage into Arkansas and southeastern Oklahoma. There is only one other living kind of alligator, and it is confined to the lower part of the Yangtze River drainage in China. These two alligators, the American and the Chinese, are closely related and no doubt of common ancestry; why do they live half a world apart?

The problem becomes more intriguing when we learn that the great paddlefishes, popularly called "spoonbills," are distributed in approximately the same fashion. There are only two living kinds of paddlefishes; one inhabits our Mississippi River drainage and the Great Lakes farther north, while the other is confined to China's Yangtze River and perhaps the Hwang Ho farther north.

A great, flattened salamander, popularly known as the hellbender, inhabits upland streams of the eastern United States and also central Missouri. It has likewise found its way into a stream on the southern edge of the Ozarks, where it looks a little different. A very similar and even larger salamander, called the giant salamander, inhabits upland streams of western China. It has also found its way into streams on the western side of the Japanese island of Honshu, where it looks a little different. The hellbenders and the giant salamanders are the only living members of their family.

As a matter of fact, faunal similarities between China and eastern United States are numerous, involving mammals, birds, snakes, lizards, frogs, salamanders, fishes, and invertebrates; and the floral similarities are even more impressive. It has been solemnly reported in the scientific literature that certain Chinese girls, sent to college in Georgia, often wept with homesickness at the familiar sight of dogwood, willow, sweet gum, hazel nut, black gum, hop hornbeam, birch, poplar, magnolia, alder, beech, elm, chestnut, maple, oak, sassafras, mulberry, witch hazel, and pine.

Even more remarkable patterns of distribution exist to plague the biogeographer. In California live three little-known kinds of salamanders, closely related and known collectively as "web-toed salamanders." One of these, called the limestone salamander, is found only in a very limited area of Mariposa County, where it lives beneath rocks; a second kind, the Shasta salamander, is confined to a restricted area of Shasta County, where it inhabits caves and crevices; and the third, the Mount Lyell salamander, is found at scattered localities along the western slopes of the central and southern Sierra Nevada, where it lives beneath rocks at high elevations. There are two other kinds of web-toed salamanders, very much like the California ones in appearance and habits but certainly not in distribution: one lives in the French and Italian Maritime Alps and in the Apennines, while the other is confined to the island of Sardinia in the Mediterranean.

A few more distributional puzzles may be cited, to show that every part of the globe offers intriguing problems. In South America there are many kinds of turtles that withdraw the head by bending the neck sideways. Appropriately, they are called side-necked turtles, and they are placed in two different groups, or families. The members of one family are known collectively as "hidden-neck turtles." They are found not only in South America but also in Africa and Madagascar. However, abundant fossil remains show us that hidden-neck turtles once lived also in Europe, Asia, and North America; so we may surmise that the living hidden-necks are but relics of an ancient group that once spread widely over the world. But what about the other family, the so-called "snake-necked turtles"? They are found not only in South America but also in Australia and New Guinea; and they have left not a single fossil outside the two far-separated portions of the family's present range. And these snake-necked turtles are strictly of the fresh water; they are not known to venture even into brackish coastal waters, much less swim nearly half way around the world.

Various other organisms are distributed somewhat like one or the other family of side-necked turtles. Thus a single frog family includes the African clawed frogs, one of which is famed for use in pregnancy diagnosis; and also includes the curiously flattened "Surinam toad" and its South American congeners. This distribution parallels that of the hidden-neck turtles. Another frog family, whose members are called "slender-toed frogs," is distributed roughly like the snake-necked turtles; for these frogs have two centers of abundance, the Australia–New Guinea region and the New World tropics, and seem to have left no fossil record outside the two portions of the modern range. (But unlike snake-necked turtles, the slender-toed frogs have been able to push northward into Central America and Mexico, and even a slight distance into the extreme southern United States.)

The pouched mammals—the marsupials—remind us of snake-necked turtles, also. Marsupials are especially abundant and diverse in Australia, the home of kangaroos, wallabies, bandicoots, koalas, wombats, and the like. Five families of marsupials inhabit the Australia–New Guinea region, with a few of the species

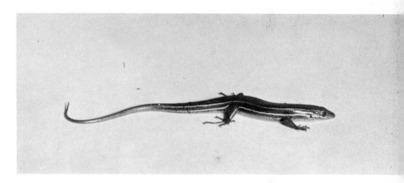

The northern tourist, visiting Florida, may see an unfamiliar lizard on the window screen: the green anole, a harmless insect-eater, shown at upper left. (Tod Swalm) Below it, here displayed in a Florida zoo, is the American alligator, which inhabits the southeastern United States. The only other living alligator, shown at top, is the Chinese, a shorter and stockier species. Below it, one of several blue-tailed skinks inhabiting the southeastern United States. Their nearest relatives are from eastern Asia. (Ross Allen's)

Side-necked turtles of the family Chelidae live in South America and the Australian region. Shown are species from the Amazon. (W. T. Neill)

ranging northward into the East Indies. There are two other living families of marsupials; they are essentially South American, although some of the species have ranged northward into Central America, and one—the Virginia opossum—has ranged far into North America. (But unlike snake-necked turtles and slender-toed frogs, the marsupials have left a fossil record outside the present centers of distribution; one marsupial family, the one that includes our Virginia opossum, lived in Europe ages ago.) We may suspect, and rightly, that there is a reason for the persistence of primitive mammals, frogs, and turtles in Australia and South America; but the reason is far from obvious.

Biogeographic problems need not involve oddly disjunct distributions. Take, for example, the toads, warty little amphibians known to almost everyone. The numerous kinds of toads, while diverse in size and habitat, are similar in that they must spend a part of their life in fresh water, first as an egg and then as a tadpole which eventually transforms into the more terrestrial toad. Neither egg, tadpole, nor adult can survive much exposure to salt water; even a narrow inlet is a formidable barrier to toads. One would think that these amphibians never had much chance of spreading very widely over the globe. It is then astonishing to learn that toads have reached every major land area of the world except Madagascar, the Australian region, and the frigid lands. Why is it that toads, seemingly not well equipped to cross natural barriers, have spread farther than most mammals or birds or reptiles?

Why does extreme southern Florida have an abundance of West Indian plants but few West Indian animals? Why does extreme southwestern Africa have a very distinctive flora but not an equally distinctive fauna? Why do the really gigantic land tortoises live only on remote little islands in the Pacific and Indian oceans? Why does temperate South America have a vast area of grass-land comparable to our North American prairie, yet have no herds of grazing animals except introduced cattle? The problems of biogeography are fascinating; their answers are even more so, as we shall discover.

From the above remarks one fact should be apparent: the distribution of a plant or animal, however puzzling, is likely to have counterparts, so that a frog may be distributed much like a turtle, an alligator much like a fish, a salamander much like a tree or a fern or an insect. In short, there are patterns of distribution, and these are of special interest to the biogeographer, who wants to know how the patterns came about. To find out, the biogeographer uses data from many

On the facing page, below, are members of the frog genus Rana, represented on all continents save Antarctica. A familiar member is the leopard frog, distributed from Canada to Nicaragua. (Ross Allen's) Above it is another member, the jungle frog, ranging from Vera Cruz to the Amazon. (James N. Layne)

Until man spread the house cat, the cat genus Felis was not as widespread as the toad genus Bufo. Toads have reached all continents save Australia and Antarctica, and also inhabit many islands. (J. K. Keen)

Giant tortoises, once widespread, survive today only on a few remote islands. Here an aggressive male of the Aldabra species tries to arouse the interest of a female from the Galapagos. (Ross Allen's)

fields of science; he is interested in modern climate and geology, in the environments of past ages, in the fossils that are dug from the earth, in oceanography, and in the biology of the animals and plants themselves.

Of course the individual biogeographer may be limited in his interest to certain kinds of organisms. As comparatively few people are interested equally in plants and animals, biogeography is sometimes divided into zoogeography, the study of animal distribution, and phytogeography, the study of plant distribution. But as we have already seen, many very diverse animals and plants may have much the same range, and the distribution of one group of organisms may provide a clue that helps to account for the distribution of some completely unrelated group. Thus a biogeographic investigation profits from breadth of interest.

The following chapters explain how various biologists determine the geographic ranges of plant and animal species, and how biogeographers analyze the patterns of distribution thus revealed. We shall discuss the patterns of the land, the fresh waters, and the oceans. Emphasis is on the plants and animals that live in the United States today, but we shall also have something to say about all the other areas of the globe, and about many organisms that have long been extinct.

2 *". . . a rustic knows plants and so perhaps does a brute beast, but neither can make anyone else the wiser."* ". . . by paying a cent each for all insects that were brought in, I obtained from the Dyaks and the Chinamen many fine locusts . . . and numbers of handsome beetles."*** **HOW THE BIOGEOGRAPHER GETS HIS DATA**

The world had to be explored, most of the animals and plants discovered, and the geographic range of each species determined at least approximately, before the distributional patterns could become evident. Even with all this accomplished, biogeography could not progress until the animals and plants were properly classified. Taxonomy, the science of classification, is basic to biogeography, and a few examples will show why this is true.

As mentioned, Australia is the home of many marsupials, primitive mammals which usually give birth to premature young and then carry them in a pouch. The higher mammals, the so-called placental mammals which give birth in more familiar fashion, have not been very successful in reaching that isolated continent. The only native Australian placentals are bats, which are strong flyers; seals, all accomplished swimmers; various whales and porpoises, which are marine; the dugong (a relative of our manatee or sea cow), also marine; quite a

---

* LINNAEUS, on his plan to name the natural world. ** ALFRED RUSSEL WALLACE, on his exploration of the Malay Archipelago, 1854–1862.

14

number of rats and mice, whose ancestors probably arrived on drifting logs; and a wild dog, the dingo, probably brought from Asia by primitive man.

But in the year 1888 the manager of Idracowa cattle station, in the barren heart of Australia, followed a peculiar trail through the sand flats and acacia scrub, and finally caught up with a small animal that seemed to be a mole. Its forelimbs were stout, flattened, and heavily clawed for digging through the ground; its eyes were useless structures hidden beneath the skin; its ears were but minute openings; its pale fur had the "moleskin" look, like fine plush. The animal's snout was protected by a horny shield. The little beast looked much like the familiar moles of Europe and North America, but even more like the golden moles of southern Africa; for these latter also have a shield on the snout and a pale fur that matches the color of the local soil.

Apparently the Idracowa mole was a distinct Australian species of a family that was otherwise South African. This was tantamount to saying that the Australian and African species were of common ancestry, and had dispersed from a single stock of moles. Moles are placental mammals, and it was astonishing to find this one in Australia. These animals are neither flyers nor seafarers; and if in ancient times it was somehow possible for moles to reach central Australia from southern Africa, why did not scores of other placental mammals follow the same route? Then someone noted that the female of the Australian "mole" had a pouch, a backwardly directed marsupial pouch that would not plug up with sand when the animal burrowed. Careful anatomical investigation proved that the little beast from Idracowa, for all its appearance and behavior, was in no way kin to moles; it was another marsupial, and so not out of place in Australia. When properly classified, the Australian burrower offered no unique problem to the biogeographer.

The similarity between the pouched "mole" and the true moles results from what is called adaptive convergence. Unrelated animals may come to resemble each other because they live in the same way and develop similar modifications of anatomy and behavior. There is adaptive convergence among plants, also. For example, the usual "cactus garden" contains at least a few plants that are not cacti at all but spurges, members of a different family. The true relationships of plants and animals must be discovered before problems of distribution can be attacked.

Classification may be almost as old as language itself. Ancient man prob-

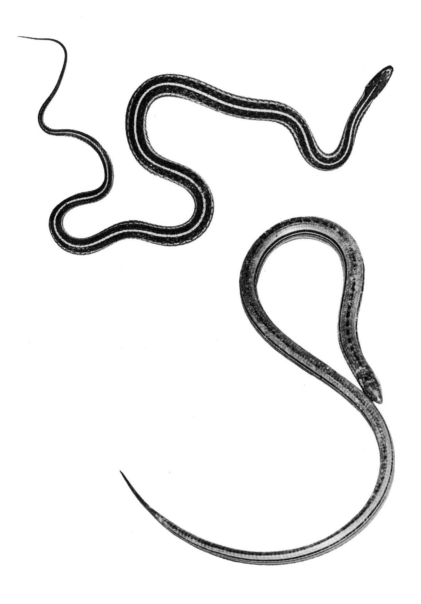

The garter snake, at top, is a true snake, widespread in North America. (Ross Allen's) The "glass snake," below it, is not a snake but a legless lizard. Early biologists were sometimes fooled by superficial resemblances between unrelated species. (Tony Stevens)

ably categorized at least the animals that were edible, or that would bite. Primitive tribes of today often have a remarkably accurate classification of the local animals and plants. However, the taxonomic scheme we now use is generally credited to a Swedish naturalist, Carolus Linnaeus, who was born in 1707. He set about the ambitious task of classifying all living organisms in a way that would indicate both differences and relationships. He saw that the common or vernacular names were useless for this purpose: one organism may have different names in different regions, the same name may be used for many organisms, and numerous plants and animals have no common name at all. Linnaeus wanted to give every kind of plant and animal a distinctive name that was either in Latin or that was Latinized from some other language, for in those days Latin was understood by scholars everywhere.

Linnaeus saw certain obvious relationships among living things. He saw, for example, that a dog is related to a wolf and to a jackal, while a cat is more related to a lion, tiger, or leopard. So he placed all of the dog-like animals in a single category called a genus; and he named this particular genus *Canis*, from the Latin word for dog. Then, to distinguish among the various members of this genus, he called one species *Canis familiaris* (the dog), another species *Canis lupus* (the European wolf), and still another species *Canis aureus* (the jackal). But the cat-like animals he placed in a different genus, *Felis*, with the species *Felis catus* (the house cat), *Felis leo* (the lion), *Felis pardus* (the leopard), and *Felis onca* (the jaguar). Linnaeus used the same system for plants. Thus he placed all buttercups in the genus *Ranunculus*, with the creeping buttercup called *Ranunculus repens*, the bulbous buttercup *Ranunculus bulbosus*, and the hairy buttercup *Ranunculus hirsutus*.

The names employed by Linnaeus were usually descriptive of the animal or plant; the Latin scholar would know that *aureus* meant "golden," that *pardus* meant "spotted," and that *repens* meant "creeping." The Latin names he used were not always coined by him; some animals and plants had been named by earlier workers. Nor did Linnaeus invent the concept of genus and species. His great contribution was the indication of genus and species in a formal and strictly binomial fashion.

Most of the names favored by earlier workers actually were lengthy descriptions in Latin, and Linnaeus himself was slow to abandon long Latin phrases. Not until the tenth edition of his *Systema Naturae* did he come around to the practice of naming all animals and plants in the aforesaid binomial way. Of course he was sometimes fooled by adaptive convergence, and we do not accept all his groupings today; but we still use the Linnaean system of genus and species. Our modern scheme of taxonomy dates officially from the appearance of Linnaeus's tenth edition, in 1758.

Needless to say, thousands of additional genera and species have been discovered and named since Linnaeus's time. Today we have rather elaborate codes of nomenclature, prescribing ways in which newly discovered organisms must receive their scientific names. We have also found it convenient to use a good many taxonomic categories in addition to genus and species. However, only one of these additional categories—the family—need concern us here. Just as related species are placed in one genus, so related genera are placed in one family. The family names of animals all end in -idae, and the family names of plants usually end in -aceae. Thus the family Canidae now includes the genus *Canis* (dog, wolf, coyote, dingo), the genus *Vulpes* (red fox, kit fox, etc.), the genus *Speothos* (South American bush dog), and several other genera of doglike animals. We may write "canid mammals" or simply "canids," meaning all the various species of the family Canidae. The hidden-neck turtles mentioned in the previous chapter make up the family Pelomedusidae, the snake-necked turtles the family Chelidae, and the slender-toed frogs the family Leptodactylidae; but we may write of pelomedusids, chelids, and leptodactylids.

A family name always has biogeographic implications. At some time in the past, all the members of a given family had a common ancestor somewhere, and we must try to discover how the living members reached the areas they now occupy. Not only must the various species be discovered and classified; their respective ranges must also be determined at least approximately. Geographic ranges are worked out mostly by people who are not especially interested in the entire science of biogeography but rather in the distribution and taxonomy of some particular group of plants or animals.

In a few countries, professors sometimes think it beneath their dignity to hunt for plants and animals, such work being left to underlings. Fortunately this attitude, vanishing everywhere, was never widespread in the United States. Our country has had a traditional interest in fieldwork and exploration by scholarly men; some of our early naturalists are said to have fought Indians with one hand while pickling snakes in rum with the other. The modern biologist knows that

At top, a cactus in New Mexico. (New Mexico State Tourist Bureau) Superficially resembling a cactus, the Stapelia (at bottom) actually belongs to the milkweed family. (W. T. Neill)

Above, a wolf at Lac au Canards, Quebec, Canada. (Canadian Government Travel Bureau) On the facing page, above, is a coyote in the Colorado Rockies. Wolf and coyote are closely related, and placed by scientists in the same genus, Canis. (Colorado Game, Fish, and Parks Department) The South American bush dog at right somewhat resembles the wolf and the coyote, but differs from them so markedly that it is placed in a different genus, Speothos. (Ross Allen's)

At top, African white pelicans in South Africa. (South African Tourist Corporation) Below them are American white pelicans in Prince Albert National Park, Saskatchewan, Canada. Other species of white pelicans inhabit Europe, Asia, and Australia. Thus a group of closely related species may be widely distributed over the globe. (Canadian Government Travel Bureau)

he can learn a great deal from observation of living organisms in a natural environment. And so botanists tramp the woods and fields, keeping a watchful eye for interesting plants. Mammalogists hide dozens of little traps to take some rare mouse or shrew; ornithologists, shotguns loaded with fine shot, stalk and bag unusual birds; herpetologists roll logs and rocks in the hope of finding a secretive frog or lizard or snake; ichthyologists drag seines through the water, or dip with nets in the shallows, or poison with rotenone, to collect the fishes of a lake or stream. Entomologists strip rotting bark for beetles, pursue butterflies with gauzy nets, or set light-traps to attract night-flying insects. Hundreds of ingenious techniques have been invented for the capture of wildlife in the interest of science.

Of course biological collecting may not be carried on just to discover the geographic ranges of various plants and animals, but it often has this salutary result. Biological expeditions to remote places are especially apt to uncover new facts about distribution. Once collected, a specimen must be preserved so that it can be studied by later workers. Plants are pressed, dried, and mounted on sheets. Mammals are skinned; the skins are stuffed and the skulls are cleaned and boxed. Birds are skinned and stuffed; reptiles, amphibians, and fishes are usually preserved entire in alcohol or formalin. Insects may be pinned and dried. Other techniques of preservation are employed as needed.

Once preserved, a specimen is tagged with certain information vital to biogeographers and taxonomists: the place and date of collection, and the collector's name. Often the collector supplies more than the bare minimum of collecting data. A finished label might read something like this: "8 miles N-NE of Port Moresby, Papua, New Guinea; 3 December, 1943; W. T. Neill, collector. In trail through *kunai* grassland near bank of Laloki River, by night." The label is attached firmly to the specimen, at least in the case of most mammals, birds, reptiles, amphibians, and fishes. Of course the needed information might be written directly onto a herbarium sheet bearing a pressed plant, or onto a cleaned and dried skull.

Even one inaccurate label can lead to a great deal of confusion. In 1905 a scientist announced the discovery of a distinct kind of snapping turtle in New Guinea. The announcement was based on one museum specimen, crudely stuffed and dried, and labeled only "Fly River, New Guinea." All the other snapping turtles were confined to the New World, from southern Canada to

Ecuador. But the Fly River country was almost unexplored; it certainly harbored one unique turtle, the pit-shell, so why not a snapper as well? Biogeographers puzzled over the curious distribution of the snapping turtle family. Chelid turtles, leptodactylid frogs, and marsupial mammals had a somewhat similar distribution, but they were all part of a very ancient South American fauna with Australian relatives; snappers rather clearly were recent invaders of north-western South America by way of the Isthmus of Panama.

No more Fly River snappers came to light, not even during World War II when many Australian, Dutch, and American servicemen collected New Guinean reptiles and sent them to various museums. Suspicion grew; and careful inspection revealed the "Fly River" turtle to be an ordinary American snapper, the same kind that is served in the restaurants of Chicago and Philadelphia. Somehow the specimen had been mislabeled, and certain of its diagnostic features had been obscured by amateurish taxidermy. Thus another biogeographic puzzle was finally laid to rest.

Having been preserved and labeled with the proper collecting data, specimens are stored until needed for scientific study. Preparation and storage are generally carried on at museums and universities, where vast collections may accumulate. Thus the United States National Museum, which is a part of the Smithsonian Institution, in 1964 housed 14,654,250 specimens of mammals, birds, reptiles, amphibians, fishes, marine invertebrates, mollusks, and worms; 16,220,460 specimens of insects, spiders, centipedes, and the like; and 3,084,624 plant specimens; to say nothing of 13,080,604 fossils. And each year this museum somehow makes room for about 1,250,000 more biological specimens, newly received from all over the world.

Museum specimens provide a concrete foundation for distributional studies. These specimens are studied chiefly by people who are not biogeographers in any broad sense, but who are interested in the taxonomic relationships and geographic ranges of certain plants or animals. Thus the investigator might be concerned only with frogs of the genus *Rana*, or butterflies of the family Nymphalidae. Alternatively, the investigator might be less restricted taxonomically and more restricted geographically, so that he wishes to examine all the birds from Colombia or all the ferns from Georgia.

In the course of his project the scientist may examine hundreds or even thousands of specimens. Eventually he publishes the results of his study in some

Biologists employ many techniques to obtain specimens. At top, a caiman, relative of the alligators, is seized by hand in an Amazonian swamp. Below it, mammals collected by an expedition to the Amazon. Skins, tagged with collecting data, will be deposited in a museum. (Bruce Mozert)

A *painted turtle found on a road in peninsular Florida, 200 miles outside the proven range of the species. This turtle is popular in the pet trade, and the isolated Florida record would probably be dismissed as the result of introduction.* (Tony Stevens)

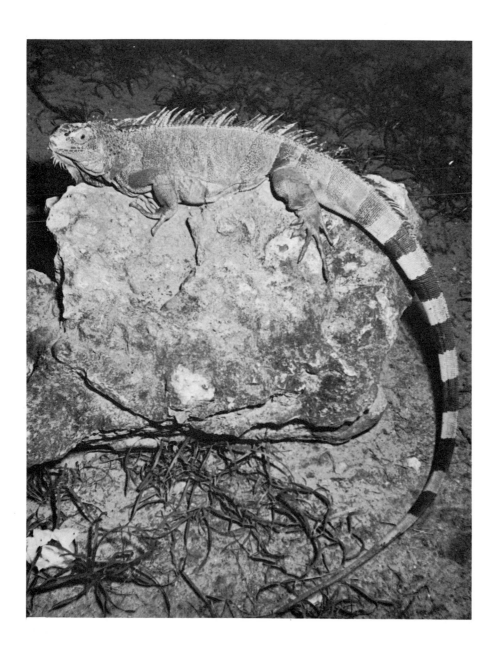

Each species of living thing is confined to certain situations. The iguana is an arboreal reptile of the New World tropics. (Tod Swalm)

technical journal. The average person never sees one of these journals, but more than 20,000 periodicals are devoted to the publication of biological discoveries, and a large percentage of these are open to the publication of new taxonomic and distributional findings.

The finished study may deal mainly with analysis of taxonomic relationships among a group of plants or animals, but even so, in most cases it will also include at least one map showing something about the distribution of the species under investigation; or in the absence of maps, distribution will be summarized verbally. Regional accounts, such as those dealing with the violets of New England or the fishes of the Great Lakes or the salamanders of New York, may include a great deal of information about distribution.

To prepare his maps or distributional summaries, the specialist must evaluate the available locality records. Recalling the "New Guinea" snapping turtle that was really American, how can he be sure that each museum specimen was correctly labeled? And might not a supposedly valid record be based upon a captive specimen that had escaped at a point, say, fifty or one hundred miles outside the actual range of the species? (For in these days of rapid mass transportation, and of innumerable pet fanciers and vivarium keepers, animals and plants certainly turn up far outside their natural ranges.)

There are some general rules of distribution that are useful in the interpretation of locality records:

First, every species of plant or animal has a definite geographic range. If this range cannot be described or mapped accurately, the failure stems from insufficient fieldwork.

Second, the geographic range of a species is somehow limited by environmental factors. It may be limited by a tangible barrier such as a mountain chain, a desert, or an ocean; but on the other hand it may be limited by more subtle factors, such as the average annual rainfall or the minimum winter temperature or the chemical and physical characteristics of the soil. Distribution in the sea may be limited by the depth or the temperature or the salinity of the water. An animal may be barred from some area by the lack of food or of nesting sites; an animal or a plant may be barred by an inability to compete with some hardier species. Several limiting factors may act together; and the factors that operate on one side of a species' range may not be the same factors that operate on the other side.

Third, unrelated species may be limited in much the same way. For example, wherever plants grow they tend to form natural groupings, or communities, in response to the local environmental factors such as rainfall, temperature, topography, soil, and fires. Thus in the eastern United States there are forests predominantly of white spruce, balsam fir, and larch; others of white pine, red pine, and hemlock; still others of sugar maple and beech; or of shortleaf pine and loblolly pine; or of various oaks and hickories. Elsewhere, different plant groupings are easily recognized: tall-grass prairie with its host of little spring-blooming flowers; sagebrush with a variety of low shrubs and bushes; chaparral; mangrove swamp; tropical savanna; and many more. Each community is characterized not only by some conspicuous plants but also by many less conspicuous ones and by various animals. An animal might depend directly on the plants of the community, say for food, or for knotholes in which to nest, or for moist leaf litter in which to burrow. Alternatively, the animal may not need any particular plant of the community, but may simply require the same conditions of climate and soil. Even in the sea, where higher plants are few, there are oft-repeated groupings of animals, and we may recognize a fauna characteristic of the coral reefs, the oyster bars, the tide pools, or the gulfweed flats.

Fourth and last, some animals and plants can thrive under a wide variety of conditions, and so may appear in various communities; but a great many are relatively specialized in their needs, and so may be able to inhabit only two or three communities and perhaps only one. In most groups of organisms some degree of habitat specialization is the rule, and very tolerant species are decidedly in the minority.

Therefore a locality record is immediately suspect if it is in a region or environment or community where the species is otherwise unknown. The out-of-place record is doubly suspect if it is based on just one specimen and has gone unverified for many years. And so a range map or distributional summary, prepared by a specialist, will seldom place a species in an area from which it is really lacking. However, it may fail to indicate the presence of a species in an area where it actually lives. This circumstance exists because biological exploration of the world is far from complete. A collector may occasionally discover a population of a plant or animal well outside its previously known range, and some finds are so interesting as to warrant a separate published report in an appropriate technical journal. The chance of making such a discovery depends on the kind of organism that is studied and the region where collecting is done.

A great many kinds of plants and animals doubtless remain to be discov-

In New Guinea a big monitor lizard is largely confined to tropical grassland. Here it eats a harmless snake. (W. T. Neill)

(Facing page) From time to time, new species are added to the scientific lists. In 1939 a strange little turtle, shown at top, was caught in the Chipola River of Florida; it proved to be a species that had hitherto eluded the collectors. (Bruce Mozert) Below it is a puzzling salamander collected in Marion County, Florida. It seems to represent a previously undiscovered species, but its formal recognition as such must await the collection of more specimens. (Tony Stevens)

ered; each year taxonomists bring to light nearly 5,000 more species of plants and about 10,000 of animals. Of course a majority of the newly discovered species are inconspicuous; they often belong to such groups as the fungi, algae, mollusks, jointed worms, roundworms, spiders, crustaceans, and especially the insects.

Some notable finds of the twentieth century include the African okapi, a forest-dwelling relative of the giraffe; the kouprey, a gigantic wild ox of central Cambodia; and the Congo peacock, whose existence was first suspected when an ornithologist saw a puzzling feather in a native's hat. Among other recently discovered species are the New Guinea crocodile, originally described on the basis of two skulls from a native longhouse; a blind, white salamander brought up from a deep artesian well in Georgia; and the large fish *Latimeria*, first taken in a trawl net off the South African coast, and belonging to a very ancient group. Among the lower animals, significant finds include the mollusks called *Neopilina*, found on the bottom of the deepest seas, and representing survivals from a remote period; the beard worms, vaguely wormlike animals of the deep sea, equipped with tentacles but not a digestive tract; and a blind, white crayfish found clinging to the ceiling of submerged caverns in Florida. Among plants, the most striking recent discovery was that of the dawn redwood, a splendid tree restricted to a few valleys of central China.

This list suggests that the really spectacular finds are likely to be made not only in remote regions but also in relatively unexplored habitats closer to home.

Biogeographers are in debt to those people who search out plants and animals around the globe; in debt, also, to preparateurs, curators, and taxonomists. Maps and distributional summaries, published in technical journals and based upon museum specimens, provide raw material for biogeographic investigation.

3 *"Let man learn that everything in Nature, even motes and feathers,*

*goes by law and not by luck."** **INTERPRETING THE RANGE OF A SINGLE SPECIES**

A great many animal and plant ranges have been mapped just approximately, not precisely. This circumstance does not prevent the biogeographer from investigating many problems. Thus the subdivision of the land into distinct biogeographic regions has been done chiefly on the basis of the vertebrates (mammals, birds, reptiles, amphibians, and fishes) and the seed-bearing plants, because these groups are the better known ones from both a taxonomic and a distributional standpoint.

Even when studying distribution in a relatively limited area, the biogeographer can make use of ranges that have been mapped only approximately. One example will clarify this statement. Numerous plants and animals, widespread in the lowlands of the southeastern United States, do not range very far into peninsular Florida, but stop short somewhere around the latitude of Gainesville or Ocala in the northern part of the state. As far as we know, the canebrake rattlesnake ranges southward to a locality about ten miles north of Gainesville, the yellow-bellied turtle and the flatwoods salamander precisely to the latitude of Gainesville, the river frog and the glossy watersnake nearly to Ocala. Further collecting might extend some of these ranges by perhaps fifteen or twenty miles,

* RALPH WALDO EMERSON

but the extensions would be of interest chiefly to the herpetologist, the student of reptiles and amphibians; the biogeographer would be concerned with the general picture, which is not altered by minor extensions of known geographic range. And when the biogeographer tries to discover what barrier has kept so many organisms out of peninsular Florida, he may not need to know how far west the flatwoods salamander ranges in Mississippi, or how far north the canebrake rattlesnake ranges in Virginia. Of course the biogeographer can proceed with more assurance when dealing with ranges that have been mapped rather accurately. Let us examine a few such ranges, and see how they lead to an important generalization.

There are thousands of species from which to choose. An especially interesting one is the four-toed salamander (*Hemidactylium scutatum*), a little amphibian that inhabits eastern North America. Scarcely three inches long, it is a secretive dweller beneath ground debris. Few people have ever heard of this salamander, and still fewer have ever seen a living individual; but the species has been of considerable interest to biologists, who have written more than a hundred technical papers about its distribution, anatomy, life history, and behavior.

The four-toed salamander was first discovered in 1838, in north-central Tennessee. As biological collecting continued, it was found to live in extreme southern Maine, extreme southern Quebec, and southern Ontario; the greater part of Michigan and Wisconsin; southern New Hampshire and Vermont as well as central and southern New York; Massachusetts, Rhode Island, Connecticut, New Jersey, Pennsylvania, Delaware, Maryland, and West Virginia; southern and eastern Ohio; northern and southern (but not central) Indiana; most of Virginia except its southern portion; northeastern Kentucky; central and eastern Tennessee; and the uplands of Alabama. This distribution sounds odd but becomes more intelligible when expressed in terms of physiography: the four-toed salamander is primarily an inhabitant of the Appalachian uplift and the Great Lakes region.

The northern limit of this range is not surprising. Unlike birds and mammals, salamanders have no internal mechanism for maintaining a constant body temperature, and so must take on the temperature of their environment. An individual four-toed salamander might be able to live, chilled and motionless, for quite a long while; but a population of these salamanders cannot exist unless the individuals can hunt for food, elude their enemies, and reproduce, as well as go through egg and larval stages. In a far northern area, the summer may be too brief to permit the carrying on of the salamander's life cycle.

The southern limit of the range does not surprise us either. The salamanders of eastern North America are mostly characteristic of cool, moist uplands; not many of them range southward even as far as central Florida. It is not always certain why this is the case; but many of the species apparently are stimulated to breed only during cold, rainy spells of winter or early spring, and so do not invade regions where winters are very mild or very dry.

But the progress of biological collecting also revealed outlying colonies of the four-toed salamander, some of them separated by more than three hundred miles from the main region of occurrence. There are isolated colonies to the south of the general range, in eastern Georgia and adjoining parts of western South Carolina, as well as in extreme southwestern Georgia and a nearby area of northern Florida. There are colonies west of the general range, in eastern Louisiana, in southwestern Arkansas, and in eastern Missouri. Far north of the main region of occurrence, there are colonies in the Canadian province of Nova Scotia.

This curiously fragmented distribution provokes a question: why is the range of this species not more nearly continuous throughout eastern North America? The answer is revealed by a study of the salamander's life history. The adult of the four-toed salamander rarely ventures into the open. It spends most of its time in hiding, under stones, fallen timbers, mossy clumps, and leaf litter— the debris of the forest floor or of nearby meadows and clearings. In such places it can simultaneously hide from many predators, avoid extremes of temperature, escape desiccation, and find tiny insects, spiders, and worms upon which to feed. When it does come forth, it is protected by camouflage: its upper surfaces are bronze and brown, like fallen leaves.

The salamander must venture abroad at least during the breeding season, when it migrates to some nearby pool, bog, or sluggish bog stream. Here the female mates and deposits her eggs. Usually the eggs are laid not directly in the water but just above it, either in thick mats of cool moss or beneath the bark of a soggy, rotten log at the water's edge. In such places the jelly-like eggs stay as wet as though they were actually in the water; yet they (and the nesting female as well) are not exposed to fishes or other aquatic predators.

The eggs hatch into active little larvae which immediately wriggle into the water and hide among rootlets or bottom debris. The larvae, too, are camouflaged, marked with fine traceries of black and brown on a mottled background of green and yellow. From time to time the larvae emerge from cover to pursue tiny aquatic organisms. Once in a while the larvae may pick up some parasites, but usually nothing very harmful to them. After about six weeks in the water, the larvae transform into miniature replicas of the adults. At transformation they migrate away from the breeding pond and take up adult life in some nearby forest or meadow.

The behavior of this amphibian does not involve intelligent selection among a number of possible alternatives, but merely results from automatic responses to stimuli coming from the environment and from the salamander's own physiological rhythms. In other words the four-toed salamander is fitted, by anatomy and behavior, for life only in areas that provide both woodlands and nearby pools—and not just any woodlands or pools but only those that fill the salamander's food and shelter requirements, as well as sufficient immunity from enemies and predators. Many regions lack such woodlands and pools, and therefore lack the four-toed salamander as well.

The above account reveals how the distribution of a species may be intimately related to its life history; but it still does not reveal how the four-toed salamander was able to reach so many localities far beyond the limits of its continuous range. A species does not evolve independently a dozen times, in a dozen different places. If the four-toed salamander occurs at two localities that are separated by an expanse of what is now unfavorable terrain, we should investigate the possibility that, at some time in the past, the intervening country was not so unfavorable as to prohibit invasion by this amphibian. Especially should we inquire whether eastern North America once had a climate that was cooler and wetter than the present one, thus permitting a wider spread of rich woods, mossy pools, and their associated fauna.

As a matter of fact, there is abundant evidence in eastern North America of a former climate that was cooler and wetter than the present one. Some of the evidence comes from palynology, a science that deals with (among other things) the identification of pollen grains that were blown into ancient bogs and swamps. Pollen is, of course, the fine dust from the cones and catkins and blossoms of the seed-bearing plants. It may be surprising to learn that microscopic examination will permit identification of many different kinds of pollen, or that the seemingly delicate grains will fossilize and persist for thousands of years; but such is the case. Pollens of various northern trees—fir, Canadian spruce, larch, and arborvitae—have been found in fossil deposits as far south as north-central Florida, eastern Texas, and northern Mexico. Today these conifers (cone-bearing trees) are widespread in the far north, but enter southeastern United States only in the moist, relatively cold environment of the higher mountains. Of course pollen grains blow about, and we could not insist that whole forests of northern conifers once blanketed eastern United States. But on the other hand these trees do not yield much pollen as compared with the broad-leaved trees such as oaks, hickories, elms, and maples; and the presence of fir and spruce pollen, at least in any abundance, is fairly good evidence that some northern conifers were present not too far away though they might not have dominated the landscape.

Vertebrate remains from Florida also suggest a cool, moist climate at some time in the past. For example, the eastern porcupine once lived in Florida. (It feeds upon the bark of trees that grow in cool, moist woods, and especially upon the bark of certain northern conifers.) It has since fallen back a thousand miles to the north. A fossil bed at St. Petersburg, Florida, yielded bones of the porcupine; of a marmot, probably the woodchuck, which today ranges southward only into the high country of northern Alabama; of a lemming mouse, whose modern relatives inhabit bogs and wet meadows as far south as the mountains of northern Georgia; and of the worm snake (*Carphophis*), which today does not live south of the rich woods of the Georgia Piedmont. Another fossil bed, near Melbourne, Florida, yielded bones of the red fox, which today does not occur (except where introduced for fox-hunting) south of the uplands of Georgia and Alabama; and a tooth probably from the wapiti or American elk, which in historic times did not live south of upper South Carolina.

A meadow vole, today found no farther south than the mid-Piedmont of Georgia, occurs as a fossil three hundred miles farther south, in Florida. The ruffed grouse, which today follows the northern coniferous forest southward down the Appalachians, also left its bones to fossilize in Florida deposits. The tree swallow now occurs as a breeding bird no farther south than Virginia, and passes over Florida only as a migrant; but its immediate ancestor nested in Florida, as shown by fossils of both juveniles and adults.

Obviously many "northern" animals and plants once reached Florida and

At left, a wapiti, often called American elk, in Montana. The wapiti once ranged far more widely in North America than it does today. (Montana Highway Commission) Above, a beaver building a dam at Loch Lomond, New Brunswick, Canada. Today the beaver ranges southward to extreme northern Florida, but in the past it occurred well down in the Florida peninsula. (Canadian Government Travel Bureau)

then fell back to the north. In fact some of the northerners were able to persist in Florida, although only in a few small areas that offered an especially cool, wet environment. Thus an impressive lot of northern plants live in or near the great ravines along the Apalachicola River not far west of Tallahassee. These ravines are kept cool and wet by numerous clear spring runs; and because of their high, steep walls, the ravines receive sunlight for only a portion of the day.

In this part of Florida one is astonished to see white cedar, beech, pallid and shagbark hickories, swamp leatherwood, big-leaved magnolia, and Fraser's magnolia, along with such smaller plants as trilliums, bloodroot, mandrake, columbine, hepatica, bellworts, Solomon's seal, baneberry, pepperroot, rue anemone, false rue anemone, swamp buttercup, and giant equisetum. A good many of these plants are isolated in the ravine country, and most are actually characteristic of the uplands from central Georgia northward. Some "northern" animals live in these Florida ravines, too: the beaver, the red-spotted newt, the eastern ribbon snake, a camel cricket not otherwise known south of the Piedmont, a crayfish belonging to a species that is otherwise Appalachian, and many more.

There is also evidence that many plants and animals, now confined to eastern United States, at one time were able to range much farther west, into regions that today are too dry for them. Fossils from the central and western states include many organisms that now are strictly eastern in distribution. Some of the species, falling back to the better-watered East, left relict colonies behind, in small areas that remained fairly damp. In the Ozark-Ouachita region, for example, one finds far-isolated colonies of otherwise "eastern" species such as the queen snake, the hellbender, the red-backed salamander, the zigzag salamander, a little fish called the streamline chub, the mountain madtom (a catfish), certain stream-dwelling crayfishes, the black locust tree, and the yellow-wood tree—as well as the four-toed salamander whose distribution prompted this discussion.

In short, a relatively cool, wet period once permitted the four-toed salamander (and its habitat) to advance far south and west. With a shift toward a warmer and drier climate it fell back to the north and east, where cool, moist woodlands persisted in and near the Appalachian uplift and the Great Lakes region; but it left behind isolated colonies wherever the local environment permitted.

Now let us consider briefly the Nova Scotian population of the four-toed salamander. The continued survival of the species this far north is not surprising; for Nova Scotia is almost an island, and its seasonal change of temperature is moderated by the surrounding ocean. The question is how the salamander ever reached the Nova Scotian peninsula. It is significant that this region harbors isolated populations not only of the four-toed salamander but also of Blanding's turtle, the eastern ribbon snake, and a white-footed mouse. Nova Scotian populations of the ringneck snake and the smooth green snake may be isolated, also. About seventy-five kinds of trees, found on Nova Scotia, do not otherwise range north of Maine or even Massachusetts. Apparently, then, various organisms reached Nova Scotia at a time when the climate of eastern Canada was warmer than it is today; and a subsequent lowering of temperature did not extirpate all of them from this near-island where climate vagaries are buffered by the ocean.

There have been past fluctuations of both temperature and rainfall, so marked and so recently as to have affected the distribution of many species that are still living today. Sometimes many fluctuations can be documented in a single fossil deposit; pollen analysis may reveal a number of distinct strata, some with plants characteristic of a cool, moist period and others with plants indicative of a warm, dry period.

Cores bored from the ocean bottom also reveal numerous fluctuations of temperature. Little organisms, called Foraminifera, live in the upper water, and fall to the bottom when they die. On the bottom their hard shells form deposits which grow thicker and thicker as the centuries go by. Many kinds of Foraminifera can live only in waters of certain temperatures; and so there are species characteristic only of arctic waters, others of subarctic waters, still others of cool temperate, or warm temperate, or tropical waters. When cores are bored into the foraminiferal sediments of the ocean bottom, it becomes obvious that the sea-water temperature has fluctuated; for at different levels in the core, cold-water Foraminifera appear, persist for a time, and then vanish, to be replaced temporarily by warm-water species. Presumably the land temperatures were fluctuating like those in the sea; and, for reasons that need not concern us here, rainfall fluctuates with temperature, cooler periods also being wetter. Through the ingenious technique of radiocarbon dating, we are sometimes able to learn the ages of the various levels found in a sea-bottom core, or of the various strata recognized during palynological analysis of a bog, and so assure ourselves that the climatic fluctuations took place in relatively recent times, geologically speaking.

From the foregoing discussion we can now derive an important generalization: the range of any organism reflects the environmental conditions that obtain today, but it may also reflect, at least to some degree, the different environmental conditions that have obtained in past ages.

4 *"An age will come after many years when the ocean*

will loose the chain of things, and a huge land lie revealed; when Tiphys

*will disclose new worlds and Thule no more be the ultimate."**

CONTINENTS IN
BIOGEOGRAPHIC VIEW

The preceding chapter dealt with considerations to be borne in mind when analyzing the distribution of a single species. Let us proceed now to patterns of distribution that are duplicated by numerous, and often unrelated, species. Land and sea must be treated separately, for the two offer vastly different environments; and we may begin with the land for it has received more study. Islands present special problems, to be discussed later; continents will be considered first.

At the outset it is well to say something about the mapping and naming of the continents. Curiously, it is impossible to obtain a base map that shows the true geographic relations of the continents, islands, and seas. This is because the world is roughly globular, and the surface of a globe cannot be portrayed on flat paper without some distortion.

In the year 1569 the Flemish cartographer Gustav Krämer, who Latinized his surname to Mercator (merchant), prepared a map in such a way that a straight line drawn on it would correspond to a single compass heading on the

* SENECA, Roman philosopher of the first century.

actual surface of the earth. This "Mercator projection" was very useful to the old merchant-seamen, and it persisted. Indeed, until the beginning of World War II the average person thought of the world only in terms of its Mercator projection, the familiar world map that hung on every schoolroom wall and that graced every geography book. It is still a commonly used projection when large areas of the world are to be portrayed; and its peculiarities must be understood.

On Mercator's projection Greenland looks bigger than South America, but it is really much smaller. Canada looks twice as big as the United States, but it is not. Extreme eastern Asia and extreme western North America usually appear at opposite sides of the map, but they are only 56 miles apart. A straight line across the tropical Pacific, say from Vietnam to the Panama Canal, looks much shorter than a great arc from Vietnam north along the coast of eastern Asia, thence across Bering Strait, and south along the west coast of the New World to Panama; but the arc is shorter on the face of the earth. (This statement may sound unbelievable, but it is easily tested by bending a wire around a scale model of the world.) These and similar deficiencies of the Mercator projection are not widely known.

There are projections other than the Mercator, but each is somehow in-accurate in depicting relative distances, or directions, or the shapes and sizes of oceans and land masses. Our geographic thinking has been conditioned not only by the distortions inherent in map projections but also by political boundaries which draw our attention away from natural features. The familiar schoolroom classification of the continents should be reviewed, and slightly amended to suit present purposes.

Many of us were taught in grammar school to name the continents Europe, Asia, Africa, North America, South America, Australia, Antarctica. But geo-graphically, Europe is merely a peninsula of Asia, and it is often desirable to lump the two as Eurasia. Asia is popularly taken to include nearby islands groups such as the Philippines or Japan; but as we shall consider islands separately, in subsequent discussion the name Asia refers to the Asiatic mainland only. Similarly, the name Africa refers only to the African mainland, and excludes the great island of Madagascar in the Indian Ocean. Australia offers no terminological problem although it has one island state, Tasmania. Antarctica may actually be two land masses capped by one sheet of ice, but the circumstance is unimportant here. The familiar political boundaries of South America coincide well enough with the physical boundaries of the continent, and we may ignore the slight extension of Colombia into the isthmus that is otherwise occupied by Panama.

Our own continent offers several problems of geographic terminology. From a standpoint of physical geography North America extends southward to South America. But we shall reject this usage, and place the southern limit of North America at the Isthmus of Tehuantepec in Mexico. The term "Middle America" will be used for all the land between the Isthmus of Tehuantepec and the Isthmus of Panama. Finally, we shall use the term "United States" to mean the forty-eight continuous states; for tropical Hawaii and far-northern Alaska generally merit separate consideration in any biogeographic discussion.

Almost all of the land in the world, except for remote oceanic islands, forms an essentially continuous system. Thus Asia is continuous with Africa by way of the Isthmus of Suez, and Europe is barely separated from Africa by the Strait of Gibraltar. Various islands, from Cyprus to the Balearics, help break the Mediterranean Sea barrier between Eurasia and Africa. Northeastern Asia (Russia's Chukotski Peninsula) and northwestern North America (Alaska's Seward Peninsula) are separated only by the narrow Bering Strait, which itself is interrupted by islands; while our Aleutians and Russia's Komandorskies lie like stepping-stones between the Old World and the New.

Baffin Island, Greenland, Iceland, the Faeroes, and the British Isles form an arc from eastern Canada to western Europe; and farther north another island arc extends from Greenland through Spitsbergen and Franz Josef Land to North Land and so to Russia's Taymyr Peninsula. Numerous islands help fill the water gap between Asia and Australia: Sumatra, Borneo, the Celebes, Java, the Lesser Sundas, the Moluccas, Timor, Ceram, New Guinea. From New Guinea hundreds of smaller islands stretch eastward into the Pacific: the Bismarck Archipelago, the Solomons, the Marshalls, the Gilberts, the Ellices, Fiji, Samoa, Tonga, the Cook Islands, the Societies, and the Tuamotu. Still farther east are Pitcairn, Easter Island, and Sala-y-Gomez; and beyond them the Chilean islands of San Felix, San Ambrosio, and the Juan Fernandez group.

North America is continuous with South America by way of Middle America; and in addition, the West Indian archipelago stretches between the two, from the Bahamas and Cuba through Hispaniola and Puerto Rico, thence through the Virgin Islands, the Leewards, and the Windwards, and so to Trinidad just off the coast of Venezuela. The Andean uplift follows the west coast of South America southward, disappears beneath the sea at the Strait of Magellan, reappears as Tierra del Fuego and then dips again beneath the sea, apparently swings around beneath the surface as the Scotia Ridge, and finally rises as the Palmer Peninsula of Antarctica.

When the continents are viewed in their proper light, as forming one great system, it becomes a little easier to see how, given enough time, a group of land animals might spread over most of the world, especially if past climates permitted invasion of areas that today are inhospitable. As a matter of fact, climate is not the only thing that has changed throughout the ages; there have also been changes in the relationship of the sea to the land. We shall discuss this topic at length in another chapter; for the present suffice it to say that land masses have moved up or down in response to stresses in the earth's crust, and that the seas have become shallower or deeper according to the amount of water bound up in the polar ice caps and according to the depression or elevation of the ocean bottom. During earth's long ages, many an isthmus has been drowned to form a strait; many a strait has been elevated to form an isthmus.

One area is of special interest in connection with submergence and emergence of the land. This is the Bering Strait, which narrowly separates Alaska from Russia. Today it is a channel 56 miles wide, joining the Chukchi Sea on the north with the Bering Sea on the south. On a clear day it is possible to stand on Cape Prince of Wales in Alaska, and look across to Cape Dezhnev in Siberia; Eskimos have often paddled their frail kayaks across, to bring back Siberian brides and sled dogs. Big Diomede, Little Diomede, and smaller islands lie in the narrowest part of the channel, with St. Lawrence Island a little to the south. The water in Bering Strait is comparatively shallow, mostly 20 to 30 fathoms, and the channel bottom is really a drowned plain. This plain is very smooth and flat to the south; but to the north it is irregular, rising to form shoals and even islands. From north to south it stretches 1,300 miles—600 miles more than the north-south length of Alaska along its Canadian border. Throughout a large part of geological time the plain has been above water, forming a broad highway, the Bering Bridge, for intercontinental dispersal.

The Isthmus of Panama is another important area that has been both elevated and submerged. At a very remote period it was above water just as now, permitting plants and animals of that day to move freely into South America. Then the Isthmus was submerged, producing a wide strait that has been dubbed the Panama Portal; and it remained submerged for long ages. During the time that the Isthmus of Panama was submerged, the Bering Bridge stood above water; for a good fifty million years it was easier for land and freshwater animals to move between North America and Asia than between North America and South America.

Turning to elevation and submergence of land in the Old World, we need mention only that overland movement between Asia and northern Africa has been possible for a very long time; but overland movement between Asia and Australia has not been, except perhaps at a remote period.

5 *". . . this division of the Archipelago into two regions characterized by*

a striking diversity in their natural productions, does not in any way

correspond to the main physical or climatal divisions of the surface."*

THE THREE GREAT
FAUNAL REALMS

For reasons to be given later, the broad outlines of plant distribution are not precisely those of animal distribution. The present chapter will consider the way animals have dispersed, in relation to the continental areas defined in the preceding chapter. The broadest patterns of animal distribution are explicable in the light of previous remarks about the spatial relationships and geologic history of the main land masses.

Animals have moved so readily between Eurasia and Africa, and between Eurasia and North America, that these continents together compose but one faunal realm. In this realm, which today includes nearly two-thirds of the world's land area, there arose the greatest diversity of animal life. Numerous well-known families and genera originated in one or another part of this realm. Here we could list paddlefishes, salmon and trout, suckers, minnows (Cyprinidae), pikes, sticklebacks, bass and sunfishes, and true perches (Percidae) among the fishes. The amphibians would include true frogs (*Rana*) and toads (Bufonidae), as well as newts, hellbenders, waterdogs, and in fact all salamanders. Among the

* ALFRED RUSSEL WALLACE, *The Malay Archipelago*, 1869.

48

reptile groups that arose in the realm are snapping turtles, musk and mud turtles, the common turtles (Emydidae), glass lizards (*Ophisaurus*), beaded lizards (Helodermatidae), rat snakes, racers, king snakes, cobras, mambas, true vipers, moccasins, and rattlesnakes. Characteristic birds include ostriches, vultures, pheasants, peacocks, chickens, grouse, turkeys, auks, woodpeckers, shrikes, wax-wings, creepers, titmice, and chickadees. Sheep, antelope, cattle, giraffes, deer, camels, hippos, peccaries, pigs, rhinos, horses, elephants, cats, skunks, weasels, pandas, bears, foxes, wolves, beavers, squirrels, rabbits, moles, shrews, and apes are among the mammals. (And this list could be greatly prolonged.)

But Australia was apart from this mainstream of vertebrate evolution. The primitive animals that had reached Australia at a very remote time were able to persist in the absence of the more advanced animals that were evolving else-where in the world. Impressive relics of Australia include a lungfish, the only living member of a family (Ceratodontidae) that was widespread in the days of the dinosaurs. Equally remarkable are the duck-billed platypus and the echidnas, exceedingly primitive mammals that are reptile-like in depositing leathery-shelled eggs and also in having but a single ventral opening for the elimination of all body wastes and for conjugation. And of course we must mention the mar-supials, such as the kangaroos, koala, wombats, bandicoots, and Tasmanian devil. Some of the marsupials developed into interesting parallels of the more ad-vanced Eurasian–African–North American types; the Australian pouched "mole" is a spectacular example, but there are others, superficially shrewlike or rabbitlike or catlike or even wolflike.

The water gap between Australia and Asia, broken by numerous islands, was not an insuperable barrier to the southward spread of every Asiatic type of ad-vanced animal. From time to time throughout the ages, various birds were able to reach Australia, where they continued their evolution. Thus the Australian bird fauna today includes some highly distinctive types, such as the huge casso-waries; and other types that are less highly modified but still distinctive, such as the lyrebirds and the birds of paradise; as well as many other types that are also represented in Asia and perhaps elsewhere.

Some animals were good at island-hopping, either by swimming or by drift-ing on trees that were blown or washed into the sea. Such animals occasionally reached Australia from the north. Good examples are the Muridae, or Old World rats and mice, the family that includes our house mouse, barn rat, and

The salmon and trout family is distributed from the Arctic southward to northern Africa, Formosa, northwestern Mexico, and the Appalachians; it has been widely introduced elsewhere. Shown above is the rainbow trout from the Chama River of New Mexico. Its natural range is from Alaska to Baja California. (New Mexico Department of Development)

(Facing page) The bass and sunfish family is strictly North American, represented from Canada to Mexico. Shown at top is a crappie from a Louisiana river. Its range is from Minnesota and Ontario southward into the Gulf states. The typical frogs, family Ranidae, are concentrated in Eurasia, Africa, and North America. A few species reach northern Australia; one reaches northern South America. Pictured at right is a bullfrog from Louisiana; its range is eastern North America, but it has been introduced elsewhere. (Louisiana Tourist Commission)

roof rat (all introduced into America from Eurasia). Australian murid rats, although less publicized than the marsupials, are strikingly diversified. Some are brushy-tailed sand hoppers resembling our kangaroo rats; some build great stick nests in the fashion of our wood rats; some store fat in the tail, for use in lean seasons; some are aquatic, much like our muskrat in appearance and habits. This diversification, this adaptive radiation into various environments, implies that ancestral murids reached Australia a very long time ago.

Australia is also noted for a diversity of venomous snakes belonging to the family Elapidae, an advanced family probably of Asiatic origin; and for a scarcity of the typical harmless snakes (Colubridae), which predominate in all other regions but which are represented in Australia only by a few species that live in trees or that enter the water frequently.

Movement of animals out of Australia seldom went farther than the nearest islands, such as Tasmania and New Guinea.

Australia, then, by virtue of long-continued isolation, forms a second faunal realm, characterized by primitive relics, some of which parallel the more advanced types; by the diversified descendants of a few higher groups of animals whose ancestors probably arrived from Asia by swimming or by accidental rafting from island to island; by flying animals that exhibit varying degrees of divergence from relatives elsewhere; and by an absence of numerous higher groups that predominate on other continents.

Up to a certain point, South America had a history much like that of Australia. At a remote time it was open to colonization by primitive animals. Then it was cut off from the rest of the world by the formation of a water gap, the Panama Portal. A portal only for marine life, it formed a barrier behind which South American land animals evolved for ages in isolation.

Some of the old South American types, isolated by the Panama Portal, have relatives in other parts of the world. Thus the South American arapaima, a 200-pound freshwater fish, has one smaller relative in South America, but its family, the Osteoglossidae, is also represented today by one species in Africa, one in southern Asia, and one in New Guinea and adjacent Queensland; while fossil osteoglossids appear early in the fossil record of North America, Eurasia, and Australia. Pelomedusid turtles, once widespread, survive in Africa as well as in South America; and, as we have already noted, chelid turtles, leptodactylid frogs, and marsupials persisted in South America just as in Australia.

The common turtles, family Emydidae, are represented in Eurasia, northern Africa, and North America; a few inhabit Central America, and one line of descent has reached South America and the West Indies. Shown are the young and adult of a box turtle, which ranges from southern Maine and Michigan southward through Florida and eastern Texas. (Bruce Mozert)

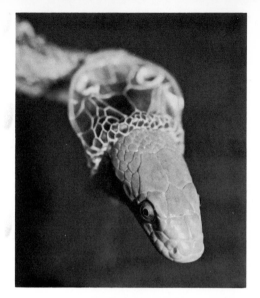

The rat snake genus, Elaphe, is present in Eurasia, North America, and Central America. A common rat snake, which ranges from southern Canada to the Florida Keys and western Texas, is illustrated at top. It is beginning to "shed its skin." The moccasin genus, Agkistrodon, is represented in Asia, North America, and Central America. Shown at bottom are brightly patterned young and drab adults of the cottonmouth moccasin, which ranges from southern Virginia and southern Illinois southward through Florida and southeastern Texas. (Ross Allen's)

At top, an ostrich in captivity in South Africa. Ostriches are now confined to Africa, but once inhabited Asia as well. (South African Tourist Corporation) Below it, the ring-necked pheasant, native to temperate Asia but widely introduced elsewhere; here shown in South Dakota. (South Dakota Department of Highways)

At top, impala antelopes in South Africa. The impala has close relatives in Africa and Asia but not in the New World. Below, Cape buffalo in Kruger National Park, South Africa. Other wild members of the cattle family are Eurasian or North American. (South African Tourist Corporation)

Especially characteristic of South America are the mammals called edentates: the armadillos, anteaters, and sloths, as well as the extinct ground sloths and tortoise armadillos. The catfishes also underwent astonishing adaptive radiation in South America where there are about a dozen families and a thousand species; these range from the tiny, semiparasitic candirú which will enter the urethra of man and larger animals, up to the enormous lau-lau which is reputed to gulp down swimmers and which carries its young in its gigantic mouth.

But as we have seen, the water gap that isolated South America was eventually bridged. A broad highway, the Isthmus of Panama, permitted interchange of the North American and South American faunas. Some of the old South American types apparently were not greatly affected by the opening of the bridge; they did not move northward into Middle America, but neither did they die out before the irruption of the more advanced northern types. Examples include the pipid frogs, some but not all of the leptodactylid frog genera, the gigantic snakes called anacondas, the chelid turtles, some of the alligator-like caimans, and some of the smaller marsupials.

Other South American types pushed northward with the bridging of the Panama Portal. Several of the colubrid snake genera and of the leptodactylid frog genera, as well as one species of caiman, pushed far up into the Middle American tropics. Some groups went even farther north. The armadillo and a few leptodactylids reached southern United States; the Virginia opossum reached northern United States; the porcupines set a record by reaching Alaska and Labrador. (But none of the South American animals reached Asia.)

Still other South American types pushed northward, even into the United States, yet eventually died out there; and also became extinct in their South American homeland. Examples are the ponderous ground sloths, some of them bulkier than an elephant; and the tortoise armadillos which, although mammalian, were covered with a bony shell. A few South American types pushed into North America only to die out there, yet persisted in South America. For example the giant, semiaquatic rodents called capybaras inhabit only South America and Panama today, although their fossil remains are not rare in Florida.

With the bridging of the portal, there was far more movement of animals into South America than out of it. From the north came various frogs and two genera of salamanders, a rattlesnake and several genera of harmless snakes, as well as mud and snapping turtles. There also came a shrew, deer, peccaries, cats, skunks, weasels, a bear, camels (the llama and its allies), rabbits, squirrels, mice, and many others. These newcomers, entering South America at its northwestern corner, found the Andean uplift in their path. Many of them, such as the camels and some of the smaller cats, were especially able to thrive in the cooler environments offered by the mountains, and so they spread southward along the Andean chain. However, some of the newcomers, such as the jaguar, were able to invade many environments, and so to spread throughout most of South America. Many of the old South American types died out in the face of this invasion from the north. Among the ones that became extinct were a variety of marsupials, ranging from superficially rodent-like species to great carnivores somewhat resembling bears and lions; and, of the placental mammals, five large groups of hoofed beasts.

It is likely, or in some cases certain, that various animals were able to cross the Panama Portal long before it was bridged. A raccoon reached South America from the north across the portal, and a South American ground sloth may have crossed in the other direction. The movement of leptodactylid frogs into Middle America from South America may have begun before the Isthmus of Panama rose, with various species accidentally drifting across the portal on floating trees. Birds, of course, reached South America on many occasions, and that continent came to develop the richest bird fauna in the world.

South America, then, forms a third faunal realm, characterized by some primitive relics; by more or less diversified descendants of groups that were able to cross the Panama Portal at various times; by numerous advanced types that arrived from the north at a relatively late date; and by many extinct animals that vanished rather abruptly with the arrival of the more advanced northerners.

The faunal realms, as defined herein, were really three main theaters of animal evolution on land and in the fresh waters; and they were centers from which, and within which, land animals dispersed. Antarctica may well have been a fourth theater of animal evolution. Today this ice-locked continent supports only a very meager land fauna; but it once had a milder climate, as shown by coal beds formed from ancient swamps, by well-preserved imprints of fossil plants, and by the trackways of what probably were reptiles. It is fascinating to speculate about what fossils will someday be found beneath the South Polar ice cap.

Here and there in the world, certain land areas do not form a part of any realm. Between two adjacent realms there must be a transition zone, where some

At top, white-lipped peccaries taken by Maya Indians in southern British Honduras. The peccary family is distributed from Texas to Argentina. It entered South America after the closing of the Panama Portal. Below, a baby raccoon from Florida. The raccoon family is widespread in North and South America, with two members (pandas) in Asia. Raccoons reached South America from the north before the Panama Portal was closed. (Ross Allen's)

At top, an agouti from Colombia. Agoutis are part of the old South American fauna, although members of the family have pushed northward into Central America and the West Indies. (Pamela Cheatham) Below it, a panther from Florida. The species, belonging to the Felidae or cat family, entered South America with the closing of the Panama Portal. (Tod Swalm)

At top, a tapir from Colombia. Tapirs, too, entered South America from the north. Below it, a brocket deer from Colombia. The deer family was yet another late entrant into South America. (Pamela Cheatham)

The emu, a flightless bird confined to Australia. (Ross Allen's)

of the animals are from one realm and some from the other. The islands east of Borneo and west of New Guinea are transitional between the Australian and the Eurasian–African–North American realms. Both Middle America and the West Indies are transitional between the South American and the Eurasian–African–North American realms. Islands far out in the ocean will usually defy classification according to realm, for they generally harbor but few land animals and these only as a matter of chance dispersal. The meager land fauna of Antarctica does not especially ally that continent with any particular realm. Some islands, such as Madagascar and New Zealand, present unique problems to be discussed later.

Following usual biogeographic practice, we have defined the faunal realms in terms of their mammals, birds, reptiles, amphibians, and freshwater fishes. These backboned animals are vastly outnumbered by invertebrates—the lower animals such as flatworms, mollusks, jointed worms, the spiders and their kin, crustaceans, the insects (over 725,000 known species!), and many more. But the majority of the invertebrate groups are uninteresting to any but a few specialists; in this book we shall have but little to say about them, except in connection with the sea, where they predominate. And when we come to the plants, we shall concentrate on the familiar seed-bearing and flowering groups, having little to say about the algae, fungi, liverworts, mosses, ferns, and other groups of lower plants.

6 *"... the trees ... cannot be explained, for their multitude ... not even the native Indians know them. ... Some have fruit and others flowers ... and one may see every stage of development at a given time and during any part of the year."** **THE CONTINENTAL TROPICS**

As regards the distribution of animals, each of the three great faunal realms is diverse within itself. Each may be divided into what are called faunal regions, and some of these latter in turn may be partitioned into subregions. The Eurasian–African–North American realm, most complex of the three, may be divided first into tropical and nontropical portions.

What, precisely, is meant by tropical? The dictionary says that tropical lands are those lying between the Tropic of Cancer and the Tropic of Capricorn. But the Tropic of Cancer is simply a line north of which, and the Tropic of Capricorn a line south of which, the sun's rays never strike the earth vertically. Obviously the temperature may vary from place to place along either of these lines; for the temperature of an area is affected not only by latitude but also by winds, ocean currents, elevation, and proximity to large bodies of water. When a biogeographer speaks of the tropics, he usually means areas where the local plants and animals are not exposed to freezing or near-freezing temperatures.

There are good reasons why many geographic ranges are limited by the

---

* GONZALO FERNÁNDEZ DE OVIEDO Y VALDÉS, on the New World rainforest, 1526.

64

possibility of such exposure. There are in the world today about a million known species of animals, and of these only about 20,000—the birds and mammals—have an internal mechanism for regulating the body temperature. All the others must take on the temperature of their environment. These others—fishes, amphibians, reptiles, and all the invertebrates—are often called ectotherms, a term signifying that they must draw on their environment for heat. Some ectotherms can briefly raise their body temperature a few degrees by muscular activity; this is particularly true of certain active, stream-dwelling fishes. Honeybees, clustered in their hive during a cold spell, can raise the temperature of the cluster by the muscular activity of wing vibration; and in the hot summer they can also lower their temperature, perhaps through increased evaporation. Various ectotherms maintain a remarkably constant body temperature, but only by restricting their activity to times when and places where this temperature prevails; such actions are especially characteristic of certain insects, lizards, and snakes. But these cases are exceptional; a great many ectotherms have no way of coping with lower temperatures, and so live only in areas that offer no danger of chilling. Most of the cold-sensitive ectotherms are harmed even by temperatures that are still well above freezing, perhaps because the animal's physiological processes are slowed or halted, or because the numbed animal is unduly exposed to predation or disease. (The sensitivity of tropical animals to cold may be best known to aquarium fanciers who keep tropical fishes: an unexpected cold wave, or a defect in the heating system on a wintry day, and certain of the tiny fishes will be floating about, belly up, even though the room temperature may still be fairly comfortable.)

Plants, too, rely on the environment for warmth. In colder regions some plants overwinter only as dormant seeds; others die back to the roots, which are protected through the winter by the warmth of the ground; still others shed the leaves and withdraw the sap to the roots, leaving only the dry, woody parts to be exposed to winter cold. But like so many ectotherms, many plants have no way of coping with lower temperatures, and so grow only in areas where there is no danger of chilling. Some tropical plants are harmed by temperatures that are still well above freezing; and most are killed by the formation of ice crystals in their tissues, at temperatures just below the freezing point of water. (The sensitivity of tropical plants to cold may be best known to Florida gardeners who set out tropical herbs and shrubs in areas not truly tropical: an unexpected cold wave

Desert in the Cape region of South Africa. Flowers spring up in the brief rainy season. (South African Tourist Corporation)

The Algodones dunes in California's Imperial Valley. Contrary to popular belief, most deserts do not resemble this expanse of shifting sand. (All-Year Club of Southern California)

and some of the plants blacken and die immediately, while others do not blacken but wilt and die later on.)

Many birds and mammals, while capable of regulating the body temperature from within, have evolved so many adaptations to tropical life that they do not invade areas offering low temperatures.

At this point it is well to clear up a common misconception about the tropics. It is widely believed that the tropics are characterized by very high temperatures. This is not true. Remarkable as it may sound, every summer the temperature probably soars higher in, say, Kansas City or St. Louis or Cincinnati than at most localities in the tropics. For example, at tropical Belém in Brazil, only 100 or so miles from the Equator and at an elevation of only 42 feet, the highest temperature recorded during an 18-year period was 95 degrees Fahrenheit, the lowest 64; yet Toronto, Canada, about 2,500 miles from the Equator at an elevation of 379 feet, has recorded an extreme high of 103, and a low of 29 below zero. A tropical area does receive more heat than a temperate one; but the heat is well distributed throughout the year, so that no month is exceedingly hot or exceedingly cold. As regards temperature, a tropical climate is equable; a "temperate" climate is highly intemperate.

It should also be noted that not all tropics are covered with rainforest or jungle. In general, and with local exceptions, rainfall is heavy in the low latitudes, permitting the growth of lush rainforest; but the rainfall progressively decreases away from the Equator, and the rainforest is eventually replaced by tropical grassland or thorn forest or scrub. Still farther beyond the Equator, rainfall is insufficient to support even these latter communities, and so deserts develop. At about the latitudes where these deserts are first met with, or at least not far beyond, the first freezing temperatures are also likely to be encountered. Then beyond the deserts the rainfall increases again. In other words rainfall increases both north and south of those deserts that lie at the edge of the tropics. Such deserts, on account of their aridity and their wide range of temperatures, will themselves be inhabited only by a very limited fauna, yet will separate a fairly large tropical fauna from a fairly large warm-temperate one. These deserts are therefore a barrier between two faunas, but they can also be thought of as accomplishing both a climatic and a faunistic transition from tropical to temperate.

It may seem novel to regard the great sand deserts as anything but fully tropical. Actually, most deserts have at least occasional frosts in some parts, and in any event the local extremes of temperature would prove intolerable for most tropical animals. For example, at Bir Milrha south of Tripoli in the Sahara, on a Christmas morn the temperature was a degree below freezing, but during the day it rose to 99 degrees! In El Goléa, a Saharan town at an elevation of 1,247 feet in north-central Algeria, during a 19-year period the temperature rose as high as 124 degrees, and fell to a low of 22. Also, in the arid lands there may be large areas that are relatively cold on account of high elevation. Thus Géryville, at 4,298 feet in semidesert of the Atlas Mountains on the northern edge of the Sahara, over a 28-year period recorded a high of 108 degrees, and a low of 1. Such areas do not provide the equability of temperature that is needed by, and that limits the distribution of, so many animals and plants.

Having defined the tropics from the standpoint of biogeography, we may now briefly review the continental areas that offer a tropical climate. The tropical portion of Asia includes the extreme southern tip of China; practically all of Indochina, Thailand, the Malay Peninsula, and Burma; and most of the peninsula occupied by India and Pakistan. The greater part of this tropical area is bounded on the north by the slopes of the Himalayan uplift and by deserts or semideserts, although the climatic and faunistic change, from tropical to temperate, is very gradual along the coast of China and in the islands that lie not far offshore. We can recognize no true tropics in most of southern Asia west of, say, Karachi. The Near East is part of a great desert belt that borders the tropics from eastern Asia to western Africa. Within southwestern Asia, only the lower part of Saudi Arabia might perhaps be listed as tropical. At Aden, on the southern tip of the Saudi Arabian peninsula, the lowest temperature recorded over a 33-year period was 61 degrees, the highest 109. The local fauna is sparse, and the tropical species are of African affinity; so if lower Saudi Arabia is considered tropical, it should be linked with tropical Africa which lies just 20 miles away across Bab al Mandab at the mouth of the Red Sea.

In Africa, the temperate lands are separated from the tropical ones by the Sahara Desert. On the African east coast, Port Sudan might be taken as the approximate northern limit of a tropical fauna, and on the west coast Dakar. The northern coast of Africa is temperate, and so to be linked with the temperate European areas to the north. (It is not generally realized that such North African cities as Tunis and Algiers are about as far north as Norfolk, Virginia, or the southern border of Kansas.) There are deserts, the Kalahari and the Karroo,

At top, wild elephants in South Africa. Elephants today inhabit both Africa and Asia. Below, rhinoceros in Umfolozi Game Reserve, South Africa. Rhinos, too, are represented in both Africa and Asia. (South African Tourist Corporation)

At top, zebras in Kruger National Park South Africa. (South African Tourist Corporation) Other wild members of the horse family inhabit Asia. The domestic horse (shown here in South Dakota) is of Asiatic ancestry, but has been spread widely by man. (South Dakota Department of Highways)

at the edge of the tropics in southern Africa; and the southwestern tip of the continent is temperate, not only by virtue of its distance from the Equator, but also because it is chilled by the cold, northwardly flowing Benguela Current just offshore. For climatological reasons this current also has a drying effect on the coast along which it passes. Many tropical animals do not range into the dry or temperate areas of southern Africa; but others, more hardy, have been able to invade these areas from the tropics farther north. The dry and temperate areas do have some characteristic animals, but these are too few to alter the essentially tropical aspect of the fauna, and the areas need not be set apart as a separate faunal region.

Thus there are two tropical Old World portions of the Eurasian–African–North American realm, one African and the other Asian. (Southern Saudi Arabia is hereafter thought of as part of the African portion.) The two are now separated by the deserts and semideserts that extend from Saudi Arabia to Pakistan. But this entire arid expanse, like many others, has seen environmental changes throughout the ages; neither deserts nor arms of the sea have prevented faunal interchange between tropical Africa and tropical Asia.

These two tropical areas have many families and genera, even some species, in common. Familiar animals, found in both areas, include elephants, rhinos, the cheetah, the leopard, the lion, horses (the African zebras and wild ass, the Asian wild horse), a family of monkeys, and the great ape family. Less familiar ones include the lorises, pangolins, bamboo rats, the Old World porcupines, hornbills, the small birds called honey guides, agamid lizards, snakes of the genus *Python*, arboreal frogs of the family Rhacophoridae, climbing perches, and several families of catfishes. Some animals that are now strictly African, such as hippos, giraffes, ostriches, and baboons, were once present in Asia but died out there. On account of these and other faunal similarities, the tropics of Africa and the tropics of Asia are placed together as a single faunal region. This region may then be divided into Asian and African subregions, as the two have faunal differences also.

Turning now to the New World, how far north do the tropics extend? Much of South America lies within the tropics, and a tropical environment extends thence northward through Central America and into Mexico. Discussion would be simplified if the tropics reached northward in Mexico only as far as the Isthmus of Tehuantepec, which we have taken as the southern limit of North

America; but this is not the case. The tropics extend farther north than the Isthmus, and so provide us with a tropical area in the southern tip of North America.

In the interior of Mexico the tropics reach only a little beyond the Isthmus, for in the state of Oaxaca the land soon begins to rise into cool highlands. Along the warm coasts, however, tropical conditions extend farther north, roughly to Tampico on the east coast and Mazatlán on the west. This tropical area in extreme southern North America is so small that it need not be set apart as a distinct region, especially since most of its tropical animals are not peculiar to it but simply enter it for a short distance, coming up from Middle America or, less frequently, South America. Also, many other animals range into this small tropical area from the temperate United States, or from the Sierra Madres which carry a temperate climate and a temperate fauna southward from the United States almost to the Isthmus of Tehuantepec.

There are two different barriers to the spread of tropical animals northward from Middle America. The Isthmus of Tehuantepec probably was a strait at a remote period. Since the rise of the modern fauna it has generally been a bridge. But today, at least, it is a so-called "filter bridge" that offers a very limited series of environments and permits only certain animals to filter through. The Isthmus is now covered mostly with grassland, scrubland, and rather open forest; it supports almost no lush rainforest, and so is a barrier to the northward spread of Middle American animals adapted only for rainforest existence. Then, a little farther north than the Isthmus, temperate conditions develop; and these block the further spread of many tropical animals that did pass through the filter bridge of Tehuantepec.

Above Mazatlán on the Mexican west coast, a desert stretches northward through the Mexican states of Sinaloa and Sonora into western Arizona, southern Nevada, and the southeastern tip of California, as well as down the peninsula of Baja California where it is broken by a mountain range. This desert has received various local names—the California portion is called the Mojave Desert—but the name Sonoran Desert may be used to designate the entire arid expanse from (roughly) Mazatlán to Las Vegas. The Sonoran Desert, like its Old World counterparts, lies at the edge of the tropics, and has a sparse fauna of animals that can withstand aridity and a wide range of temperatures.

Above Tampico on the Mexican east coast, east of the Sierra Madre Oriental,

Macaques, at top, are represented in both Asia and Africa. (Ross Allen's) Below them, baboons in Kruger National Park, South Africa. Baboons are mainly African, but formerly inhabited southern Asia as well. (South African Tourist Corporation)

At top, giraffes in Kruger National Park, South Africa. Now strictly African, giraffes once lived in Asia also. (South African Tourist Corporation) The true chameleons (lower photo) are distributed from Africa and Madagascar to southern Asia and southern Europe. (Ross Allen's)

the country is dry but not a desert. It has patches of thorn forest to the south; and beyond these, an area of grassland and mesquite extends northward into southern New Mexico and southwestern Texas. In the United States just east of the Rocky Mountains, and also between the Sierra Nevada and the Rockies, there are large areas that are dry and barren enough to be called deserts; but they are relatively cool deserts, developed in the lee of the mountains which block off rain-bearing winds from the Pacific. We need not consider such deserts at this point.

The lower keys of Florida are nearly frost-free. Key West, at the tip of the keys, often claims the title of the only frost-free city in the United States. Over a 40-year period this island community recorded temperatures up to 100 and down to 41 degrees. But the land fauna of the keys is largely of temperate origin, not strikingly different from the fauna of, say, Ocala in north-central Florida, where the temperature over a 39-year period varied between the extremes of 105 and 12 degrees. Thus from a standpoint of animal distribution, southern Florida is appended to the temperate lands farther north.

Most of South America is tropical, but portions are not. Tierra del Fuego, at the southern tip of the continent, has a climate much like that of Arctic tundra, and a temperate climate extends northward into southern Argentina and southern Chile. Tundra and cold woodlands follow the crest of the Andean uplift for almost its full length. Just east of the Andes in Argentina there is a strip of desert, which passes eastward into semidesert and northward into thorn forest and grassland. The coastal strip west of the Andes is also relatively cold for its latitude, being chilled by the cold, northwardly flowing Peru Current; and this current, like Africa's Benguela Current, results in coastal aridity, so that desert conditions follow the Peruvian coast northward to within about three hunded miles of the Equator. But the nontropical portions of South America were populated chiefly by tropical animals, the ones that could withstand aridity and extremes of temperature. There is no need to set apart southern or Andean or west coastal South America as a distinct region, in spite of a few characteristic animals.

Australia is barely tropical in the north and temperate elsewhere, with a great central desert; but the curious relict fauna dominates all the environments, and we need recognize but one faunal region in the Australian realm.

7 *". . . there is a considerable degree of resemblance in the volcanic nature of the soil, in the climate, height, and size of the islands, between the Galapagos and Cape Verde Archipelagoes: but what an entire and absolute difference in their inhabitants!"** **ARCTIC, TEMPERATE, AND INSULAR LANDS**

With tropical areas set apart as needed, there remains the task of partitioning the nontropical portion of the Eurasian–African–North American realm; and then of considering the islands of the world, with a view toward adding them, where feasible, to the appropriate realms and regions.

As previously noted, Bering Strait was once a bridge with a milder temperature than at present. Nevertheless the strait now is, and for a long while has been, a sizable water gap in a cold area. Quite a few groups of animals never made the crossing between Siberia and Alaska; and so some biogeographers would draw a line through Bering Strait, separating the nontropical portion of the Eurasian–African–North American realm into Old World and New World regions. There is certainly justification for this procedure, but an alternative arrangement may have more to recommend it.

As regards both fauna and environment, the nontropical portion of this realm is most homogeneous at the north, in the lands that border the Arctic

* CHARLES DARWIN, *The Origin of Species*, 1859.

Ocean. Furthermore, the circumpolar Arctic environment is unique within the realm, what with its brief summer, long and very cold winter, frozen subsoil, relatively permanent snowfields and glaciers, long days and nights, and absence of trees. Much of the Arctic animal life is also unique. It is well to retain all the nontropical portions of the Eurasian–African–North American realm as a single region, and to set off the Arctic as a subregion thereof.

The southern limits of the Arctic subregion do not correspond to the Arctic Circle. Owing to the tilt of the earth's axis, at the North Pole the sun remains above the horizon for six months and then below the horizon for six months; this effect diminishes southward, and along a line about 1,600 miles south of the Pole there is only one day of the year when the sun does not rise and only one day when it does not set. This line is the Arctic Circle, and of course temperature will vary from place to place along it. From a biogeographic standpoint, the southern limit of the Arctic subregion is the northern limit of tree growth.

With the setting apart of this circumpolar subregion, there remain two temperate subregions. One includes extreme northern Africa, all of Europe except the Arctic lands of northern Scandinavia and northern Russia, and all of Asia north of the tropics and south of the Russian Arctic; the other includes all of North America below the Arctic. (As mentioned, we may ignore a small area with tropical climate and some tropical animals in the extreme southern tip of North America.)

As for islands, some of them were formerly connected to a mainland in relatively recent times. Such islands consequently harbor a good portion of the modern mainland fauna—the portion that can live in the relatively limited series of environments usually offered by islands. There is no objection to including the so-called "continental" islands—those with former mainland connections—in the appropriate realms and regions. An exception may arise, however, in the case of an island that was connected with a mainland only at a very remote date; for the island fauna may be made up of a few oddities evolved locally during long isolation, and perhaps some relict types no longer extant on the mainland. An island, whether continental or not, could be added to a given region if it received a sizable fauna mostly from that region.

Sumatra, Java, Borneo, and Ceylon were formerly connected with the nearby mainland, and may be added to the subregion that embraces the Asian tropics. Formosa, like islands generally, has a more equable temperature than prevails at the same latitude on the mainland, and so belongs to the tropical Asian subregion even though nearby mainland areas have been considered climatically and faunistically transitional between tropical and temperate.

Japan may be added to the subregion that includes temperate Eurasia. With Japan temperate and Formosa tropical, the Ryu Kyus between them may be considered climatically and faunistically transitional. New Guinea has been connected with Australia, and supports a basically similar relict fauna. However, the island lies closer to the Equator, farther into the tropical belt of high rainfall. New Guinea has an enormous mountain range with peaks about 16,000 feet high; a vast tropical river, the Fly; and sizable expanses of lush rainforest as well as grassland. It offers some tropical environments that are poorly represented in Australia, or not at all; and it has some unique animals such as the pit-shell turtle which is the only known member of its family (the Carettochelydidae), and an echidna unlike the mainland ones. New Guinea should be added to the Australian region, but as a distinct subregion thereof. Tasmania belongs to the Australian subregion.

A few New Guinean marsupials have spread eastward as far as New Ireland; one of them, and a few New Guinean rodents, reach the Solomons, as do some New Guinean frogs. A small boa and several lizards have spread even farther eastward, but perhaps as stowaways in the thatching of native vessels. The islands east of New Guinea, from the Bismarcks to Samoa, may be included in the New Guinean subregion. Excluded, however, are those islands from New Caledonia and Fiji southward to New Zealand, and the numerous little islands of the South Pacific east of Samoa. For reasons to be given later, these islands are not to be classified.

In the New World, Tierra del Fuego and the Falkland Islands may be appended to the southern tip of South America. Trinidad was formerly annectent with Venezuela, and has a fairly large and typical South American fauna. Newfoundland may be added to the temperate North American subregion, while Novaya Zemlya and ice-covered Greenland belong with the Arctic subregion. Spitsbergen and Franz Josef Land, while perhaps not continental, have received their meager fauna from the Arctic, in which subregion they may be included.

Small islands lying just off a mainland (such as Hainan off the South China coast), or lying just off a continental island (such as Bali at the eastern tip of Java), are too numerous to mention individually; they may be added to the appropriate realms and regions.

Certain islands merit special comment. As already noted, the Pacific islands east of Borneo and west of New Guinea are transitional between the Eurasian–African–North American realm and the Australian realm, having been populated by animals coming from both directions. These islands are not continental. The largest of them, Celebes, is quite old, and has evolved some characteristic animals with Asiatic affinities. These include the "black ape" (really a monkey), a curious four-tusked hog called the babirusa, and the anoa or pygmy water buffalo, as well as a pygmy elephant which has become extinct.

The Philippines have not been a part of the Asiatic mainland, although a more westerly island of the group, Palawan, may once have been temporarily connected with Borneo. At various times, certain of the Philippine Islands probably were connected among themselves in a fashion unlike the present one. There is reason to believe that at times the water gaps in the Philippines were larger and more numerous than at present, and some of the islands smaller. Speaking generally, the Philippines were populated by animals that came from the tropical Asian mainland, especially by way of Borneo, and that crossed several water gaps. Accordingly, there is no objection to placing the Philippines in the same subregion with tropical Asia.

The British Isles are continental. Ireland became separated from Great Britain before the latter became separated from mainland Europe. But the fauna of the islands was several times wiped out, or nearly so, by glaciers that formed during the so-called Ice Age. Shortly after the last retreat of the glaciers, there was still continuous land from the mainland to Great Britain but not from the latter to Ireland; and a few cold-hardy European species crossed into Great Britain. As the postglacial climate moderated, the English Channel formed, restricting further immigration to species that could cross a barrier of salt water. Ireland, wiped clean by glaciation, was repopulated mostly by species that could cross both the English Channel and the Irish Sea; such species were few. (The absence of snakes from Ireland is attributable to geologic history, not the good St. Patrick!) The British Isles may be added to the subregion that includes temperate Europe.

The islands of the West Indies (except Trinidad) are not continental.

There is not much evidence that they ever were interconnected among themselves to a significantly greater degree than at present, although there is evidence that some of them were once larger and environmentally more diverse than at present. They were populated by animals coming across water from North, Middle, and South America. The North American influence is weakest, and is evident mostly to the north, in Cuba and the Bahamas; the South American influence is evident mostly to the south, in the Lesser Antilles. While the West Indian fauna has many distinctive elements, its strongest affinities are with Middle America. We have considered Middle America to be transitional between the South American realm and the Eurasian–African–North American one; and it would be consistent to look upon the West Indies as similarly transitional.

The Bermuda Islands lie in the Atlantic Ocean about 600 miles east of the United States coast. They lie farther north than is generally realized, at about the latitude of Montgomery, Alabama; but their temperature is moderated by the surrounding ocean. Over a period of 32 years, Bermudan temperatures varied between the extremes of 94 and 39. The native land animals show no West Indian affinities. Land birds of Bermuda—those actually living on the islands and not merely passing through on migration—are all North American: the Florida gallinule, ground dove, crow, catbird, bluebird, white-eyed vireo, and cardinal. There is one native lizard, a little skink of the genus *Eumeces*; other species of the genus are found in Africa, Asia, North America, and Middle America, but not the West Indies. Modern man has lately introduced half a dozen West Indian frogs and small lizards to Bermuda; but if these be ignored, the island group may be included in the temperate North American subregion.

Previously it was stated that certain islands are to be excluded from the classification of the land into faunal subdivisions. In general, unclassifiable islands are those that lie far from any mainland, and that have had no mainland connection at least since the rise of the modern fauna. Some of the unclassifiable islands are small, and monotonous of environment. They may be low islets of sand and coral, or they may be steep, rocky upthrusts; but in any event they offer so little diversity of environment that they could support but a few species. The ones they do support need not have arrived from the nearest mainland or from any one mainland; for winds and ocean currents often determine the directions from which such islands receive their limited fauna. The island

The tuatara, below, is a primitive reptile surviving only on New Zealand. At right, another of New Zealand's animal oddities, the Kiwi on North Island, New Zealand. (New Zealand Information Service)

species are likely to be adapted for life on or near beaches, and for dispersal on floating logs and trees. Unfortunately, such species—for example mice, and some of the lizards called skinks and geckos—are also likely to stow away on the vessels of primitive and modern man; seldom can we be sure that a small, remote island received its meager fauna without the help of man.

Among the small, remote, and unclassifiable islands are many in the Southwest Pacific east of Samoa, such as the Line Islands and the Tuamotus; many in the Southwest Pacific east of the Philippines and north of New Guinea, such as the Carolines, Marianas, and Marshalls; a few in the Southeast Pacific, such as Easter Island, Sala-y-Gomez, and the Juan Fernandez group; and the Atlantic islands from the Azores and the Cape Verde group through Ascension and St. Helena to Tristan da Cunha and the South Sandwich Islands. Antarctica, although large, is like these small islands in supporting only a meager land fauna; and like them, it goes unclassified.

Other unclassifiable islands are geographically situated in such a way that they have been populated to an equal degree from different faunal subdivisions. Iceland, for example, was wiped clean of its land fauna by glaciations. With the final retreat of the ice this island, lying between Greenland and Norway, was repopulated by some animals from the Arctic and some from temperate Europe. The Galapagos Islands, about 600 miles off the west coast of Ecuador, received their fauna partly from South America and partly from Middle America; but a few of the Galapagos species, notably two big iguanas and some finches, now have no close relatives on the mainland. The islands were named for, and are especially famous for, their giant tortoises, called *galápagos* in Spanish; but these reptiles are living remnants of a group that was formerly widespread on all continents except Australia.

Some islands are unclassifiable because their fauna is made up largely of so-called endemics—species, genera, families, or higher groups not represented anywhere else in the world. Among such islands are those in the Indian Ocean: Madagascar, the Comoros, the Mascarenes, and the Seychelles.

Madagascar lies about 260 miles off the east coast of Africa, but its fauna is not African. It has no soft-shell turtles or common turtles; no agamid, lacertid, or monitor lizards; no pythons, cobras, mambas, or vipers; no ostriches, hornbills, woodpeckers, honey guides, or finches. There are no monkeys or apes; no rabbits, squirrels, or Old World porcupines; no foxes, weasels, hyenas, or cats; no elephants, rhinos, giraffes, or antelope. The horses and cattle are domestic imports only. The native Madagascan mammals include an endemic family (Tenrecidae) of large insectivores, a variety of rats and mice, several endemic genera of small carnivores belonging to the mongoose and palm civet family (Viverridae), and three families of lemurs. These last are primates, members of the group that includes the monkeys and the apes; but they are more primitive than any monkey, and today live only on Madagascar and the nearby Comoro Islands. Other Madagascan oddities include two endemic genera of lizards belonging to the Iguanidae, a family otherwise represented by one species on Fiji and Tonga and several hundred more species in the New World; and several turtles of the family Pelomedusidae, one of them belonging to an ancient genus (*Podocnemis*) that has also persisted in South America.

Madagascar once had a pygmy hippo, but it died out, as did the elephant bird. This latter was a flightless bird nine feet tall, weighing about 1,000 pounds. It lived on the island so recently that its great eggs, each holding about two gallons, were commonly used by the natives as water containers. (And these eggs, carried north by Arab seamen, bolstered the legend of the roc, the fabulous bird that made off with Sindbad.)

The Mascarene Islands, lying 400 to 850 miles east of Madagascar, were noted for endemic flightless birds of other kinds. Mauritius had the dodo, flightless rails, and a supposedly flightless parrot; Reúnion had the dodo-like solitaire, a flightless rail, and an almost flightless heron; Rodriguez had the white dodo, a flightless rail, and a supposedly flightless heron. All these birds, along with some giant tortoises, were killed out by man within historic times. Some curious boa-like snakes live on Mauritius and nearby Round Island. The islands also have various African and Asian animals, some of them perhaps introduced.

The Seychelles, about 600 miles northeast of Madagascar, have some legless amphibians called caecilians, two endemic genera of frogs belonging to the family Ranidae, and some giant tortoises; but no flightless birds. There are also a few African and Asian animals on the Seychelles, some perhaps introduced.

Even more puzzling than the islands of the Indian Ocean are the unclassifiable islands east of Australia, from Fiji and New Caledonia to New Zealand. New Zealand, two great islands about 1,000 miles east-southeast of Australia, harbors three species of frogs belonging to the family Leiopelmidae. This is the most primitive family of frogs, and its only other member inhabits certain

Above, a giant anole, an iguanid lizard species confined to Cuba. (Tod Swalm)
Below, the iguanid lizard genus Cyclura is restricted to the West Indies. Shown is
the rhinoceros iguana, a Cyclura of Hispaniola. (C. M. Binder, Jr.)

uplands of California, Oregon, Washington, British Columbia, Idaho, and Montana. Also on New Zealand is a reptile known by its native name of tuatara. Although superficially lizard-like, it is not a lizard but one of the so-called beakheads. Apparently all the other beakheads died out even before the dinosaurs did. New Zealand once had large, flightless birds, the moas; these may have been exterminated by prehistoric man. Another endemic flightless bird of New Zealand, the kiwi, survives today. Several other New Zealand birds had nearly or quite lost the power of flight.

New Caledonia, about 800 miles east of Australia, was reached only by birds and lizards. Both groups have evolved distinctive types on the island. Most interesting, perhaps, is a cranelike bird, the kagu, placed in a family of its own; it has well developed wings but does not fly. Walpole Island, near New Caledonia, once was inhabited by an almost unbelievable land turtle, a gigantic reptile that could not withdraw the head into the shell, and that was horned like a bull. It belonged to a family (Meiolaniidae) that once inhabited Australia and South America, a distribution roughly paralleling that of the marsupials, chelid turtles, and leptodactylid frogs.

Lord Howe Island, 300 miles east of Australia, has only birds and lizards. One of the lizards and several of the birds are distinctive; and two of the birds— a rail and a gallinule—are flightless. The island once had a meiolaniid turtle, also. Fiji, about 750 miles northeast of New Caledonia, might almost be included in the New Guinean subregion, but its small land fauna includes a lizard of the family Iguanidae. This same lizard also lives on Tonga. The iguanids are the dominant lizards of the New World, and it is astonishing to find a member of the family on these remote little islands of the Old World.

8 *"A new scientific truth does not triumph by convincing its*

adversaries, but because its opponents die, and a new generation grows up

that is familiar with it."* "He said true things, but called them by wrong

names."** **REFINING THE**

ZOOGEOGRAPHIC SCHEME

Before leaving the topic of faunal realms and regions, it seems proper to give credit to a few of the people who helped to develop the zoogeographic scheme. They were remarkable individuals, some of them; and certain of their ideas will help us to draw up a meaningful summary of animal distribution.

Biogeography began as a descriptive science, concerned with defining the major patterns of animal and plant distribution. Of course the early zoogeographers could not have been expected to do more than map out faunal subdivisions, in the days when geology, paleontology, archeology, and ecology were but infant sciences. Indeed, until the latter nineteenth century, the philosophical framework of the times was scarcely conducive to explanation of distributional patterns. John Lightfoot, vice-chancellor of Cambridge University, had announced that the world was created on September 17 in the year 3928 B.C., at nine o'clock in the morning; Archbishop James Ussher of Ireland later corrected him, pronouncing the date to have been October 23 in the year 4004 B.C. It was hard to explain how certain snails, slugs, and dung beetles made their way from

* MAX PLANCK   ** ROBERT BROWNING, *Bishop Blougram's Apology.*

California to southern Europe (or the reverse) in less than 6,000 years, and still harder to explain why they should have made this trip and not some other.

One zoogeographer, a little more than a century ago, saw clearly that animal species had wandered, but he also felt that the wandering of every species must have begun from a certain twin-peaked mountain, Büyük Agri Dag, in northeastern Turkey near the Armenian border; for this mountain, better known as Ararat, was the traditional resting place of Noah's ark!

Also, at one time it was widely believed that numbers had a significance beyond that of mere enumeration. Certain numbers were thought somehow to be better or luckier or more fundamental than others. (And the superstition is still widespread.) Several early zoogeographers accordingly felt that there should be a certain number of major taxonomic groups and a corresponding number of faunal realms; a "quinary system," based on the number five, was especially favored.

Toward the end of the seventeenth century, the English naturalist John Ray, often called the founder of zoology, thought it well to indicate each animal's country of origin; but otherwise he had little to say about zoogeography. Carolus Linnaeus, founder of our taxonomic system, was similarly casual about distribution; in the various editions of his *Systema Naturae* he restricted biogeographic observations to occasional Latin statements such as "*Habitat in America*" or "*Habitat in Indiis.*"

William S. Macleay, a Scottish naturalist, in 1819 proposed the aforesaid quinary system. In this he was followed by such men as the English physician William Leach and the Austrian zoologist Leopold Fitzinger. William Swainson, an English naturalist and writer who studied birds, mammals, and insects, was so taken with the quinary system that in 1835 he divided the world into five continents: Europe, Asia, Africa, Australia, and America.

Charles Lucien Bonaparte, nephew of Napoleon, was a French naturalist who spent six years in the United States. In 1827 and 1837 he published a comparative list of European and North American birds. He realized that there was a correlation between anatomy and function, as regards structural characters employed by taxonomists to distinguish among various animal groups. The realization later proved useful in anatomy, taxonomy, and even biogeography.

In 1841, Hermann Pompper of Leipzig divided the continents according to their temperature zones, listing the known mammals, birds, and reptiles of each zone. His work was important in showing that faunal subdivisions need not correspond with continental outlines.

Johann Wagner, a Bavarian, in 1844–1846 summarized the distribution of the mammals, holding Mount Ararat to have been the center of dispersal for all life. Wagner, incidentally, was later privileged to describe an extraordinary fossil found near his home; it was a fine specimen of *Archaeopteryx*, a link between the reptiles and the birds. (How astonished he would have been to learn that the reptile-birds vanished 135 million years before man appeared on earth!)

Edward Forbes, a British marine zoologist, in 1844 advanced the concept of former land bridges, and tried to explain faunal distribution in the British Isles as the result of land connections that subsequently vanished. This concept of a changing earth was very important, although many later workers, in the absence of geological data, relied too heavily on supposed sunken bridges to explain disjunct distributions.

In 1853 the Austrian biologist Ludwig Schmarda discussed the geographic distribution of animals, in a great volume of 755 pages. He opened with remarks about the effect of climate on distribution, and subdivided both land and sea into numerous faunal "realms." Some of his subdivisions were well founded, others not. Schmarda did recognize the fundamental unity of North America and Eurasia in the Arctic, and saw that the African shore of the Mediterranean should be grouped with the European one. He also segregated the islands between Asia and Australia, placing them with neither continent.

In 1858 the Englishman P. L. Sclater reviewed the distribution of birds throughout the world. He erected six well-chosen faunal subdivisions, all of them coinciding approximately with regions or subregions recognized today. Sclater was a man of unusual versatility: lawyer, Oxford scholar, director of the London Zoo, editor of various journals, and the author of 1,287 technical papers. He seems almost to have established a tradition of versatility among zoogeographers, for several of his immediate successors were comparably gifted in many fields.

The most interesting figure in early zoogeography was Alfred Russel Wallace, an English naturalist, sociologist, author, and explorer. He traveled in two of the world's richest faunal areas: the Amazon of South America and then the Southwest Pacific from the Malay Peninsula to New Guinea. His travels among the East Indian islands took eight years, during five of which he was accompanied only by a few natives. From this trip he brought back 125,660 specimens of mammals, birds, reptiles, mollusks, butterflies, beetles, and other insects. The collection included hundreds of species new to science.

While investigating the differences between island species and their mainland relatives, Wallace concluded that the individuals of a given species vary somewhat among themselves, some variations being harmful and others helpful to the individual. Individuals exhibiting helpful variations are more likely to survive and reproduce, while the ones exhibiting harmful variations are more likely to succumb to predators or other hazards. A species therefore changes in the direction of greater fitness to its environment, through the continual weeding out of the less fit individuals.

Curiously, another English naturalist and traveler independently arrived at the same conclusions, and similarly from a firsthand study of island life. He was Charles Darwin, and he has generally received the credit for this concept of evolution through natural selection, although the concept was first announced in a joint paper by Wallace and Darwin, read before the Linnaean Society on July 1, 1858.

Even though ill with some tropical disease acquired during his exploration of the East Indies, Wallace wrote numerous scientific papers based upon his field work, and in 1869 published a popular account of his East Indian travels. This latter work, *The Malay Archipelago*, went through ten editions in Wallace's lifetime, and was reissued in facsimile edition as late as 1962. But Wallace's crowning work appeared in 1876: a two-volume set entitled *The Geographic Distribution of Animals*. Wallace's nomenclature and his scheme of faunal regions have been widely used ever since, with only minor emendations.

Something more must also be said about Wallace's illustrious colleague, Charles Robert Darwin. The young Darwin set out to be a physician, like his father and grandfather before him; but—as he later wrote in his autobiography—he found the classroom lectures "something fearful . . . intolerably dull." Changing his university, Darwin abandoned medicine and decided to become a clergyman, an ironic choice in view of the impact his discoveries were to have upon the theology of that day.

Immediately after his graduation, Darwin obtained a post as naturalist with an exploring expedition supported by the British government. On the now-famous *Beagle* he circumnavigated the world. The voyage lasted five years, much of the time being spent along South American coasts. The experiences of this trip not only led Darwin to formulate the principle of natural selection; they also stimulated him to investigate such diverse fields as geology, paleontology, ecology, genetics, marine biology, and animal behavior—sciences then in a formative stage.

On the voyage, Darwin spent much time ashore in South America, exploring both slopes of the southern Andes. Far inland he found marine deposits, rich with seashells; found, too, the fossilized remains of trees. The local geology convinced him that a grove of fine trees once overlooked an ocean—the Atlantic Ocean now seven hundred miles away. The land had then sunk beneath the waves, carrying the trees with it; but later the sea bottom had been pushed up into a high mountain range, bringing to light the mineralized remnants of the ancient forest. Darwin surmised that this parched and barren land had once supported a tropical forest, in the days before the rise of a mountain range had blocked off all the warm, moisture-laden winds. He did not realize the vast age of the deposits he had found, but he reached conclusions of prime significance in biogeography: even great mountain ranges have not always been in existence, there have been profound changes in the relationship of sea to land, climates have changed, and living things have not always been distributed in the way they are today.

Although experiences of the voyage led Darwin to the principle of natural selection, he worked for twenty years more on the book that would document this discovery. Just when he was almost ready to publish the results of his labor, he received from his friend Wallace a manuscript expressing almost identical conclusions; it had been sent from the East Indies where Wallace was still exploring. Highly ethical, Darwin was moved to withhold his own work; but Wallace would not hear of any such thing, and the two agreed on a joint presentation.

After Wallace and Darwin, the next major contribution to zoogeography was that of a Hungarian-American, Angelo Heilprin. As versatile as Sclater or Wallace, Heilprin was a scientist and author, a traveler and mountain climber, a teacher of geology and paleontology, and leader of the Peary relief expedition to the North Pole. He was one of the first scientists to explore southwestern Florida and the Everglades, not long after Seminole Indian hostilities had ceased there. In 1887 Heilprin published *The Geographic and Geological Distribution of Animals,* a work that touched upon paleontology as well as upon modern distribution. This work accordingly ushered in a new era in zoogeography. Heilprin also modified Wallace's arrangement by the inclusion of North America and temperate Eurasia in a single region, and by recognition of the Sonoran Desert as faunistically transitional between warm-temperate and tropical.

In 1884 the American naturalist Theodore Gill presented his arrangement of faunal subdivisions. His work was important because it pointed out the usefulness of freshwater fishes in defining realms and regions. There were other men who helped in one way or another to define the faunal subdivisions: Albert Günther, Thomas Henry Huxley, W. T. Blanford, Richard Lydekker, Ernst Haeckel, Christian L. Brehm, Spencer F. Baird, Edward D. Cope, and many more; but their work need not be detailed here.

After Wallace and Heilprin, zoogeographic interest began to center on the paleontological finds that were being made almost daily. It was exciting to learn, for example, that the beautiful birds called trogons, today nearly confined to the tropics of Asia, Africa, and the New World, seemingly originated in France. Zoogeographers lost interest in realms and regions, becoming concerned instead with "faunal analysis." This latter dealt with the place of origin of the animals found in any given area.

In 1954 an American zoologist, Karl P. Schmidt, pointed out that both regional classification and faunal analysis were valid approaches to zoogeography. Schmidt also slightly revised the regional classification, employing paleontological data to choose between alternative arrangements.

In 1957 the American zoologist P. J. Darlington published a book, *Zoogeography: The Geographical Distribution of Animals,* treating the distribution of vertebrates that live on land or in fresh water. A major contribution in the modern tradition, this book also analyzed the faunal realms, regions, and subregions. Darlington pointed out that the subdivisions erected by Wallace hold up quite well in the light of modern knowledge. Of course there have been many other twentieth-century contributions to zoogeography; but most of them exemplify various paleontological or ecological approaches, to be considered in other chapters.

Apparently the early zoogeographers found it easier to define the faunal subdivisions than to name them aptly. Since the time of Ludwig Schmarda, each realm and region has received several names. But fortunately, the nomenclatural muddle is readily simplified by two rules:

First, capitalize the words "realm," "region," "subregion," and "transition" when they are used for a formally defined zoogeographic subdivision; but do not capitalize adjectives such as Arctic unless they relate directly to a subdivision. Then there will be no danger of confusing a Region (formal subdivision) with a region (area or general vicinity); no hesitancy in ascribing an "arctic climate" to a mountain peak outside the Arctic. Second, name the subdivisions in the familiar terms of geography and climate. These precepts are followed in the summary below, and throughout the remainder of the text.

Summary of Faunal Realms, Regions, and Subregions

The South American Realm includes the continent of South America, along with the nearby islands of Trinidad, Tierra del Fuego, and the Falklands. This Realm has only one Region, the South American Region, and this latter has only one Subregion, the South American Subregion. North of the South American Realm is the Middle American Transition (from the Isthmus of Panama to the Isthmus of Tehuantepec), and the West Indian Transition (the Lesser Antilles, the Greater Antilles, and the Bahamas). The two together may be referred to as the Caribbean Transitions.

The Australian Realm includes continental Australia, Tasmania, New Guinea, and small islands east of New Guinea from the Bismarcks to Samoa. This Realm has but one Region, the Australian Region; but the latter includes two Subregions. (1) The Australian Subregion is made up of continental Australia and Tasmania; (2) the New Guinean Subregion includes the islands from New Guinea to Samoa. To the west of the Australian Realm is the Celebesian Transition. It includes the islands west of New Guinea and east of Borneo.

The other continents together make up the Eurasian–African–North American Realm. Within this Realm there are two Regions, one mostly tropical and the other temperate or arctic. The African-Asian Tropical Region includes (1) the African Tropical Subregion, which is Africa south of the Sahara, along with the southern tip of Saudi Arabia; and (2) the Asian Tropical Subregion, which extends from extreme southern China through most of the Indian peninsula, and which also includes Sumatra, Java, Borneo, the Philippines, Ceylon, and Formosa.

The African Tropical Subregion is bounded on the north by the Saharan and Saudi Arabian deserts, in which is accomplished a climatic and faunistic change from tropical to temperate; and the Asian Tropical Subregion is bounded on the north by deserts and by the southern slopes of the Himalayan uplift, these areas also accomplishing a climatic and faunistic change from tropical to temperate. The Asian area of climatic and faunistic change, from tropical to temperate, may be placed a little farther north in the offshore islands than on the mainland, passing through the Ryukyus. There is no need to apply any formal name to areas in which the climate and fauna change from tropical to temperate, and such areas do not belong to either of the adjoining subdivisions.

A portion of the Eurasian–African–North American Realm, that lying north of the tropics, is the Arctic-Temperate Region; no other Region has both Arctic and temperate portions. This Region has three Subregions. (1) The Arctic Subregion is circumpolar in the lands north of tree growth. This Subregion includes Greenland, Spitsbergen, Franz Josef Land, and Novaya Zemlya. (2) The North American Temperate Subregion includes the remainder of North America southward to the Isthmus of Tehuantepec. Newfoundland and the Bermuda Islands belong to this Subregion, also. (3) The Eurasian–African Temperate Subregion includes northern Africa, north of the Sahara; and that portion of Europe below the Arctic but above the deserts or mountain slopes that form the northern boundary of the Asian Tropical Subregion. Japan and the British Isles also belong to the Eurasian-African Temperate Subregion.

The South American Subregion, largely tropical, has a temperate or colder area in the far south; the African Tropical Subregion has a temperate area in the far south; and the North American Temperate Subregion has a tropical area in the far south. Each small, far-southern area has some animals not present elsewhere in the Subregion, but need not be set apart as a separate zoogeographic subdivision.

The Celebesian Transition, the Middle American Transition, and the West Indian Transition have all been secondary theaters of animal evolution. Unclassified lands include the islands of the Indian Ocean; the islands of the Southwest Pacific from Fiji and New Caledonia to New Zealand; small islands of the Southwest Pacific lying north of New Guinea and east of the Philippines; small islands east of Samoa in the Southwest and Southeast Pacific; the Galapagos; Antarctica and nearby islands; Iceland; and small islands well out in the North Atlantic or in the South Atlantic.

Subregions have been divided into smaller units. Rather than do so at this point, we shall pass on to plant distribution, and comment briefly on the animal life that exists in each phytogeographically distinctive area.

9 *"... we traversed 4,000 leagues in the ocean .... We did not discover*

any land, except two desert islands; on these we saw nothing but birds and

trees, for which reason we named them the Unfortunate Islands."*

PRELIMINARY REMARKS
ON PLANT DISTRIBUTION

There are several different ways of looking at plant distribution, but at this point we shall consider only one: a subdivision of the world, based upon the distribution of plant species, genera, and families. This phytogeographic approach parallels the zoogeographic approach of Sclater, Wallace, Heilprin, and Schmidt, although the floral subdivisions do not always correspond to the faunal ones.

Before defining the floral subdivisions of the world, it is well to explain why they are not identical with faunal Realms and Regions. In discussing this topic it should be remembered that not all plant groups are distributed in patterns that would be unfamiliar to a zoogeographer. The faunal subdivisions reflect present and past geographic relationships among the continents, and the existence of climatic zones; such major factors of the environment have influenced plant distribution as well as animal. In later discussion it will be seen that various floral subdivisions, like the faunal ones, are partly explicable in terms of the long isolation of Australia and South America, the bridging of the Panama Portal and Bering Strait, and the differences between tropical, temperate, and arctic en-

* FERDINAND MAGELLAN, in his journal, 1519.

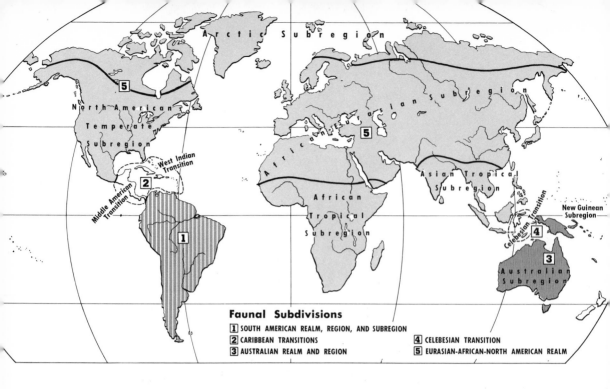

Faunal Subdivisions

1 SOUTH AMERICAN REALM, REGION, AND SUBREGION
2 CARIBBEAN TRANSITIONS
3 AUSTRALIAN REALM AND REGION
4 CELEBESIAN TRANSITION
5 EURASIAN-AFRICAN-NORTH AMERICAN REALM

vironments. But when comparing plant distribution with animal, we must take into account some additional considerations, heretofore ignored.

Just as faunal Realms, Regions, and Subregions are based on the distribution of the higher animals, so the floral subdivisions are based on the distribution of the higher plants. When comparing the faunal with the floral subdivisions, we are really comparing the geography of the vertebrates, or backboned animals, with that of the spermatophytes, or seed-bearing plants; and of course we are limiting ourselves to the living species, genera, and families of both animals and plants. This is not to say that the biogeographer always ignores invertebrates, lower plants, or extinct organisms; but these do not usually enter into the delimitation of the major faunal or floral subdivisions.

Modern vertebrates and modern spermatophytes, two groups so remote taxonomically, are vastly different in both their age and in their means of dispersal over the globe. The differences are obviously significant in biogeography. The older an organism, the longer time it has had to spread; the better it is at crossing barriers such as mountains, deserts, or seas, the greater chance it will have to reach all areas suitable for its continued existence.

Which are older, the modern plants or the modern animals? This question

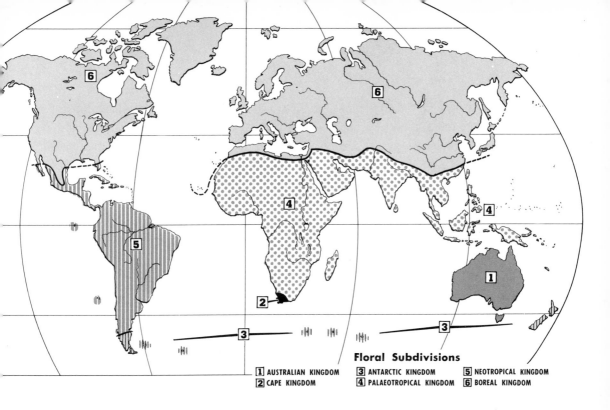

Floral Subdivisions

1 AUSTRALIAN KINGDOM	**3** ANTARCTIC KINGDOM	**5** NEOTROPICAL KINGDOM
2 CAPE KINGDOM	**4** PALAEOTROPICAL KINGDOM	**6** BOREAL KINGDOM

was strikingly answered by excavations into fossil beds that crop out along the Powder River Basin in eastern Wyoming. These beds have been explored by paleontologists—students of ancient life—for more than seventy-five years, and a great deal has been learned about the conditions under which they were deposited. At a very remote time, when dinosaurs still ruled the land, a great arm of the sea split North America into eastern and western portions, and the Wyoming beds were laid down along the shore of this sea as it slowly retreated. Some of the beds were formed in shallow sea water, or about brackish river mouths and estuaries, while others were deposited in freshwater swamps and stream valleys not far inland of an ancient seashore.

The fishes of the Wyoming beds mostly belonged to extinct genera if not families, although a sturgeon and a gar represented genera still living today. The frogs and salamanders represented extinct genera and sometimes families, except for one frog which perhaps has a living relative in the Philippines. Snakes, lizards, and turtles similarly were members of vanished genera and often families, with the exception of a lizard possibly belonging to a genus still present in North America. The alligators and crocodiles of the beds all belonged to extinct genera, the birds all to extinct families. The mammals were few, small, and

primitive; there were only marsupials, insectivores, and some strange little beasts —the so-called multituberculates—that left no descendants. Scattered throughout the deposits were the teeth and bones of dinosaurs, ancient and sometimes gigantic reptiles that had already dominated the world for 165 million years.

These beds were so old as to shed but little light on the zoogeography of modern vertebrates, yet their plants were largely of modern aspect. There were cattail, water lettuce, willow, lotus, grape, ash, sequoia, and black haw, along with close relatives of the poplar, magnolia, basswood, and dogwood, to name but a few familiar kinds.

When the fearsome *Tyrannosaurus,* greatest of the meat-eating dinosaurs, stalked through the old Wyoming forest in search of prey, it trampled herbs and shrubs and saplings that we could easily recognize today. When the duck-billed *Anatosaurus* emerged dripping from a lake, its scaly hide festooned with water lilies, it plowed its way through sedges, cattails, cannas, and willow bushes that would not seem out of place around a millpond or a park lake. The Wyoming beds were laid down near the end of dinosaurian reign, and the meek little mammals would soon take over the earth; but for a while longer, when the earth shook to the tread of reptilian feet, the insectivores and marsupials would cower beneath fallen timbers that we would probably call sycamores, magnolias, ashes, and oaks.

In short, the higher plants average much older than the higher animals, and have had far more time to disperse. They have had millions of extra years in which to bypass barriers such as mountain ranges, deserts, and seas. Indeed, some living plant genera and families date back to a time before most features of the modern landscape had even developed. Also, numerous higher plants apparently disperse across barriers more readily than do most higher animals.

Biogeographers have had chance to compare the dispersal ability of higher plants with that of higher animals through studies made on Krakatoa, a volcanic island lying between Java and Sumatra. In 1883 the island blew up with a force vastly exceeding that of any hydrogen bomb, a blast in the 10,000-megaton range. Six cubic miles of earth, with rock, trees, wild beasts, houses, and people, were blown into the upper atmosphere and drifted as dust for years, providing the world with unusually colorful sunsets. All that was left of Krakatoa was a peak covered with lava, pumice, and ash; life was exterminated there, with the possible exception of deeply buried roots and soil microbes. Other islands, miles

Pasque flower, state flower of South Dakota. Its genus, Anemone, *is widespread in temperate lands and on mountains in the tropics. (South Dakota Department of Highways)*

away, were also sterilized by the volcanic explosion; the nearest unsterilized land lay 25 miles distant. Biologists thereafter kept watch on Krakatoa to see how it was repopulated by plants and animals.

Nine months after the explosion a lone spider was found on the island, but in the absence of prey it was doomed. Three years after the blast, 11 species of ferns and 15 species of flowering plants had reached Krakatoa, and in another ten years the island was covered with greenery. There were coconut trees on the beach, stands of wild sugarcane farther inland, clumps of figs and papayas, and four species of orchids in the fast-developing forest.

The higher animals were far less successful in reaching Krakatoa. Twenty-five years after the eruption, the island harbored 263 species of animal life, but most of these were lower animals, especially insects. Four species of land snails had also arrived. The only vertebrates were 16 kinds of birds and two reptiles: a python and a monitor lizard, both excellent swimmers. In less than fifty years the island supported a fine tropical forest, and now harbored 47 species of vertebrates; but most of these latter were birds and bats, the only land mammals being two kinds of rats.

Biologists, noting the rate at which land vertebrates arrived, estimated that two or three million years might elapse before the island was repopulated by all the Indonesian animals it could support. Of course a stretch of ocean is but one of many possible barriers to distribution; but it is a relatively effective one, and the studies on Krakatoa permit us to generalize about the dispersal ability of the higher plants as compared with that of the higher animals.

10 *". . . it seems settled that inside of the Australian border-belt one finds many deserts and in spots a climate which nothing can stand except a few of the hardier kinds of rocks."* "Not a tree was seen, nor a shrub. . . . The only vegetation we met with was a coarse, strong-bladed grass growing in tufts, wild burnet, and a plant like moss, which sprung from the rocks."*** FLORAL SUBDIVISIONS OF AUSTRALIA, SOUTH-TEMPERATE AFRICA, AND THE FAR-SOUTHERN LANDS*

Modern phytogeographers divide the world into large floral units called Kingdoms, and Kingdoms into smaller units called Provinces. (These words may well be capitalized where they relate to formally defined areas.) The present chapter and several that follow list the Kingdoms and Provinces of the world, and describe some of the plants that characterize each. Also, attention is called to

* MARK TWAIN, *Following the Equator.* ** CAPTAIN JAMES COOK, on his discovery of South Georgia Island, 1775.

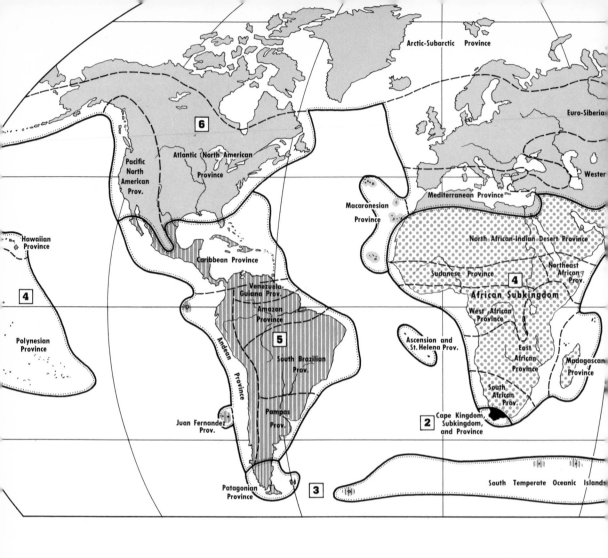

significant correspondences, or noncorrespondences, of plant distribution with animal.

There is an Australian Kingdom whose existence in part reflects the long separation of Australia from the remainder of the world. Surprisingly, this Kingdom, unlike the Australian Realm, does not include nearby New Guinea even though this island was once connected with the mainland. As we have seen, New Guinea has relatively heavy rainfall and a variety of tropical environments, while most of Australia has scanty to moderate rainfall and a series of temperate environments. New Guinea has been populated chiefly by plants coming from tropical Asia via Indonesia, a route not open to many land or freshwater vertebrates.

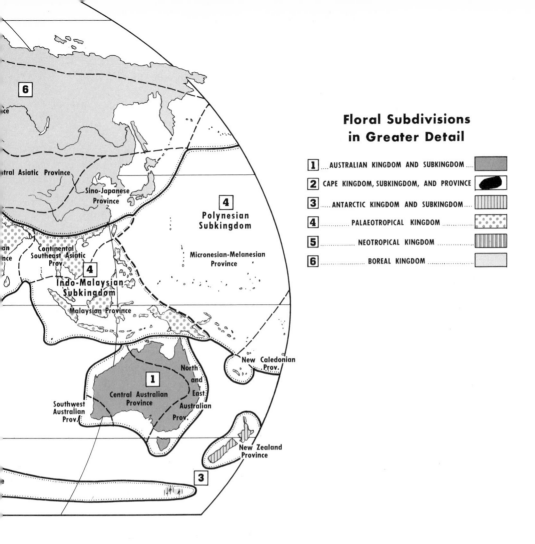

**Floral Subdivisions
in Greater Detail**

1 .... AUSTRALIAN KINGDOM AND SUBKINGDOM ....

2 CAPE KINGDOM, SUBKINGDOM, AND PROVINCE

3 ..... ANTARCTIC KINGDOM AND SUBKINGDOM ....

4 .... PALAEOTROPICAL KINGDOM ....

5 .... NEOTROPICAL KINGDOM ....

6 .... BOREAL KINGDOM ....

Also excluded from the Australian Kingdom are the smaller islands lying north and east of New Guinea, for they likewise support a flora that is essentially of Asian rather than Australian affinity. There has been some mingling of Asian and Australian plants in the East Indies, but there is no phytogeographic counterpart of the Celebesian Transition. Tasmania, just south of the Australian mainland, is the only sizable island that belongs to the Australian Kingdom. This Kingdom is therefore coextensive with the Australian Subregion of the zoogeographers.

The Australian Kingdom need not be split into Subkingdoms, but it has three distinct Provinces. The North and East Australian Province includes a strip of country roughly 200 to 300 miles wide, stretching eastward from Derby through Wyndham, Darwin, and the peninsulas that border the Gulf of Carpentaria, and thence southward through Cairns, Charters Towers, Brisbane, and Sydney to about Melbourne or Ballarat. Tasmania also belongs to this Province. Although including some scrubland and grassland, the North and East Australian Province encompasses most of the forests in Australia. Few areas of the Province support lush rainforest; most support a drier and more open forest, with trees whose hard, coarse leaves are resistant to desiccation and wind damage.

The Southwest Australian Province is small and includes only the southwestern tip of the continent, another area with a relatively dry and open forest. The rest of the continent, the driest part, forms the Central Australian Province. Here the vegetation includes grassland, with or without scattered trees, and various scrublands such as those locally known as mallee and mulga. (Australians apply the name "scrub" to some of their finest forests, but we shall use the word to mean a dry area with coarse shrubs and stunted trees.) Parts of this Province are almost bare of plants: the Nullarbor Plain supports only saltbush and bluebush; the flat, red, pebbly "gibber plains," widespread in the heart of the continent, are vegetated only along gullies where a few hardy trees may grow.

The Australian Provinces have not been erected because two are largely forested and one is not. A Kingdom, or any subdivision thereof, is recognized because various plant families, genera, and species are peculiar to it; not because the characteristic plants happen to group themselves into forests or grasslands or scrublands. The general nature of the vegetation is mentioned in order to give an idea of the local landscape, and to provide a clue as to why some animals should be confined to a floral subdivision. Also, the provincially restricted plant groups are not as interesting as the ones that range widely over the continent, giving the Kingdom its unity.

Plants of the Australian Kingdom have such engaging names as bimbil, wandoo, cootaminda, red tingle, willow geebung, tuckeroo, lillypilly, and bastard box; but these, along with most of the characteristic Australian plants, are not well known outside their own country, and need not be discussed. However, many trees and shrubs of this Kingdom have been carried to other lands as ornamentals, and a few of these may be cited.

The most typical plants of this Kingdom are trees of the genus *Eucalyptus*. Three out of every four individual trees in Australia belong to this genus, which includes more than 600 species, and which predominates in 95 per cent of

Australian forests. Within the genus, the diversity of species is impressive. Thus the coolabah, made famous by the Australian folksong "Waltzing Matilda," is a big *Eucalyptus* that provides welcome shade in the dry interior of the country; it withstands not only drought but also a range of temperatures from 120 degrees Fahrenheit to below freezing, and its dense wood weighs 89 pounds per cubic foot. In contrast, the snow gum is a *Eucalyptus* that grows on mountain slopes up to 5,000 feet, and that may have to withstand a hundred hard freezes each winter. The mountain ash is a *Eucalyptus* that grows to a height of 300 feet and a diameter of nine feet; it is the tallest hardwood tree in the world, our taller redwoods and Douglas firs being conifers. "Mallee" is a scrubland dominated by dwarf *Eucalyptus*, the individual trees being mostly below 20 feet in height, and sprouting amidst tussocks of coarse porcupine grass. All the species of *Eucalyptus* shed their bark. One species, the manna gum, sheds in such quantities that an acre of these trees may be littered with fifty tons of fallen bark.

A few species of *Eucalyptus* range northward into New Guinea or beyond, but only one of these has been able to invade the rainforest; this one has spread all the way to the Philippines. (Here one might draw a parallel with the arboreal marsupials called cuscuses. Adapted to rainforest life, they have spread farther into the East Indies than other marsupials, being found from northern Australia and the island of New Guinea to the Moluccas and the Celebes.)

Many species of *Eucalyptus* are grown in the United States, in parks, yards, and roadside plantings. This genus belongs to the myrtle family (Myrtaceae), which also includes the bottlebrushes and the cajeput tree, brought from Australia to enhance our gardens. The myrtle family has two centers of abundance, beyond which only a few genera range. These centers are Australia and the tropics of South America. This distribution calls to mind the marsupials, leptodactylid frogs, and chelid turtles, as well as the extinct horned turtles of the family Meiolaniidae. It is significant that the most characteristic Australian plants, the myrtles, are distributed about like the most characteristic Australian mammals, amphibians, and reptiles.

The acacias, with sunny yellow flowers and fernlike leaves, are familiar in the United States, but most of them are imports from Australia. The commonest plant community in Australia is the so-called mulga, a scrubland dominated by dwarf acacias; mulga covers about 30 per cent of the continent. Acacias belong to the genus *Acacia*, of the plant family Mimosaceae; there are about 600 species of the genus in Australia, while another 200 grow in other countries, especially Africa but also southern Asia, South America, and the warmer parts of North America. Some animal groups are distributed much like the acacias; for example the snake family Elapidae. A majority of the elapid genera are confined to Australia or nearby islands, but there are other genera in Africa as well as a few in southern Asia, South America, and the warmer parts of North America.

Australia harbors some floral oddities. Among them are the so-called beef-woods, horsetail trees, or "Australian pines." The last name is inappropriate for they are not pines at all; their branches are covered with leafless green stems which superficially resemble long pine needles. These curious trees belong to the genus *Casuarina*, a genus so distinctive that it is placed in a family of its own, the Casuarinaceae. A majority of the *Casuarina* species are confined to Australia, but some grow in the islands that lie between this continent and Asia. One readily takes root on beach dunes, and has made its way from Australia eastward into Polynesia and northward into Burma.

This latter species has also been spread widely by man, for use as an ornamental and for windbreaks. It now grows along miles of the Tamiami Trail in southern Florida; in parks and yards of San Diego; and along the once-popular Varadero Beach, east of Havana on the north coast of Cuba. It may be seen on Jamaican dunes where old Port Royal sunk beneath the waves; in plantings on the dry Caribbean island of Aruba; and about South American yards, parks, beaches, or botanical gardens, from Caracas and Georgetown to Recife and Rio de Janeiro.

Turning now to Africa, we have already stated that the southwestern tip of that continent is cooled and dried by the northwardly flowing Benguela Current, but animals really characteristic of the cool, dry south are so few that the area need not be recognized as a separate faunal subdivision. However, about 2,500 species of higher plants grow on or near the Cape of Good Hope, many of them being very distinctive endemics. The area, for all its small size, forms a Kingdom of its own. This Kingdom is generally called the South African Kingdom, but the name is a confusing one; not only does the phrase "South African" have several different geographic and political connotations, but there is also a South African Province belonging to a different Kingdom. It would be better to refer to a Cape Kingdom, extending from Capetown northward about to Clanwilliam and eastward to Port Elizabeth.

Left, a Protea, national flower of South Africa. The genus is restricted to tropical and southern Africa. Right, pink Watsonia at Caledon, South Africa. The species is confined to the Cape Kingdom. (South African Tourist Corporation)

The Cape Kingdom is bounded by deserts that develop north of Clanwilliam; and by tropical grasslands and forests that develop above Port Elizabeth, where a southwardly flowing current warms and moistens the land. Prior to industrialization, the Cape landscape was reminiscent of the southwestern United States, with dunes and deserts, high mesas, long and jagged ridges, and the dry beds of intermittent streams. There are rocky canyons where drought-resistant trees have gained a foothold, and parched grasslands that suddenly come to flower-starred life after the infrequent rains.

Floral similarities between the Cape Kingdom and Australia are greater than would be expected from their geographic separation. Many plants of the Cape Kingdom are familiar to us at least from illustrations in seed catalogues: Agapanthus or "blue lily of the Nile;" Anchusa, called summer forget-me-not; Arctotis, with its daisy-like flowers of cream, bronze, and violet; and Asparagus plumosus, miscalled a fern, whose delicate fronds go into the making of most bouquets. There is Dimorphotheca or star-of-the-veldt, with petals of white or salmon; and the Cape honeysuckle, draped with orange-red blossoms. Other garden plants from this Kingdom include Freesia, delightfully scented and blossoming in our winter; Gazania, an old-time favorite belonging to the daisy family; Gerbera or Transvaal daisy, a popular cut flower; some of the Geraniums; and Plumbago or leadwort. One of the most important plants of the florist trade, Gladiolus, is from the Cape Kingdom; and the garden plant commonly called Montbretia is a man-made hybrid between two kinds of Gladiolus from the Cape.

In the Cape Kingdom not only is rainfall scanty, but most of it is confined to the period from October to April. Many plants of the Cape must exist as underground bulbs or tubers during the drier parts of the year, sprouting or flowering only when the weather is favorably damp. Florists speak of the "Cape bulbs;" they include various species of Amaryllis, Ixia, Strelitzia, and Watsonia, as well as the hybrid Montbretia, the "Mexican" lily, the Kaffir lily, and the callas that are used as funerary offerings. (It should be borne in mind that many garden plants are called by wholly inappropriate names that obscure true relationships and geographic origins; that cultivated plants may exhibit characteristics not found in their wild prototypes; and that many showy flowers are man-made hybrids between species from widely separated areas.)

Although the Cape Kingdom has no counterpart in zoogeography, a few interesting animals are confined to the Cape and nearby country; and some of them, like the Cape plants, are strikingly modified for life where rainfall is scant and seasonal. Thus several burrowing frogs of the family Microhylidae live on or near the Cape; rarely seen above ground, they have no aquatic tadpole stage and cannot even swim. The frog family Heleophrynidae is restricted to more southerly parts of Africa; one of its species, the ghost frog, lives only about rocky brooks on Table Mountain at Capetown.

The southernmost floral subdivision of the world is the Antarctic Kingdom. It is named not for the continent of Antarctica but for its "anti-Arctic" position on the globe. It is surprisingly fragmented. Its New Zealand Province includes the two main islands of New Zealand, as well as the Kermadec, Chatham, Auckland, and Campbell islands; its Patagonian Province embraces Tierra del Fuego, southern Argentina, the southern Andes of Chile, and the Falkland Islands; its South Temperate Oceanic Islands Province includes small, little-known islands such as South Georgia, Marion, Tristan da Cunha, the Crozets, the Kerguelen Archipelago, Heard, McDonald, St. Paul, Amsterdam, and Macquarie. Antarctica and the South Shetlands are sometimes added to this last Province; they support but a few spermatophytes.

Plants of the Antarctic Kingdom are mostly unfamiliar to the average person, although New Zealand has provided our gardens with a Clianthus or glory-vine, a Fuchsia, several Veronicas or speedwells, the Olearia or daisy tree, and a Senecio or groundsel. Of special interest to phytogeographers are the trees of the genus *Nothofagus*. Called southern beeches, and related to our northern beech, they dominate the landscape in much of the Patagonian Province and again in New Zealand. They also occur as fossils on Antarctica, which today is too cold for them. It may be that some plants of the southern hemisphere spread widely by way of Antarctica, at a time before that continent had developed its mantle of ice. Southern beeches have also spread into Australia, New Guinea, and New Caledonia.

Characteristic of the Antarctic Kingdom are tussock-forming grasses of the genus *Poa*. These grasses dominate certain plant communities not only on New Zealand and Tierra del Fuego but also on many smaller islands of the Kingdom. A curious plant of this Kingdom is the so-called Kerguelen cabbage, *Pringlea antiscorbutica*. Its leaves and its thick rootstock are edible; and the plant has been a welcome addition to the diet of seal-hunters in southern waters. It be-

Above, a south-temperate
landscape reminiscent of
north-temperate ones: Lake
Wakatipu in The Remark-
ables, a mountain range
near Queenstown, South
Island, New Zealand. At
left, the Celmesia or mountain daisy from New Zea-
land. The genus links Australia and New Zealand.
(New Zealand Information
Service)

longs to the family (Cruciferae) that includes our familiar cabbage, mustard, turnip, and radish. The Kerguelen cabbage is found only on a few small, widely separated islands: Kerguelen, Heard, Marion, and the Crozets. Three higher plants are able to grow on Antarctica itself; two are grasses and one is a member of the daisy family. These plants are not endemic to Antarctica, but range also to various islands and into the Patagonian Province.

The Antarctic Kingdom is not even roughly equivalent to any faunal sub-division. Nevertheless, among the land and freshwater animals there are a few groups distributed about like certain Antarctic plants. For example, a troutlike fish, *Galaxias attenuatus*, inhabits streams of southern Australia, New Zealand, and southern South America; its family, the Galaxiidae, is represented in south-ern Australia, New Zealand, New Caledonia, Tierra del Fuego, the southern Andes, the Falklands, and the southern tip of Africa. A wingless fly, like a few higher plants, ranges from Antarctica to southern South America.

In a comparison of floral with faunal distribution, it is worth noting that Antarctica proper, with three land spermatophytes, harbors no true land verte-brates. The Antarctic birds and seals are essentially marine, deriving their sus-tenance from the sea. Rats, mice, and cats, introduced by man, have survived on some of the islands of the southernmost Kingdom, but not on Antarctica itself. Sledge dogs, abandoned by a Japanese expedition at Showa Station on the Antarctic coast, soon died off; and Mongolian ponies, brought by the Scott party, could never have survived unaided in the general absence of vegetation.

11 *"... it is all forest on the northern slopes of the mountain—open glade and miles of forest; ground at present all sloppy, oozes full and overflowing, feet constantly wet. . . . Rivulets rush with clear water . . . they flow northwards and westwards to the Chambezi."* "There is one island called Kankorogo. . . . The soil of the shores is highly ferruginous in colour, and, except in the vicinity of the villages, produces only euphorbia, thorny gum, acacia, and aloetic plants."*** **PLANT DISTRIBUTION IN AND NEAR TROPICAL AFRICA**

The next Kingdom to be considered is called the Palaeotropical. (Phytogeographers prefer the classical spelling "palaeo-" to the modernized "paleo-" favored by zoogeographers.) It includes the tropics of Asia and Africa, and deserts bordering these tropics, as well as certain islands lying in the South Pacific, the Indian, and the South Atlantic oceans.

The vast Palaeotropical Kingdom is split into three Subkingdoms: the African, the Indo-Malaysian, and the Polynesian. The African Subkingdom corresponds fairly well to the African Tropical Subregion of zoogeography, but is

---

* DAVID LIVINGSTONE, in his journal, 1867.   ** HENRY M. STANLEY, on his search for Livingstone, 1876.

114

larger; it includes not only the tropics of Africa proper but also the islands of the
Indian Ocean, as well as the deserts that lie at the edge of the tropics in both
Africa and Asia. The Indo-Malaysian Subkingdom approximately equals the
Asian Tropical Subregion plus the Celebesian Transition. These two Subkingdoms together are therefore to be compared with, but are more extensive than,
the African-Asian Tropical Region. The Polynesian Subkingdom has no counterpart in zoogeography, for most of its islands are small, with but few land vertebrates.

The African Subkingdom is in turn made up of eight Provinces:

The North African–Indian Desert Province includes the Sahara and the
deserts that stretch from Saudi Arabia to West Pakistan. As we have already
seen, these deserts serve as barriers between faunas. However, in plant distribution their role, or at least their major role, is somewhat different; they harbor a
variety of plants adapted to arid conditions, and so are listed as a Province in
their own right.

Toward the south, the climate of the Sahara becomes progressively less
harsh, and in this direction the desert vegetation gradually passes into grassland
or scrubland, and eventually into forest. Some plant groups, coming from the
true tropics of Africa, invade the desert for varying distances, depending on their
ability to withstand drought. Thus the Sahara is usually considered a part of the
African Subkingdom. (The Mediterranean lands of extreme northern Africa
belong to a different Kingdom, to be described later.)

The Red Sea and the Persian Gulf have not been important barriers to the
eastward or westward spread of desert plants, and so the North African–Indian
Desert Province includes arid lands in both Africa and Asia. The flora of this
Province is sparse. The few trees grow mostly around oases or along the valleys
of intermittent streams. Best known tree of the region is the date palm. There
are acacias in the desert, and a mistletoe lives parasitically upon them. One
hardy oak is found in the eastern part of the Province.

The Sudanese Park Steppe Province extends in a band across Africa south
of the Sahara, from Senegambia eastward through Mali to the lands of the
upper Nile. The Province might be thought of as a southern border to the
Sahara, with sufficient rainfall to lift the region out of the desert category and
into that of grassland. This Province is a land of open spaces and big game, with
acacias, cassias, grasses, and palms. The vegetation, naturally sparse, has been

rendered even more so by certain activities of man, especially burning and cattle-raising.

The Northeast African Highland and Steppe Province embraces most of Somalia and Ethiopia, as well as the island of Socotra; it also includes the southern tip of Saudi Arabia. This last area, in phytogeography as well as in zoogeography, is linked more closely with Africa than with the remainder of the Saudi Arabian peninsula. Only a few plants are really characteristic of this Province, and most of these are confined to the mountainous sections, or else to the island of Socotra off the tip of Somalia. The montane plants include some acacias, an aloe, and a *Euphorbia*; the insular ones include a tree-sized Dracaena and a familiar, blue-flowered plant known both to gardeners and to botanists as *Exacum affine*. But the most famous plant of the Province is coffee. Native to the mountains of Ethiopia, it was brought to Arabia in the sixteenth century, to Ceylon and the East Indies by the seventeenth century, and later to many parts of the New World.

The East African Steppe Province is a large one, including Kenya, Uganda, Tanzania (Tanganyika), Rhodesia, Zambia (Northern Rhodesia), and Mozambique, as well as parts of Angola and Southwest Africa. The Province is mostly a plateau with savanna vegetation, but there are also high mountains with other plant communities. Both the plateau and the mountains are of exceptional interest.

The savannas of this Province are the richest game fields of the world, and their richness is made possible by the plant life. The grasses which dominate the plateau country, the scattered trees, the roots and tubers beneath the ground—these support elephants, rhinos, zebras, giraffes, buffalos, and a variety of antelopes. Lions, leopards, cheetahs, and hunting dogs prey upon the grazers and browsers; hyenas, jackals, and vultures pick the bones of the dead. Native peoples aspire to live in an old way, or perhaps in a new; but neither way especially compatible with persistence of the game fields. There is no study more fascinating than the interrelationships of climate, soil, topography, plants, animals, and people in the East African Steppe Province.

The mountains of this Province are interesting for a different reason. In most tropical lands there are mountains whose peaks offer a temperate or even an arctic environment; but such mountains usually are parts of long ranges that extend into temperate regions, and it is not surprising to find a temperate flora on peaks of this kind even though the nearest lowland may be tropical. But the mountains of the East African Steppe Province are somewhat isolated from all other ranges. Ancient and formerly volcanic, they rise above a tilted and dissected plateau; and their highest peak, Kilimanjaro, reaches 19,340 feet. Yet on the upper slopes of these mountains the flora includes many plants that we would call "northern": sanicles, pimpernels, thistles, globe thistles, strawflowers, bedstraws, Saint-John's-worts, buttercups, clovers, stonecrops, heaths, and geraniums. A remarkably similar flora might be found growing in, say, parts of New England or southeastern Quebec. The upper slopes of these African mountains are also noteworthy for their tree groundsels, tree lobelias, and tree heaths. Groundsels and lobelias are widely distributed throughout the world, but most of them are small herbs; and the heaths, also widely distributed, are usually low shrubs. Yet on these peaks all three groups are represented by tree-sized species, and botanists have not agreed on why this should be the case.

The biogeographer is impressed by the fact that a number of plant families, otherwise made up of low-growing herbs or shrubs, are represented by tree-sized giants on various remote islands, as well as on mountain peaks in several tropical lands. The cold upper slopes of these mountains are also remote islands in a sense, for they are isolated from other areas of comparable climate by wide stretches of tropical lowland. Both the peaks and the oceanic islands could support a variety of trees, but are so isolated that they have not been reached by many trees of more usual kinds; in such places the role of trees is left vacant for the taking. The biogeographer suggests that herbs and shrubs will then take over the role, evolving into treelike species; just as on islands that harbor no land mammals, the birds may become large and flightless, taking over the otherwise vacant mammalian role. At any rate, there is no stranger landscape on earth than the upper slopes of these African mountains, where tree groundsels sprout like a forest of lances, and tree lobelias rear their cabbage-like tops. Soggy moss covers the rocks and the tree trunks; cold mists drift ghostlike, changing sometimes to snow. There is no bird call, no chirp of insect, no voice of man—only the incessant drip of icy water, forever condensing upon the plants, the boulders, and the ground. This East African Steppe Province has no counterpart in zoogeography. It harbors a few characteristic animals, but these live mostly in the mountains; the beasts of the plateau usually range far to the north and to the south, wherever there is extensive grassland.

The West African Rainforest Province extends from the Congo Basin westward to Cameroon; and from the latter it reaches still farther west, but only as a narrow coastal strip, to Guinea. The narrow western extension of this Province is not of especial significance in phytogeography, but it has been of great importance in human history: it is the old Slave Coast, the Ivory Coast, the Gold Coast, the Grain Coast. The early attention given to this wooded shore may perhaps account for the widespread but erroneous belief that Africa is mostly rainforest or jungle. Actually, there is but little African rainforest outside this Province; and there are far greater expanses of rainforest in Asia and in the New World.

This Province, like rainforest regions generally, has many trees that are valued for timber or for rubber. Better known plants of the Province include Liberian coffee, the African oil palm, a raffia, Senegal mahogany, and a glory-bower. A curious fruit, *Blighia sapida,* named for Captain Bligh of *Bounty* fame, is native to this Province; but it is now more widely grown in the West Indies, where it is called the akee. A fine flowering maple, which is not a maple but a mallow, comes from offshore islands of the West African Rainforest Province.

Many plant groups, widespread throughout the tropical forests of the world, are rather poorly represented in Africa; notable examples are the orchids, the bamboos, and the palms, as well as the so-called aroids or members of the arum family. In fact the whole tropical flora of Africa is unexpectedly scanty, the rainforest component especially so. This is in sharp contrast with the temperate flora of the Cape Kingdom, which is extremely rich in proportion to the small size of the Kingdom. It seems likely that, in fairly recent times, drastic shifts of climate have exterminated many plant species of tropical Africa; and that the Sahara, vast and inhospitable, has slowed reinvasion by plants—especially rainforest plants—coming from the Asiatic tropics.

Some African mammals, birds, reptiles, and amphibians live only in rainforest, and so are characteristic of the West African Rainforest Province. Many of these are adapted for life high in the treetops, or else in the leaf litter of the forest floor. Certain forest animals, largely restricted to this Province, live also on wooded mountain slopes in eastern Africa; and this widely disjunct distribution supports the belief that lush forests were once more widespread in Africa than they are today.

The South African Province includes the "high veldt"—the high and nearly treeless grasslands—of Orange Free State and the Transvaal, as well as the Kalahari and Karroo deserts. The Province is not sharply distinct. It has received some plants from the East African Steppe Province and others from the Cape Kingdom. The flora of the South African Province is not especially rich. Plants characteristic of this Province are, in the main, unfamiliar to any but the specialist. However, one typically South African genus is widely known; this is *Mesembryanthemum,* whose numerous species are called fig marigolds and ice plants. They are often grown indoors in the United States, and do well outdoors in Florida and California. Other greenhouse or garden plants coming originally from this Province include the blue-flowered *Heliophila,* which belongs to the mustard family; *Gasteria* and *Lachenalia,* of the lily family; and *Eriocephalus,* a shrubby, evergreen member of the daisy family. The silver tree, *Leucadendron,* whose dried leaves are used for decoration, is another native of the South African Province; it belongs to the interesting family Proteaceae, which, although widely distributed, is best represented in southern Africa and eastern Australia.

But the most remarkable plants of this Province come from the Karroo Desert. Some of these, like the stone plants (*Lithops*), resemble pebbles scattered over the desert sand. Others have fleshy stems, often protected by sharp spines. The Karroo plants are commonly grown in "cactus" gardens, but they are not cacti; they belong to other families such as the spurges, the milkweeds, the carpetweeds, and the orpines. Even the lily family and the daisy family have produced cactus-like species in the Karroo. The peculiarities of these desert plants are simply adaptations for storing water and protecting it from thirsty animals. As so many Karroo plants have evolved extreme adaptations to desert existence, one supposes that this region has been arid for a very great length of time.

Few animals are confined to the South African Province, but some are limited to an area roughly equalling this Province plus the Cape Kingdom. For example, the frog family Heleophrynidae, already mentioned in connection with the latter Kingdom, is distributed only from Capetown to northern Transvaal. Several other frog groups are similarly restricted.

The Madagascar Province includes the islands of the Indian Ocean: Madagascar, the Comoros, the Seychelles, and the Mascarenes. They went unclassified in zoogeography, for their fauna was an odd mixture of African and Asian groups, along with other groups that were endemic—strictly confined—

At left, the red-billed hornbill of southern Africa. Right, the lilac-breasted roller, a beautiful species of southern Africa. (South African Tourist Corporation)

to one or more of the islands. Their flora is an even more puzzling mixture. Seven families and about two hundred genera of plants are endemic to these islands. Many other families and genera, although not confined to the Province, nevertheless are represented therein by endemic species. In fact about 85 per cent of the local plant species do not range outside the Province. However, many of these have close relatives in other parts of the world, not only in Africa but also in Asia and Polynesia, even Australia and the Americas.

The flora of the Madagascar Province is rich, made up of about 6,000 species. Nearly half of these belong to families and genera that are widespread in both the Old World and the New. Such plants do not ally the islands with any particular continent. Roughly 1,500 species of the Province either range into Africa, or else have their closest relatives there. On these grounds the Province is appended to the African Subkingdom. These may seem scant grounds, but the islands at least have more floral affinities with Africa than with any other region.

We have mentioned that continental Africa has surprisingly few bamboos, orchids, aroids, and palms. But the bamboos are well represented in the Madagascar Province; orchids are fairly well represented, and many of the local species have Asian relatives; some of the Madagascan aroids have relatives only in the New World. The palm family is amazingly diverse on Madagascar and even on the smaller islands; and palms of the Province all find their nearest relatives in Asia or Australia, not Africa. One palm of the Province, the so-called "double coconut" (*Lodoicea*), is extraordinary in both appearance and distribution. It reaches a height of one hundred feet, and its nut—really a seed, the largest produced by any plant—resembles two gigantic coconuts grown together. These curious nuts, rotting and incapable of germination, have drifted even to Asia; they were well known, and called "coco de mer," long before anyone discovered that the tree itself grew only on the Seychelles.

A better known floral oddity of Madagascar is the traveler's-tree, a relative of the banana. Its great leaves hold rainwater in basal receptacles, and the liquid is said to be a boon to thirsty travelers. A climbing frog lives only in these receptacles, a circumstance to intrigue the biogeographers but not, perhaps, the travelers. The traveler's-tree belongs to a genus (*Ravenala*) with only two species; one is Madagascan, the other South American. This distribution seems strange but is not particularly significant. A great many plant families and genera have made their way to all the tropics of the world, perhaps even ranging into adjacent temperate regions as well. Some of these groups, such as the yams, morning glories, Eugenias, rose mallows, Peperomias, strychnines, and vanilla orchids, still live in every tropical land. But many other groups, having reached all the tropics, then died out from several regions, leaving just a few species here and there. Perhaps *Ravenala* had such a history.

The family Didiereaceae, endemic to Madagascar, is remarkable in that its members are cactus-like in appearance. And there are some cactus-like milkweeds on Madagascar, just as in the South African Province. Although most plants of the Madagascar Province are unfamiliar to us, the Madagascar jasmine (*Stephanotis*) is widely grown in Florida. It is not really a jasmine but a climbing member of the milkweed family. The royal poinciana (*Delonix*) may have come from Madagascar; but this beautiful flowering tree was spread so widely by man that its original homeland cannot be determined with certainty.

We come now to the last, and the smallest, of the Provinces that make up the African Subkingdom. This is the Ascension and St. Helena Province, named for two little islands far out in the South Atlantic. These islands are unimportant in zoogeography; their only native land vertebrates were a few birds. Ascension never harbored many plants, but St. Helena supported a fairly rich flora in proportion to the small size of the island. Both the animals and the plants of this Province were mostly exterminated by man, or by the asses, pigs, and goats that he introduced. A century ago, the botanist Joseph Hooker estimated that one hundred species of plants had vanished from St. Helena before they could be studied. The remaining species, about forty in number, mostly belong to genera that are also represented in southern Africa. There are, however, five endemic genera on the island, and only one of these has African affinity; the other four find their nearest relatives in South America. Three of these endemic genera belong to the daisy family. This family, of worldwide distribution, is made up largely of herbs, with a scattering of vines and shrubs; but the species of the St. Helena genera are woody and tree-sized. This island is another remote spot where the role of trees has been taken over by members of a predominantly herbaceous family.

Today on St. Helena, slim Araucaria trees grow in the shaded glen where the exiled Napoleon once meditated. The three peaks of the island—Actaeon, Cuckold, and Diana the Huntress—are again clothed with vegetation; small birds flit through the grass and underbrush. On Ascension, a great volcanic cone at last merits its name of Green Mountain, for its cinder-soil has come to support a considerable growth of plants. But most plants and animals now living in this Province were imported, intentionally or accidentally, from other parts of the world.

The foregoing account has mentioned plants characteristic of various Provinces within the African Subkingdom. It must not be forgotten that the Subkingdom is unified by many plants that range into most or all of these Provinces. Notable examples of widespread African plants are a Celosia, a Clematis, a calla, the scarlet-flowered Kalanchoe, and the familiar castor bean. Accepting a variety of environmental conditions, these plants not only range widely in nature but have also become garden favorites in many lands. One would not often expect a single species of plant to grow in all the different environments of the African Subkingdom; but a genus of plants might include a number of species, each adapted to somewhat different conditions of soil, topography, and climate. Genera, and of course families, are therefore likely to be widespread. Numerous genera unify the African Subkingdom; and some other genera, like the well-known *Coleus* and *Sorghum*, are represented in the tropics of both Africa and Asia, thus helping to unify the Palaeotropical Kingdom. Plants really widespread throughout the Palaeotropical Kingdom are, in the main, kinds that can withstand drought; for most of these plants had to pass through the arid North African–Indian Desert Province in order to attain a Kingdom-wide distribution. This Province has served as a filter barrier between the African and the Indo-Malaysian Subkingdoms, hindering the passage of any but drought-resistant species.

12 *"After go men by sea to the land of Lomb. In that land groweth the pepper in the forest that men call Comba. . . . For it is a good country and a plentiful, but there is overmuch passing heat."** **PLANT DISTRIBUTION IN AND NEAR TROPICAL ASIA**

Two Subkingdoms of the Palaeotropical Kingdom remain to be discussed: the Indo-Malaysian and the Polynesian. The Indo-Malaysian Subkingdom is smaller than the African, and is divided into but three Provinces; nevertheless it has a richer flora than the African, in proportion to its geographic extent. The Subkingdom includes the greater part of India, and most of the large peninsula occupied by Burma, Thailand, Indochina, and Malaya, as well as a coastal strip along southern China. This continental portion of the Subkingdom corresponds rather well to the Asian Tropical Subregion of zoogeography. However, the Subkingdom encompasses far more islands than does the Subregion: not only Ceylon, Sumatra, Java, Borneo, the Philippines, Hainan, and Formosa, but also the islands of the Celebesian Transition from the Celebes through New Guinea. Little islands off the coast of India—the Maldive, Laccadive, and Chagos archipelagos—also belong to the Indo-Malaysian Subkingdom. The Ryukyus, lying between tropical Formosa and temperate Japan, in phytogeography are grouped with the former.

Some moderately well known plants, widespread over much or all of the

* "SIR JOHN MANDEVILLE," probably from the fourteenth-century account of Friar Odoric.
124

Indo-Malaysian Subkingdom, include the East Indian ironwood, source of the supposedly antirheumatic nagkassar oil; the valuable timber tree called teak; and the mango, a popular fruit in the tropics. Also familiar are the elephant's ears, prized by us for the ornamental leaves but by Asiatics mainly for the starchy tubers; and the trees of the genus *Hydnocarpus,* whose poisonous fruits yield an oil used in Asia for the treatment of skin diseases. The northeastern boundary of the Subkingdom is not sharply defined; for along the eastern coast of China there is a relatively gradual climatic transition, and a correspondingly gradual floral transition, from tropical to warm-temperate. A number of plants are largely confined to the Indo-Malaysian Subkingdom, but also range some-what farther north into China, and often into the Japanese islands. Among such plants are the woody vines known as herald's trumpets; the loquats, grown both as ornamentals and for their fruits; and the garden shrubs called Daphniphyllum. Others include the Michelias or banana-shrubs, related to the magnolias; the paper mulberries, widely planted in the United States; and the camellias that are stocked by so many nurseries.

The Indo-Malaysian Subkingdom lies at the heart of the Palaeotropical Kingdom. Many plant groups of this Subkingdom are also found to the east, in Polynesia, and to the west, in Africa. Such groups unify the entire Kingdom. Examples include the genus *Asparagus,* some species of which are grown as ornamentals and others as vegetables; the genus *Boswellia,* one species of which yields frankincense; and a genus of mistletoes (not the genus represented in the United States, but a member of the same family.) The Indo-Malaysian Sub-kingdom is diverse within itself, and is clearly divisible into three parts. One is essentially Indian; a second includes most of the Asiatic mainland tropics east of India; the third is made up chiefly of tropical islands from Sumatra and the Philippines to Timor and New Guinea.

More specifically, the Indian Province of the Indo-Malaysian Subkingdom includes southern India and the Malabar Coast, the Deccan and the Ganges Plain, the southern flanks of the Himalayan uplift, the island of Ceylon, and the little islands from the Laccadives through the Chagos Archipelago. The Indian Province is not sharply distinct; it has been invaded by plants coming from the north, the west, and the east. Endemic genera are comparatively few, and mostly clustered at opposite ends of the Province: either the Himalayan slopes or the island of Ceylon. The flora of this Province is not especially rich.

Early studies such as that of Joseph Hooker, on the plant life of "British India," might suggest a very rich flora; but British India was long taken to include Burma and the Malay Peninsula, which do harbor a large number of plant species.

It is surprising how many characteristic plants of the Indian Province have proven to be of commercial value, and so have been carried to other lands. It may be that the people of this crowded region were unusually inclined to experiment with the possibilities of the local plant life. At any rate, India has given the world jute, a major fiber; a Crotalaria valued as a cover crop; cardamon, both a spice and a medicine; and a fruit yielding chaulmoogra oil, once used in the treatment of Hansen's disease. It has given us indigo, a dye plant formerly grown in the southeastern United States; black pepper and long pepper; Indian millet; and sesame. Locally valued, if less familiar to us, are the loofah, whose fibrous fruit-skins may be used as dishrags; the curry leaf, a prized ingredient of some curries; a small herb whose extract is used to perfume the hair; lingoum, a dyewood; and a grain called eleusine. Also characteristically Indian is the toddy palm or wine palm; its flower-sheaths can be tapped to yield as much as 12 gallons of liquid in 24 hours, a liquid that is potable before or after fermentation.

Rice, dhall, horse gram, asparagus bean, rice bean, amaranth, eggplant, rat-tailed radish, taro yam, cucumber, mango, tree cotton, hemp, and gum arabic, if not originally endemic to India, at least were first brought under cultivation there. A few plants of the Indian Province are often raised in gardens or greenhouses in the United States. Here may be listed the musk mallow, a Datura, "Spanish" jessamine, a Vitex, and a fine, large morning glory. Three popular orchids—a Cymbidium, a Vanda, and a Dendrobium—are also from India. The fig genus, *Ficus*, has two interesting species that are characteristic of the Indian Province. One of these is the great banyan tree, a floral oddity often mentioned but seldom seen. The other is the best known of all indoor plants, the rubber plant; in the United States it grows in pots along halls and around lobbies, but in India it is a forest tree reaching a height of one hundred feet. (House plants needing little sun are likely to be from forests where the ground is deeply shaded.)

The inclusion of Ceylon in the Indian Province is somewhat dubious. The local flora shows a strong affinity to that of the Indonesian islands. It is true, however, that the more common and conspicuous plants of Ceylon grow also in India, or else have close relatives there. The island has many endemic species, the most famous being the hard, black wood called ebony. The plant family Dipterocarpaceae, widespread in the Old World tropics, is represented by several endemic genera on Ceylon. Some of the dipterocarps (the name means "two-winged seed") are typical trees of the rainforest. The trunk, supported by great buttresses, may rise one hundred feet before branching into a leafy crown. The friut, borne high in the air, releases winged seeds that drift like moths; many seeds will fall to earth at some distant point, where the seedling will not have to grow in competition with the parent tree. The smaller islands of the Indian Province, from the Laccadives through the Chagos Archipelago, are low coral atolls, so monotonous of environment as to support but few plants.

Some families of animals live only in the Indian Province. Among them are the shield-tailed snakes (Uropeltidae) and a gigantic crocodilian called the gharial, placed in a family of its own (Gavialidae). Certain more widespread families—for example the cichlid fishes, the caeciliid amphibians, the colubrid snakes, the platanistid dolphins, and the lorisid primates—have endemic genera in India. It is curious that India proper does not harbor an assemblage of highly distinctive plants, comparable to the vertebrate oddities. In this connection it is significant that the more striking animal endemics are largely restricted to relatively inaccessible habitats. Thus the caeciliids, legless and superficially worm-like, tunnel in damp ground or humus; the uropeltid snakes burrow through ground debris in mountain forests; the slender loris prowls only at night, and then in the treetops; the gharial and the Ganges dolphin swim in large rivers. This restriction of the animal endemics and the scarcity of plant endemics suggest that a part of the flora and fauna has been exterminated by the activities of man; for the country has been long and heavily settled, the landscape altered by cutting, burning, and intensive agriculture of several kinds. India is in fact one of the most habitable portions of Asia; it supports about a fifth of the world's population, in an area roughly half the size of the United States.

India is geographically important—it is often called a subcontinent—and merits additional description. To a remarkable degree it is isolated by natural barriers: the Himalayas to the north; other steep mountain chains to the east, toward Burma; high mountain walls to the west, in West Pakistan and Afghanistan; and the ocean elsewhere. (Barrier regions are excluded from the Indian Province, but are conveniently discussed here.) The hills and mountains of the

The gharial, a distinctive reptile confined to India. (Harold R. Piel)

east, and the eastern part of the Himalayas, are rainy and heavily forested. The greater part of the Himalayas does not belong in the Palaeotropical Kingdom; for tropical climate and vegetation do not climb very far up the flanks of the towering ranges. The Himalayan flora and fauna have their affinities toward central China and Japan. The western Himalayas are drier than the eastern. The uplifts of West Pakistan and Afghanistan are very dry, sparsely vegetated; they are linked both florally and faunally with Saudi Arabia and the Sahara, not with the Asiatic tropics.

A distinctive part of India, lying within the Indo-Malaysian Subkingdom, is the great plain that constitutes the northern part of the country. It bears the broad valleys of the Indus, the Ganges, and the Brahmaputra river systems. This area is well watered on the east, in Assam and the lower Ganges Valley, but becomes progressively drier toward the west. The rivers are fed primarily by meltwater from the Himalayan snows. In the spring the rivers overflow their banks, to deposit mud and silt; much of the plain is floored by alluvium—water-deposited soil—hundreds of feet thick. But when the snows have finally melted, most of the streams will go dry or nearly so, at least in their upper reaches. Such environmental conditions make for a comparatively sparse flora and fauna.

The remainder of the country is the plateau that forms the Indian peninsula. Lying well inside the boundary of the Indo-Malaysian Subkingdom, it is very diverse within itself. Of interest are the Satpura Mountains, which trend from the northeast to the southwest. Forested and well watered, they have been a highway whereby plants and animals of the wet tropics have been able to move across India. There is a wet, forested area, hilly in part, along the west coast of the peninsula also. The northern portion of the plateau has a moderate rainfall and some forests; but the high southern portion is comparatively dry and barren. The peninsula is the only part of India lying below the Tropic of Cancer; but the configuration of the land, the proximity of a warm sea, and the direction of the winds all help to bring tropical or nearly tropical conditions northward to the upper Ganges and the eastern Punjab. The tropics extend a little farther north in India than anywhere else in the world. Off the south coast of India is the island of Ceylon. Its fauna, unlike its flora, is typically Indian, without Indonesian affinity.

The vegetation of India is geographically diverse just as is the climate. Thus there is a dry, warm desert in the lower Indus Valley; a dry, warm short-grass prairie in the upper Indus; a savanna, with grasses and scattered trees, over much of the plateau; a tropical forest along the west coast from Bombay southward, as well as in eastern Bengal and southern Ceylon; and a mangrove swamp along much of the coast.

Indian animals are also distributed in spotty fashion. To consider some of the nonendemic species: the tiger characterizes the forests of the Sundarbans, the Terai, central India, and the Western Ghats. The sloth bear is a forest animal; the sun bear is Himalayan, living mostly above 5,000 feet. The wild dog inhabits Assamese forests. The one-horned rhino is found in the northern alluvial plain, the two-horned rhino in Assam and thence southward. Various monkeys and deer are mostly forest animals, and so is the leopard; but the last also dwells in the canebrakes along the Ganges. Wild elephants still live in the forests of eastern Bengal and Assam, in Ceylon, and (very sparingly) in the peninsula. The gaur, a huge member of the cattle family, is met with in the hill forests of southern India, the wild buffalo in Assam. In the open grasslands and semi-deserts are to be found the cheetah, wolf, hyena, jackal, and wild hog. An antelope lives on the coastal flats of Gujarat and Orissa, a wild ass in Sind and Cutch. The high, dry country of the west harbors ibex, wild sheep, and wild goat.

Indian birds are similarly restricted to certain vegetation types; but the most familiar kinds—several parrots, the mynas, the peacock, pheasants, and the jungle fowl—all come from the forests. In general, India's nonendemic animals of the dry habitats are distributed across southwestern Asia and sometimes into northern Africa; those of the damper habitats, especially rainforest, are distributed toward Indochina, the Malay Peninsula, and sometimes the larger islands of northern Indonesia.

Now let us move to another floral Province, the Continental Southeast Asiatic, which includes most of the continental Asiatic tropics lying east of the Ganges delta, say from about Dacca in East Pakistan to Yungkia (Wenchow) in China. Hainan, Formosa, the Ryukyus, the Andamans, and the Nicobars also belong to this Province. However, the southern half of the Malay Peninsula is excluded, its floral affinities being with the Indonesian islands. This Province is not sharply differentiated, for a floral transition takes place within its borders, from temperate China to the tropical islands of Indonesia. However, it is clearly distinct from the Indian Province.

Endemic families of the Continental Southeast Asiatic Province are few, endemic genera not especially striking or familiar. Most species of the Province belong to genera represented also to the north and to the south, less frequently to the west.

A good many garden or crop plants probably originated in or near the Province: purple Bauhinia, the camphor tree, Begonia rex, "Cape" jasmine, a ginger lily, scarlet Ixora, crepe myrtle, red banana, "Rangoon" creeper, "Ceylon" Sansevieria, and a Torenia or wishbone flower. (Unfortunately, it is not always possible to determine the exact original range of a cultivated plant; many kinds were spread widely by man in prehistoric times, and some—for example rice and bananas—probably were brought under cultivation at several different times and places.)

In southeastern Asia, the flora of the mainland tropics is far less interesting than that of the Indonesian islands. The reverse situation obtains with the fauna. The Asiatic tropical fauna is rich on the mainland, somewhat less so on the nearer islands, and much reduced on the more distant islands. This difference between floral and faunal distribution should come as no surprise; other Kingdoms have already shown us that plants have a special ability to reach islands, and to produce distinctive endemics thereon.

The Malaysian Province includes the tropical islands of the Southwest Pacific, from the Philippines southward through Timor, and from Sumatra eastward through New Guinea. Appended to this Province is the southern portion of the Malay Peninsula, northward about to the latitude of Kota Bharu. A strip along the north coast of New Guinea, roughly from Hollandia to Madang, might perhaps be removed from the Indo-Malaysian Subkingdom and added to the Polynesian; but the circumstance is not important here. There are extensive grasslands, usually with *Eucalyptus* or other trees of the myrtle family, in Timor and drier parts of New Guinea. Otherwise, most of the Malaysian Province is heavily forested.

More than 2,000 plant genera have been reported from the Province, and about 40 per cent of these are endemic or nearly so. The total number of species is uncertain, but may be near 40,000. New Guinea alone harbors about 9,000, with 90 per cent of them endemic. Borneo and the Philippines also have large floras, with much endemism. The orchid family is especially noted for endemism in this Province; it includes more than 1,000 species confined to the Philippines, and more than 2,500 confined to New Guinea.

Among the endemics of the Malaysian Province are the durian tree, whose spiny fruit is relished by man and beast in spite of its foul odor; the breadfruit, now widely grown in several tropical lands; and a rattan palm whose dried resin is known as dragon's blood. Other endemics include taro, whose sprouts and starchy tubers are edible; sago, likewise yielding starch; the mangosteen, a delightful fruit; and an ornamental Acalypha. We might also list a scarlet-flowered glory-bower; a shrub yielding the purgative croton oil; a giant bamboo; "Chinese" hibiscus; and Codiaeum (miscalled "croton"), prized for its variegated leaves. Originally endemic to this Province were cloves, nutmeg and mace (both from the same fruit), and ginger.

A famous endemic genus of the Malaysian Province is *Rafflesia*. One of its species, *Rafflesia arnoldi*, produces the largest flower in the world, a fleshy blossom three feet across, weighing fifteen pounds. Not only is the plant a floral oddity; it also illustrates how the distribution of a species may be limited by a complicated life history. The Rafflesia seeds are sticky, and are transported unintentionally by birds and mammals; they have been found adhering to the feet of elephants and to the beaks of seed-eating birds. Somewhere in the forest, a Rafflesia seed is rubbed off against a partially exposed root. The seed adheres to the root and eventually forms a parasitic union with it. Not all roots are equally acceptable as hosts; most individual Rafflesia plants are parasitic upon large grapevines. The great flowers of the Rafflesia open directly from the root of the host plant; they smell like rotting flesh, and are pollinated by flies and carrion beetles. Thus the distribution of this Rafflesia is limited not only by temperature and rainfall, but also by the presence of suitable hosts, seed transporters, and pollinators.

Some islands of the Malaysian Province have high peaks, which offer a temperate environment on their upper slopes. In such places are to be found plants of temperate affinity, especially blackberries, buttercups, and violets. Among both the plants and the animals of the Province, continental affinities are strongest in islands nearest the mainland. Otherwise, faunal distribution does not especially parallel floral distribution within the Province. New Guinea has many endemic animals, but most of them have relations to the south, in Australia; while the plants of New Guinea, endemic or not, generally have their closest relations in some other part of the Indo-Malaysian Subkingdom. The central islands of the Malaysian Province are somewhat transitional between eastern and western portions of the Indonesian archipelago; but they are not transitional between Australia and Asia, as they are in zoogeography.

Even ignoring problems of faunal transition, animal distribution is complex in the Malaysian Province. A few islands are inhabited by distinctive animals with no close relatives anywhere in the world. Among them are the pit-shell turtle, the only member of a family (Carettochelydidae) confined to New Guinea; and the earless monitor lizard, only member of a family (Lanthanotidae) restricted to Borneo. The cobegos, strange gliding mammals miscalled "flying lemurs," are found in Sumatra, Java, Borneo, and the Philippines; but one species ranges somewhat beyond the confines of the Province, into the upper Malay Peninsula and Thailand. Some animal families, present on the Asian mainland and perhaps elsewhere, are represented in the islands by endemic species or even genera. Reptilian examples include the Komodo "dragon," a monitor lizard of the Lesser Sundas; and the sail-tailed lizard, an agamid which ranges from the Philippines through the Celebes and the Moluccas to New Guinea. Birds include the great argus pheasant of Sumatra, Borneo, and the Malay Peninsula; also the black cockatoo and the crowned pigeons of New Guinea; and the maleo fowl of the Celebes. Among mammals, the black "ape," anoa, and babirusa are confined to the Celebes; a dwarf water buffalo and several rats to the Philippines; the proboscis monkey and giant forest hog to Borneo. Various tree shrews are restricted to Borneo, Sumatra, and the Philippines; the siamang, a small ape, to Sumatra; and the orangutan to Borneo and Sumatra. Finally, many animals range over parts of the Asiatic mainland and also reach some of the East Indian islands. Thus the Indian elephant and Malayan tapir reach Sumatra; one rhino ranges to Sumatra and Borneo, another to Sumatra and Java; the tiger reaches Sumatra, Java, and Bali.

The complexities of animal distribution in the Malaysian Province have largely resulted from the geological history of the region. A strong faunal similarity from one land mass to another usually reflects a former continuity of land; and conversely, a marked faunal difference usually reflects the long existence of a saltwater barrier. But in contrast, plants have dispersed readily across the seas and straits of the Province; and the degree of floral similarity from one place to another is likely to be correlated with the degree of environmental similarity.

13 *"Here and there you come upon a bank of sand exceeding fine and white, and these parts are the least productive. The plants (such as they are) spring from and love the broken coral, whence they grow with that wonderful verdancy that makes the beauty of the atoll from the sea. The cocoa-palm in particular luxuriates . . . ."** **PLANT DISTRIBUTION IN AND NEAR POLYNESIA**

The Polynesian Subkingdom is the third and last major component of the Palaeotropical Kingdom. This Subkingdom extends westward through the Palaus, the Bismarck Archipelago, the Solomons, and New Caledonia; eastward through Easter Island and Sala-y-Gomez; northward through the Hawaiian group; and southward through Norfolk Island and Lord Howe (but not including New Zealand or the Kermadecs). The Subkingdom is, accordingly, very extensive, but its actual land area is relatively small. A great many islands of the Subkingdom are low atolls with a limited flora mostly of widespread, salt-tolerant plants. Other islands of the region are made up in part of former coral reefs, now elevated well above the sea; and such islands may support a larger flora than the atolls. Yet other islands are craters and rocky upthrusts of volcanic origin; these offer a greater diversity of environments and harbor a greater variety of plants.

* ROBERT LOUIS STEVENSON, *In the South Seas.*

134

The Subkingdom is made up of four Provinces: the Hawaiian, the New Caledonian, the Polynesian, and the Micronesian-Melanesian. (Phytogeographers know two of these Provinces as, respectively, the "Region of New Caledonia" and the "Region of Micronesia and Melanesia;" but a slight alteration of this terminology is called for.) Few plant genera are to be found in all four Provinces. In this Subkingdom, made up of widely scattered islands, locally endemic genera and species considerably outnumber the more widely distributed ones. The palm, madder, ginseng, rue, Euphorbia, and dogbane families, especially, are represented by genera endemic to just one or a few islands. Within this Subkingdom, a genus may be represented in two, and occasionally three, of the Provinces. For example, one genus is present in New Caledonia and the New Hebrides; a second in New Caledonia, Fiji, and Polynesia proper; a third in Hawaii and the Societies; a fourth in Hawaii and Fiji. The Subkingdom is unified in this fashion, rather than by any large number of widespread genera.

The Micronesian-Melanesian Province includes the Carolines, the Marianas, the Bismarck Archipelago, the Solomons, the New Hebrides, Fiji, and the tiny islands from the Marshalls to the Ellice group. Plants of this Province are also conspicuous along the north coast of New Guinea, from about Hollandia to Madang; and so a small bit of New Guinea may be bracketed with the nearby Bismarck Archipelago. The fine, large islands along the southwestern border of this Province, islands such as New Britain and Bougainville, have not been intensively studied. Further botanical collecting may modify opinion on just where and how the Province passes into the adjoining Malaysian one. Of islands lying farther east in the Micronesian-Melanesian Province, the most interesting is Fiji. It has an endemic plant family, the only such in the Province; and it also has about a dozen endemic genera of more widespread families, especially the palms and the madders. (The palm family is notable for its production of very distinctive endemics on remarkably small islands.)

The Polynesian Province includes the islands from Samoa through the Societies and the Marquesas group, and thence to Easter Island and Sala-y-Gomez. These are mostly small, low islands without much diversity of habitat. The flora is correspondingly limited, with less endemism than in any other Province of the Subkingdom, and with a preponderance of species that can grow on or near beaches. The daisy family has produced a genus of large, woody species in Polynesia proper; called *Fitchia*, it is represented on Tahiti, Rarotonga, Raiatea, Mangareva, and Rapa. This last island has several interesting endemic species, among them a blueberry, a Coprosma, and a Corokia. Easter Island, near the eastern border of the Province, has about thirty native species of plants, only one or two of them endemic.

The New Caledonian Province includes New Caledonia, Lord Howe Island, and Norfolk Island. New Caledonia supports one of the most remarkable island floras in the world. The madder, myrtle, orchid, and ginseng families are especially diverse on New Caledonia; but the daisy and grass families, usually conspicuous on islands, are poorly represented. Trees and woody shrubs are abundant, herbaceous plants few. Over one hundred genera are endemic to the island, some so distinct as to constitute endemic families. Norfolk and Lord Howe are both small, and offer a limited series of environments as compared with New Caledonia. Nevertheless, each of these smaller islands has a sizable flora with interesting endemics. Both Norfolk and Lord Howe have some floral relationship with Australia and New Zealand, but not especially with New Caledonia.

The inclusion of Norfolk, Lord Howe, and New Caledonia in one Province is more a matter of convenience than of plant distribution. The three islands are similar in that each seems to be quite old and long isolated; each has, in proportion to its size, a large flora with marked endemism. Most plants of the New Caledonian Province are not well known. The Norfolk Island "pine," a 200-foot tree in its homeland, is often grown as a pot plant in the United States; its genus, *Araucaria*, is represented by additional species in Australia, New Guinea, and Chile. A gigantic species of fig is also endemic to Norfolk. Two palms from Lord Howe Island are widely used as ornamentals; known commonly as Kentias, they actually belong to the genus *Howea*. The genus is endemic to Lord Howe.

The Hawaiian Province is of special interest, and not simply because Hawaii has become a state of the Union. The Hawaiian islands are diverse. Some are small and low, others large and mountainous. Except in the lee of high mountains, rainfall is heavy; temperatures are mild and equable. Conditions would seem ideal for the development of a rich and complex flora, were not the archipelago so remote: it lies in mid-Pacific, about 2,400 miles from California and 3,900 miles from Japan. The Hawaiian islands are, in fact, the most isolated ones in the world; what plants grow upon them?

Coconut trees abound on the coast of rugged Tahiti, in the heart of the Polynesian Province. (Matson Lines)

Tree-sized ferns are conspicuous in parts of Hawaii. However, they are lower plants, and we are here concerned only with spermatophytes.

Some of the commonest Hawaiian plants belong to genera that are widely distributed over many lands. For example, the official flower of Hawaii, the ohia, belongs to a genus (*Metrosideros*) that is also represented in Australia, New Guinea, New Zealand, New Caledonia, Indonesia, various Polynesian islands, and the Cape of Good Hope. The kolea, once used as a source of red dye, belongs to a genus (*Suttonia*) that is also present in New Zealand. The ape-ape, with great, umbrella-like leaves, belongs to a genus (*Gunnera*) that is found not only in Hawaii but also in South America, New Zealand, Tasmania, Indonesia, Madagascar, and continental Africa. The pukiawe bush, with berries of red or white, belongs to a genus (*Styphelia*) that is predominantly Australian.

Gardeners may know the genus *Acaena*, of the rose family; one of its species, the New Zealand bur, is grown not for its flowers but for its colorful, spiny fruits. This genus, present in Hawaii, is likewise found in South America, New Zealand, the South Temperate Oceanic Islands, New Guinea, and California—an amazingly disjunct distribution. The genus *Pittosporum*, which includes several common garden shrubs, is distributed almost as strangely as *Acaena*. Species of *Pittosporum* are found in many tropical and temperate portions of the Old World, and on such islands as New Zealand, New Caledonia, New Guinea, Madagascar, the Philippines, and Madeira in the Atlantic, as well as on Hawaii. The cigar flowers are also well known to gardeners; their genus (*Cuphaea*) is largely confined to North and South America, but with one outlying species that ranges to the Galapagos and Hawaii. The sandalwood genus (*Santalum*), with seven species on Hawaii, is known also from Australia, Indonesia, Polynesia, New Caledonia, and the Juan Fernandez group off the coast of Chile.

Hawaii has no endemic plant families; but about forty genera, belonging mostly to widespread families, are confined to the islands. The daisy family has produced three Hawaiian genera of relatively large, woody plants. One of these genera, *Argyroxiphium*, includes the silversword, perhaps the most spectacular shrub of Hawaii. The whole silversword plant is covered with silvery hairs, except for a great stalk of yellow blossoms. This species grows only in the vast crater of Haleakala, a 10,000-foot volcanic cone on Maui; but additional species of the genus inhabit other islands of the Hawaiian group. The lobelia family is well represented in Hawaii, where four endemic genera are found. Some of the

The Hawaiian islands offer a diversity of environments. Above, a coastal scene near Waipio Valley on Oahu. Below, the silversword, Hawaii's most spectacular shrub. It belongs to the daisy family. (Hawaii Visitors Bureau)

Hawaiian lobelias are woody shrubs rather than tender herbs. Among other families, the violets, the campions, and the Geraniums have all produced woody giants in these islands.

Hawaii has high peaks, some of them snow capped. (Mauna Loa and Mauna Kea tower nearly 14,000 feet above sea level.) The upper slopes of the mountains support a few plants belonging to families and genera characteristic of temperate climates. Many plant groups, widespread in the tropics and perhaps elsewhere, are poorly represented in Hawaii, if indeed they are present at all. Examples include the banana family, the orchid family, and the lily family. (Of course we are not concerned here with introductions by modern man.) Plants that grow on or near beaches are likely to reach far-flung islands; but such plants are not especially numerous on Hawaii.

Now let us try to explain the peculiarities of the Hawaiian flora. The Hawaiian islands, of volcanic origin, are not extremely ancient, at least not as geologists and biogeographers would reckon age. Plants have had no vast length of time to reach this remote archipelago. The ones that did reach there belonged, in many cases, to genera whose members were especially well adapted for wide dispersal. Once established in Hawaii, these plants gradually evolved certain adjustments to the local environment, altering in size or leaf shape or flower color. In other words, over the millennia these plant immigrants often evolved into endemic species. Some of the immigrants, perhaps the earliest arrivals, became so modified that botanists would now place them in genera of their own. It is of course possible that some endemic genera of the Hawaiian archipelago were once widespread, but died out everywhere except on these islands where they were protected from certain more vigorous competitors.

Some genera diversified within the archipelago, a circumstance not surprising when one recalls that the islands are scattered over a distance of 1,600 miles, and that they differ among themselves in available habitats. The plant species of Hawaii, roughly 1,800 in number, could have been derived from no more than 275 ancestral stocks. The flora of the islands could easily be accounted for by postulating the arrival of just one plant immigrant every 20,000 years! Needless to say, the actual arrivals were not neatly spaced throughout the millennia; but it does seem clear that chance dispersal, by wind or water, satisfactorily explains the present existence of a fairly large flora in Hawaii.

Which figured more prominently, wind or water, in bringing seeds and other viable plant parts? As mentioned, plants characteristic of beaches, or of coastal dunes, are not numerous on Hawaii; and this circumstance suggests that ocean currents were not especially important in carrying plants to these islands. Typhoons may have blown far more plant seeds to Hawaii than ever the seas washed up. Whence came the original plant immigrants? Only a few of the Hawaiian plants show affinities to the north. There are affinities toward Polynesia and New Zealand, more toward the warmer parts of the New World, and still more toward the lands of the Southwest Pacific and the Indian Ocean. Thus the Hawaiian flora may have accumulated from several directions.

Within the Polynesian Subkingdom, land or freshwater vertebrates are few, and so in this region there are few really well-marked patterns of animal distribution. The patterns that are discernible usually transgress the outlines of the plant Provinces, and are better discussed without reference to these floral subdivisions.

Animals of Australian affinity are numerous on New Guinea, whence they range eastward in rapidly decreasing numbers. For example, New Guinea has fifteen genera of marsupials, thirteen of them present also in Australia; of the New Guinean genera, four range eastward to New Britain and three of these latter to New Ireland. Of the New Ireland genera, one reaches eastward to the Solomons, and none beyond. In like fashion other groups of New Guinean animals straggle eastward. Samoa is the easternmost island to harbor at least a few animals clearly of New Guinean derivation; these include insectivorous and fruit-eating bats, a scarlet bird called the cardinal honey eater, and a small boa. The tooth-billed pigeon of Samoa belongs to an endemic genus, and is not closely related to other pigeons.

From a standpoint of zoogeography, the more southerly islands of the Polynesian Subkingdom, from Fiji and Tonga through New Caledonia to Norfolk and Lord Howe, might be considered together. Not that these islands have species in common; rather, the general area harbors a small but puzzling assemblage of animals. There are a few faunal affinities toward South America, best exemplified by the iguana of Fiji and Tonga. A galaxiid fish on New Caledonia has relatives in South America and New Zealand. The kagu of New Caledonia may find its nearest kin in the sun bittern of South America. Fossil remains of terrestrial horned turtles, discovered on Lord Howe and Walpole, hint that the surviving oddities of the region are relics of a larger fauna.

There are also tantalizing hints of former land connections involving the more southerly islands of the Polynesian Subkingdom. The composition of the New Caledonian flora is about what one might expect in a continental area; and a volcanic rock called andesite, thought by geologists to originate from continents and not from sea bottoms, crops out from the Bismarcks and the Solomons through Fiji and Tonga to New Caledonia, the Kermadecs, and New Zealand. But this fascinating topic of former land connections in the Southern Hemisphere will not be considered at length here; for it involves many regions and not just a few islands of the Polynesian Subkingdom. Let us turn instead to the fauna of the Hawaiian archipelago.

The native vertebrates of Hawaii, other than essentially marine species, include only a bat and some land birds. The bat belongs to an American genus, *Lasiurus*. The land birds include some probably of American ancestry: a thrush, a goose, a crow, a hawk, a duck, an owl, a stilt, a gallinule, and a heron. A flycatcher is of Polynesian ancestry; some honey eaters have relatives scattered from New Zealand through New Guinea to Bali. A Hawaiian rail is of uncertain affinity. There is also a family of birds, the Drepaniidae, endemic to Hawaii; called Hawaiian honey creepers, they probably evolved from some ancestor of American origin. These drepaniids, so distinctive and diversified in Hawaii, may be descended from the first bird stock ever to reach the islands. Evidently birds, like plants, reached Hawaii from several directions. It is interesting to recall that plant seeds have been found adhering to the beak, feet, or plumage of birds. A few seeds may have reached Hawaii in such a way. Of course birds commonly spread plant seeds in their droppings or in regurgitated pellets; but the avian digestive process is so rapid (about a half hour to four hours) that seeds must rarely cross a wide expanse of ocean in the stomach or intestine of a bird.

As for Micronesia, its fauna, limited at best, is richer toward the west. This is not surprising, for the Palaus, at the western edge of Micronesia, are separated by only 500 miles of water from New Guinea, or from the Moluccas, or from the Philippines. The Palaus have one ranid frog, very similar to a New Guinean species. This island group also has an abundance of land birds. A bittern, a gallinule, an owl, a nightjar, and a warbler are of Asian derivation; while an osprey, an incubator bird, two rails, a pigeon, a kingfisher, a cuckoo shrike, and a whistler came from New Guinea, either directly or by way of the Philippines. Four Palau birds—two more pigeons, a starling, and a honey eater—are widespread among the islands of the Polynesian Subkingdom. East of the Palaus, the land and freshwater bird fauna dwindles rapidly. The Palaus have about 32 species, the Marianas 21, Ponape 18, Kusaie 10, the Marshalls only 3.

Several widespread bird genera have produced endemic species in western Micronesia, and a few birds of these islands are so distinctive that each merits a genus of its own. Some of the most distinctive endemics are confined to the Palaus (a warbler), the Marianas (a honey eater), Truk (a flycatcher), and Kusaie (a rail). The spectacle birds (Zosteropidae), widespread in the Old World tropics, are represented in western Micronesia by an endemic genus; this genus includes one species each on Palau, Yap, Truk, and Ponape. When winter comes to temperate Asia, some of the Asiatic birds migrate southward, remaining for months on the islands of western Micronesia before flying north again. Other Asiatic birds simply pass through western Micronesia on their migration to lands farther south.

Within the Polynesian Subkingdom, the poorest land fauna is that of the small islands lying east of Samoa. Only a few land birds range as far eastward as the Societies. One that does so is the familiar barn owl, a virtually cosmopolitan species. No land or freshwater bird reaches Easter Island. Seabirds, notably terns, albatrosses, shearwaters, petrels, storm petrels, tropic birds, boobies, and frigate birds, nest on various islands of the Polynesian Subkingdom; but they derive their sustenance from the ocean, and their distribution is better studied in connection with marine zoogeography.

Shorebirds, such as plovers, curlews, godwits, sandpipers, stilts, and phalaropes, are often conspicuous about islands of the Polynesian Subkingdom. However, most of these birds are migrants, moving from northern Asia or Alaska to some wintering ground in Australia or New Zealand, and utilizing the islands as way-stops. No shorebird nests in Micronesia, and not many anywhere in the Polynesian Subkingdom. (Hawaii has an endemic stilt; a New Guinean plover nests also on New Ireland; and a thick-knee, widespread in the Malaysian Province, lives also on New Caledonia, the Bismarcks, and the Solomons.) Various rats and mice, and some small lizards, are to be expected almost anywhere in the Polynesian Subkingdom. No doubt they have at times drifted from island to island on logs and on rafts of floating vegetation. However, they have also been spread widely in the vessels of seafaring Pacific islanders; and of course they are carried about by modern shipping. It is hard to determine the original range of these little stowaways.

14 *"Delight is a weak term to express the feelings of a naturalist who for the first time has wandered by himself in a Brazilian forest. The beauty of the flowers, the glossy green of the foliage, but above all the luxuriance of the vegetation . . . ."** **PLANT DISTRIBUTION IN AND NEAR BRAZIL**

The next Kingdom to be considered is the Neotropical, that of the New World tropics and nearby warm-temperate areas. The Neotropical Kingdom includes all of South America except the Patagonian Province, which belongs to the Antarctic Kingdom. From South America the Neotropical Kingdom extends northward to embrace the tropics of Central America and Mexico, as well as the West Indian islands. It also includes southern Florida, the Bermudas, the Galapagos Islands, the Juan Fernandez group, the southern half of the Baja California peninsula, and the Mexican east coast northward about to the Texas border.

The Neotropical Kingdom corresponds roughly to the South American Realm plus the Caribbean Transitions; but the Kingdom extends farther northward in North America, while the Realm extends farther southward in South America. About 3,000 plant genera are not represented outside the American

* CHARLES DARWIN, in his journal, 1832.

144

tropics. Roughly a third of these are distributed throughout the Neotropical Kingdom, and help to unify it.

Widespread Neotropical genera, at least moderately familiar, include *Caryocar*, whose species yield timber, souari nuts, and a vegetable oil; *Lycaste*, ground-growing orchids; and *Maranta*, its species raised for colorful leaves or gathered for starchy tubers. Also familiar are *Monstera*, ornamental climbers; *Ochroma*, with but a single species, source of the very light balsa wood; *Oncidium*, tree-growing orchids; *Tecoma*, the trumpet creepers; and *Theobroma*, one species of which yields cocoa. Almost as widespread as the above are the genera *Cattleya*, *Laelia*, and *Miltonia*, which include some of the finest orchids.

Cecropia, an interesting genus of the mulberry family, is widespread in the Neotropical Kingdom. The Cecropia trunk is hollow, and cavities extend also into some of the branches. Certain ants live only in the hollows of a Cecropia tree. Here they find a secure home, and they may also feed upon an exudate from the plant. The fiercely stinging ants defend the tree against various enemies; most importantly, perhaps, against leaf-cutter ants which might otherwise defoliate the branches. Plant collectors in the New World tropics, accustomed to climbing for orchids and air plants, soon learn that Cecropias and fiery ants always live together, be it in the West Indies, Mexico, Central America, or South America!

The Neotropical Kingdom is made up of seven Provinces, and the vast equatorial forest of South America is parceled out over three of them. These three are the Amazon, the Venezuela-Guiana, and the South Brazilian Provinces. The three are not sharply distinct from each other; they have many plant families, genera, and species in common. The Amazon Province does not include the entire Amazon Basin, with its far-reaching tributaries; rather, it is a strip about 500 to 700 miles wide, following the Amazon valley all the way from the eastern slopes of the Andes to the Atlantic coast. To the north of this Province is the Venezuela-Guiana Province, which includes the Venezuelan uplands and the Orinoco Basin. To the south of the Amazon Province, the South Brazilian Province includes the east coast, eastern highlands, and central uplands of Brazil, as well as the Gran Chaco of Paraguay and northern Argentina.

The Amazon Province is dominated by the great river for which it is named. The magnitude of this stream is staggering. It is 3,700 miles long; twelve of its tributaries are each at least 1,000 miles long. There is 14 times as much water in the Amazon system as in the Mississippi; in fact, one-fifth of all the running fresh water in the world is in the Amazon. An island in the mouth of this river is larger than the combined states of Massachusetts, Connecticut, and Rhode Island; and the Amazonian flood is recognizable in the Atlantic Ocean 200 miles offshore. Even 2,000 river miles upstream, the Amazon is two miles wide and 500 feet deep, flowing at a rate of five to seven miles per hour. The Amazon Province is the only Province, anywhere in the world, to have developed along a single river.

The great river affects plant distribution in several ways. The Amazon system is subject to flooding, different parts of it at different times of year. The water may rise 40 feet or more, inundating vast areas. Consequently the flora of the Province includes a large number of plants that are aquatic, or else that are able to withstand periodic floodings. The distinctiveness of the Province is due largely to the relative abundance of aquatic or flood-tolerant species. It must not be assumed, however, that the entire Province is subject to flooding. High areas are scattered throughout the Amazon Basin, and some of them are quite large. Many are forested but some support a savanna vegetation, a grassland dotted with palms. Toward the eastern half of the Province, especially, there are large ranges of hills covered with savanna.

All along the river, large sections of bank may be undermined by storms and floodwaters; often whole trees are borne swiftly downstream. Usually the drifting tree carries a burden of vines, orchids, air plants, and ferns, with perhaps some arboreal animals as well. The Amazon also may bear great, matted rafts of floating vegetation, accumulated in lakes and bogs but later washed into the river. The rush of water, from the Andes to the ocean, must constantly disseminate seeds and other viable plant parts, thus helping to unify the flora of the Amazon Basin. (But this is not to say that the plants of the Amazonian lowland are disseminated only by flood waters.)

Characteristic plants of the Amazon Province include the cacao, a small tree whose seeds yield the cocoa and chocolate of commerce; Pará rubber, belonging to an endemic genus (*Hevea*); and the Brazil nut, a common large tree of the forest. Locally important are a *Paullinia* vine, whose seeds are rich in caffeine; several catbriers of the genus *Smilax*; a species of *Couma*, whose milky latex is not made into rubber but is drunk as a beverage; and the ubauba, an edible fruit of the genus *Pourouma*. Among the aquatic plants is the largest of water lilies, *Victoria regia*, whose great, floating leaves will support the weight of a small child. A variety of palms also characterize the Province. Among them are an *Astrocaryum* and an *Attalea*, fiber-yielding palms with edible fruits; a species of *Leopoldinia*, from whose ashes a salt substitute is extracted; and the bussu palm, whose flower-sheaths furnish material for a coarse cloth. Of exceptional value to the Indians is the ita palm; it yields firewood, ornaments, starch, fiber, edible fruit, leaves for basketry, and sap for fermentation into wine.

The above list of typical Amazonian plants is short, but not because the Province is lacking in endemics. While not many plant genera are confined to the Amazon valley, at least 3,000 species are so restricted. Most of these, however, are familiar only to botanists. (And to Amazonian Indians, who rely on the forest for food, drink, seasonings, weapons, ornaments, clothing, shelter, timber, firewood, canoes, musical instruments, household utensils, paints, oils, glue, pottery glazes, varnishes, soap, rubber, cosmetics, insecticides, arrow poisons, fish stupefiers, basketry, cordage, fiber, intoxicants, perfumes, medicinal herbs, unguents, aphrodisiacs, and contraceptives.)

Turning now to faunal distribution, a great many animals, like numerous plants, are widely distributed over the Venezuela-Guiana, the Amazon, and the South Brazilian Provinces. There are some animals, for example a dolphin with saltwater kin, that are confined to the Amazon Province. However, the most interesting animals of the Province—the large, aquatic ones—usually range also into the Orinoco drainage, and less frequently into the Rio de la Plata. Such distribution may seem odd, but a detailed map will reveal certain tributaries of the Amazon to interdigitate closely with those of the Rio de la Plata; while the Amazon and Orinoco drainages are actually connected by the Casiquiare River, which periodically reverses its flow between the two.

Aquatic animals inhabiting both the Amazon and the Orinoco include a freshwater dolphin without marine kin; a manatee or sea cow; and the black caiman, a huge relative of the alligator. Various remarkable fishes are distributed in this way: some stingrays, without close kin in the sea; the lau-lau, a thirteen-foot catfish reputedly dangerous to man; the arapaima, a harmless fish nearly as large as the lau-lau, with a rasplike tongue used by the Indians to grate tubers; and the nine-foot electric eel, which can deliver 500-volt shocks at a frequency of 400 per second. Amazonian–Orinocan reptiles include the giant anaconda, bulkiest of snakes, a predator upon caimans; and the social river turtle, a pelomedusid whose eggs are devoured by man and beast alike.

Some aquatic animals of the Amazon range not only into the Orinoco but also into the Rio de la Plata. Among them are the ariranha, a giant river otter; and piranhas, notoriously carnivorous fishes some of which are dreaded by bathers. On the other hand, the little catfishes called candirús, even more dreaded for their habit of entering the urinary tract, are found in the Amazon and the upper Rio de la Plata, but not the Orinoco. A species of crocodile is confined to the Orinoco; while the Rio de la Plata has an endemic catfish larger than the lau-lau, and a small dolphin that may occasionally venture out to sea.

Many large Amazonian tributaries are whitish and turbid, with a burden of clay from the uplands where they arise. Other tributaries, such as the Rio Negro, arise in swamplands and are limpid, but stained by plant acids to the color of strong tea. There are also smaller tributaries, some rushing out of the hills and others meandering through the lowlands. A network of channels winds through the flooded forest; and there are also lakes, backwaters, lagoons, overflow pools, swamps, marshes, bogs—the greatest diversity of aquatic habitats in the world. And so within the Amazon Province is to be found the most impressive assemblage of freshwater animals in the world, although a majority of the species range also into at least one adjoining Province.

Returning now to plant distribution, the Venezuela-Guiana Province has not been thoroughly explored from a botanical standpoint; nor are many of its known endemics very familiar to us. The Amazonian forests extend into the southern part of this Province, but in a northwardly direction the forest becomes more and more frequently interrupted by expanses of savanna. The northern part of the Province is mostly a grassland dotted with small palms. Many endemic species of the Province are either grasses or palms.

The most fascinating part of the Venezuela-Guiana Province lies in southeastern Venezuela, near the borders of both British Guiana and Brazil. During the early nineteenth century, explorers in this region glimpsed from afar a fantastic mountain of pink rock, rising straight out of the forest and vanishing in the clouds. Streams dropping from unseen heights above made thin, vertical lines of silver over the cliff face. The Arecuna Indians, who lived not far away, called the mountain Roraima. As time went by it became evident that the top of Roraima, more than 9,000 feet above sea level, was a flat expanse of perhaps 25 square miles. A tall, slender pinnacle, Töwashing, was barely separated from the southern tip of Roraima. Nearby was another mountain, Kukenam, with a smaller top than Roraima but equally high. Other table mountains were scat-

Truly a freshwater whale: Inia, *the white dolphin of the Amazon and Orinoco.*
(Pamela Cheatham)

tered over southeastern Venezuela, all rising abruptly out of forest or grassland. Mount Duida had a flat top 250 square miles in extent, and Chimentá-tepuí was even larger. All the mountains were difficult of access, and most of them appeared to be unscalable, although Roraima had a promising ledge that sloped upward from the bottom of the mountain to the mist-wrapped top.

These mountains were of hard sandstone, and stood upon even harder pedestals of volcanic rock. Some of the mountains rose from rolling plains, others from a much higher hill country; yet the flat tops of the mountains were all at very nearly the same elevation above sea level. Geologists thought that the region had once been a high plateau. Over the ages this plateau had eroded down by thousands of feet, but the harder parts had remained as table mountains. In short, the tops of these mountains represented all that was left of an extremely ancient land surface.

The mountains, or rather their inaccessible tops, captured the popular fancy. Arthur Conan Doyle used Roraima as a setting for *The Lost World*. In this book he populated the mountain top with living dinosaurs; a winged reptile roosted atop Töwashing. (And the irascible professor, George Edward Challenger, who guided Doyle's fictional expedition to Roraima, was a caricature of Edward Drinker Cope, the eminent anatomist, paleontologist, and zoogeographer.) William Henry Hudson, born and raised in South America, wrote of this same region in *Green Mansions*. The sylphlike wild girl of Hudson's story was an allegorical figure; but her homeland of "Riolama," somewhere in the mountains of southeastern Venezuela, was Roraima. (Hudson, incidentally, was a naturalist as well as a perceptive novelist; and his books, fiction or otherwise, provide unexcelled descriptions of South America.)

Eventually, scientific expeditions reached the top of Roraima, Duida, Chimantá-tepuí, and other table mountains of Venezuela. The scientists found no dinosaurs, and if they found lovely wild girls they kept the matter quiet; but they did find an unearthly landscape, and plants belonging to previously unknown genera. Today we know that the top of each table mountain is cold and damp. (A night temperature of 47 degrees Fahrenheit has been recorded atop Chimantá-tepuí.) The surface is heavily fissured, littered with boulders, sculptured by wind and rain into miniature tablelands. Deeply carved hollows are almost cavelike, thickly overgrown with ferns and mosses. Low herbaceous plants cluster around ponds and bogs; shrubs and stunted trees grow here and there.

Many plants of the Lost World are dwarfed and strangely shaped, with unusual leaves and odd flowers. Often the leaves are covered with gray or brown fur. Much of the conspicuous vegetation belongs to the daisy, madder, melastome, or heath families; but other groups are also represented, among them the orchid, sedge, bladderwort, Rapatea, pipewort, yellow eyes, Saint-John's-wort, Bonnetia, and ginseng families. One member of the daisy family is woody and tree-sized, reminiscent of an African tree groundsel. Several species of pitcher plants trap and eat the local insects. Cherry, holly, huckleberry, black haw, Saint-John's-wort, and *Carex* sedges lend a "northern" aspect to the flora, although similar plants grow also in the high Andes.

The really striking endemics grow mostly on the flat tops of the mountains, but even the sides and bases have endemic species of plants. The Lost World has also yielded some interesting club mosses, ferns, insects, millipedes, centipedes, spiders, earthworms, snails, lizards, snakes, birds, and small mammals. There was even a catfish in a river atop Chimantá-tepuí, and a species of venomous lancehead snake was found to live only about this mountain. A black frog, found crawling over wet rocks atop Roraima, proved to represent an endemic genus of the family Dendrobatidae.

Finally we come to the South Brazilian Province, a large subdivision and one offering a variety of environments. Tongues of Amazonian forest enter the Province from the north, mostly along river valleys. There is also a narrow coastal strip of rainforest, roughly from Recife to Florianópolis. Just back of the coast are forested mountains, with some peaks exceeding 9,000 feet. At high elevations the forest is dominated by conifers, the Araucarias, rather than by hardwoods. Inland of the mountains is a vast savanna. The northeast corner of the Province is covered with scrubland, as is the Gran Chaco region.

Endemism is not marked in the South Brazilian Province, a majority of the local genera being represented elsewhere in South America. Of the genera that are endemic to the Province, only one is especially familiar; this is *Arachis*, whose members are called groundnuts. One species of *Arachis* is the peanut, an important crop plant. A good many widespread genera do have distinctive species within the South Brazilian Province. Some of these endemic species are well known: pineapple, ipecac, wax palm, yerba maté or Paraguay tea, the fiber called "Pará" piassaba, passion fruit, quassia or bitterwood, and the pineapple guava. Garden plants from the Province include an Aphelandra, a dwarf Begonia, the Bougainvillea vine, a large Datura, and a scarlet Fuchsia. Also familiar are a passion-flower vine, scarlet sage, *Sinningia speciosa* (the Gloxinia of florists), the purple-flowered Tibouchina, and the Allamanda. We might also add a Bauhinia or orchid tree, the coral tree, a vervain, a feather palm, a Geonoma palm, the black rosewood, and several orchids. Many greenhouse plants of the Bromelia family come from this Province; among them are species of *Aechmea*, *Billbergia*, and *Cryptanthus*.

Some Atlantic islands are appended to the South Brazilian Province. These include small, hilly Fernando de Noronha, and the high, mountainous Trinidade (not Trinidad off the Venezuelan coast). Neither island harbors endemic genera, but each has a few endemic species of plants, the most interesting being a rock-dwelling Combretum of Fernando de Noronha. More remote are the St. Peter and St. Paul Rocks, and the three islets of Martin Vaz; but these are barren upthrusts without higher plants.

The animals of the South Brazilian Province, like the plants, usually range into other parts of South America. From a zoogeographic standpoint, the most interesting part of the Province is the mountainous region near Rio de Janeiro; it has yielded a number of distinctive frogs and other animals. Animals of this Province, adapted for life only in savanna or scrubland, may be limited in their northward spread by the forest of the Amazon Province; and similarly, animals of the Venezuela-Guiana savanna may be limited in their southward spread by these same forests. (Even the Amazon itself is a formidable barrier to cross, what with its mighty current and its extraordinary assemblage of aquatic carnivores.) In other words the great Amazonian river system has been a highway for the dispersal of animals that live in forests, swamps, and aquatic situations; yet the same system has been a barrier to the dispersal of animals restricted to drier habitats.

The general region of the Amazon drainage may once have been an even more important barrier separating the northern from the central part of South America; for the present river system is thought to have developed in the gradually uplifted basin of an inland sea. Some zoologists believe that the ancestors of the Amazonian stingrays, and of the Amazonian dolphin with marine kin, lived in this sea, adapting to decreasing salinity as the basin was slowly elevated.

The more remote islands of the South Brazilian Province have few land vertebrates and need not be discussed. Queimada Grande, a small continental island twenty miles offshore, has a sizable fauna, including an endemic lance-head snake and an African gecko. In South America, Mexico, and the West Indies this gecko, *Hemidactylus mabouia,* usually lives beneath the overlapping central leaves of certain large agaves (*Furcraea*). One supposes that the lizard reached the New World from Africa quite a long while ago, to have developed such a close association with a Neotropical genus of plants. In the West Indies are two more African geckos that probably reached the New World by natural means, and not as stowaways on modern vessels. Like these geckos, a few plants may have crossed the Atlantic from Africa to the New World tropics. Thus the orpine genus *Kalanchoe,* familiar to gardeners, is widespread in the Asian and African tropics, yet has one oddly misplaced species in Brazil.

15 *"Its lands . . . are most beautiful . . . and filled with trees of a thousand kinds and tall . . . and I am told that they never lose their foliage."** **SOUTHERN, WESTERN, AND NORTHERN PARTS OF THE NEOTROPICAL KINGDOM**

The Venezuela-Guiana, the Amazon, and the South Brazilian Provinces, discussed in the previous chapter, lie at the heart of the Neotropical Kingdom, with other Provinces to the south, the west, and the north. The Pampas Province extends southward from the South Brazilian Province to the northern border of the Antarctic Kingdom. Included are Uruguay and the southeastern corner of Brazil, the Argentine pampas, and western Argentina.

This Province is forested only toward its northern edge. The eastern part of the Province is a virtually treeless grassland, the so-called pampas. ("Pampa" is a word borrowed by the Spanish from a Peruvian Indian language; it means a plain.) The western part of the Province is a scrubland reminiscent of the mesquite country along the Texas-Mexico border. The climate of the Pampas Province is not fully tropical. Killing frosts, moving northward out of Patagonia, may sweep across the land.

The Province is not sharply distinct. It has been invaded by plants coming

---

* COLUMBUS, off the coast of Haiti, 1492.

154

from the more thoroughly tropical lands to the north, and from the cold Patagonian lands to the south. Endemism is not marked. The pampas country is one of the largest areas of grassland in the world, but most of its plants are not confined to the Province. One endemic genus of grass, *Cortaderia*, includes the "pampas grass" widely grown as an ornamental; and among garden flowers, the Petunias came originally from this Province. Also characteristic are a creosote bush and a mesquite, both with relatives in Mexico and the southwestern United States. The western part of the Pampas Province harbors some cacti belonging to endemic genera.

Some of the most conspicuous plants of the Pampas Province were imported from other lands. The tree known locally as ombu, a relative of our familiar pokeweed, came originally from some more tropical part of South America; the cardoon, a thistle-like plant with edible stalks and roots, is native to the Mediterranean region. The milk thistle, both a vegetable and an ornamental in its European homeland, has become a weed in the Pampas Province (and in California). In the Pampas Province, cardoon and milk thistle often reach a very large size and cover extensive tracts with an almost impenetrable growth.

A fairly distinctive fauna is centered in the Pampas Province, although many of its components also range northward toward the Gran Chaco or southward into Patagonia. This fauna includes several kinds of armadillos; the restless cavy, a species of guinea pig; the mara, a fast-running rodent of the open grassland; the tuco-tuco, a colonial burrowing rodent of the Argentine pampas; and the plains viscacha, a larger burrowing rodent. The common rhea, a flightless bird superficially resembling an ostrich, is also characteristic of grasslands in southern South America.

The Venezuela-Guiana, Amazon, South Brazilian, and Pampas Provinces are all bordered on the west by the Andean Province, which extends along the South American west coast from southern Chile to northern Colombia. The Province includes the slopes and peaks of the Andes, and the land to the west thereof. The Galapagos Islands are appended to this subdivision, also. A cold, northwardly flowing current, of Antarctic origin, passes offshore of the Andean Province. Called the Peru or Humboldt Current, it serves to chill and dry the coastal lands. Only the northern part of this elongated Province is tropical; most of Chile is temperate or colder.

The tropical, mountainous, northern part of the Andean Province has a rich flora and much endemism. The highly localized endemics are mostly plants of the mountain slopes; species that can live on the crest of the range are likely to follow the Andean uplift for the greater part of its length. Among the tropical endemics of the Province is the genus *Desfontainea*. One of its species, popularly called by the generic name, is a garden shrub; its holly-like leaves are green all year, and its showy flowers, scarlet and yellow, are produced over a long season. Another endemic genus is *Eccremocarpus*; one of its species is the glory-flower, a climbing shrub with orange-red blossoms.

Many plant genera, not confined to the Andean Province, are represented therein by endemic species. Plant species, characteristic of the more northerly part of the Province, include the cherimoya, an edible fruit; *Carludovica palmata*, a superficially palmlike plant from which "Panama" hats are made; a wax palm; and quinoa, whose seeds are ground into meal. Others include cinchona, yielding the antimalarial drug quinine; the tree tomato; coca, source of the narcotic cocaine; and ratany, whose rootstock has been used medicinally and to color wines. Still other endemic species are the kidney bean; the soapbark tree; the "California" pepper tree; and melluco, with an edible root. Three species, originally confined to the northern part of the Andean Province, are today so familiar and widespread as to merit separate comment. These are the tomato, the "Irish" potato, and tobacco. All belong to the nightshade family (Solanaceae), a family otherwise noted for species with highly poisonous leaves or fruit.

Several plant species, originally confined to the more northerly Andes, have become popular garden flowers. Among these are the Browallia; the orange-ball butterfly bush; an Episcia; and the Ercilla, a climbing shrub. Others include a Eucharis with fine, white blossoms; and a Fittonia, grown for its interestingly veined leaves. Here, too, we could mention heliotrope; a Mutisia, a climber belonging to the daisy family; the garden Nasturtium and its relative the canary-bird flower; and several orchids. The puya, a gigantic, terrestrial member of the air plant family, is occasionally seen in conservatories; its 30-foot flowering stalk may bear 8,000 blossoms.

The Galapagos Islands are of volcanic origin. Relatively arid, they support patches of grassland, but more rocky scrubland with cacti and coarse bushes. The local flora includes a fairly high percentage of endemic species, but only a few endemic genera. The nonendemic genera are largely of Andean relationship.

South of the tropical mountains, the Andean Province includes a narrow coastal strip that is desert to the north and temperate scrubland to the south. Just inland of this strip the Andes rise abruptly, supporting cold steppe, cold deserts, or arctic tundra. The southern part of this Province has an interesting lot of endemics. One such is the genus *Jubaea*. It includes but a single species, the coquito palm, which will grow handsomely in California but not in Florida. Chileans make syrup from its sap. Originally confined to a small area around Concepción in Chile, the coquito is one of the few large palms to grow only outside the tropics. Other plant genera, confined to the southern part of the Andean Province, include *Lapageria*, one species of which is called Chile bells; and *Schizanthus*, the butterfly flowers or poor man's orchids.

Many widespread genera are represented in the southern part of the Andean Province by endemic species. Among these are the yellow herb lily; Darwin's barberry; a Calceolaria or pouch flower; and *Cantua buxifolia*, a shrubby relative of the Phlox. Also in this category are an Escallonia, used as a hedge plant; a pink Fuchsia; a scarlet Avens; a Lippia, the lemon verbena; and apple of Peru. Still others include a Salpiglossis or painted tongue; *Schizopetalon walkeri*, often grown as a border plant; a climbing Nasturtium with scarlet flowers; and a Verbena.

As seen from the above lists, a great many garden plants were originally endemic to the Andean Province. This does not mean that the Province is especially rich in species with showy blossoms. Rather, much of the Province being temperate, the flora is relatively rich in species that can live outdoors in temperate Europe or temperate North America. Quite a few plant genera, widespread in temperate North America, live also in temperate South America; alders and gooseberries are examples. This distribution is less surprising when one notes that mountains are nearly continuous from Alaska to Tierra del Fuego. These montains form a highway whereby temperate species have been able to pass through the tropics.

It might be asked how one can justify the inclusion of cold mountains, especially those with a temperate or an arctic flora, in a Province that also supports a tropical flora. Actually, in the tropics only about 1½ per cent of the land is covered by arctic vegetation, and another 4 per cent by temperate vegetation. The tropical plant Provinces have been erected with little reference to such aberrant, high-altitude areas.

Many animal groups are largely confined to the Andean Province. There are animals characteristic of the arctic mountain crests and of the arid coastal deserts, just as there are of the less severe Andean environments. In an earlier chapter we noted that many North American animals invaded South America with the closing of the Panama Portal, and followed the Andean uplift southward. Andean animals of North American derivation include the llama and its allies, the spectacled bear, the Andean fox, some small cats, the mountain tapir, and the little deer called pudus. However, some of the most interesting animals of the Andean Province represent original South American stocks and not North American invaders.

Old South American stocks, now Andean in distribution, include several genera of leptodactylid frogs. The genus *Telmatobius* is confined to the high Andes; one of its species lives in the icy waters of Lake Titicaca at 12,500 feet. Another geographically important genus of the Leptodactylidae is *Eupsophus*. Its species are distributed from Chiloé Island, off the coast of southern Chile, northward into the Brazilian tropics; but the more northerly species live about ledges where cold springs emerge, where steady drip and constant evaporation provide locally cool temperatures in a warm land.

The lizard family Iguanidae, an old South American one, includes several genera that are essentially Andean. Most interesting, perhaps, is *Liolaemus*, one species of which lives at 15,000 feet in Peru. During the local "summer" these lizards bask on rocks as soon as the morning sun has melted the night's snow away. The lizards' eggs are in little danger of freezing, for they are not laid; they are retained in the female's body until they hatch. Such adjustments to the montane environment imply long residence in cold lands.

The old South American rodent stocks also invaded the Andes. Among the Andean rodents is a cavy, first domesticated by the Inca and later by us; it is the familiar guinea pig of the laboratory. The chinchilla, prized for its soft fur, is another Andean rodent with a long ancestry in South America.

The seed snipes (Thinocoridae) are interesting birds of the Andean and Patagonian Provinces. They look almost like little quail, but are related to shorebirds such as the plovers. On Tierra del Fuego and the Falkland Islands the seed snipes live at sea level in a tundra-like vegetation; but the farther north they range, the higher the elevation they keep to, following the same kind of vegetation. At the northern end of their range, in the Ecuadorean Andes, they live at 14,000 feet.

The deserts of the Andean Province support a limited fauna. (They are the driest deserts of the world, and may go for years without rain.) Perhaps the most characteristic animal group is the lizard genus *Dicrodon*, belonging to the old South American family Teiidae. These lizards live in burrows, venturing forth only when the temperature is favorable. One species burrows only in the scant shade of a mesquite tree and feeds mainly upon the mesquite pods. The pods provide the only source of moisture; the lizards do not drink water. Such adjustments to the desert environment imply long residence in arid lands.

The northern, tropical end of the Andes has the world's greatest concentration of bird species, and an abundance of other vertebrates. A geographically important mammal of Ecuador and western Colombia is the mountain porcupine; it is not closely related to the prehensile-tailed tree porcupines of the Amazonian forests, and may be a remnant of the South American stock from which our North American porcupines were derived.

As mentioned elsewhere, the Galapagos Islands received their fauna partly from South America and partly from Middle America, some of the immigrants evolving into very distinctive endemics. It is interesting to note that the cold Humboldt Current, which so affects climate and distribution on the South American mainland, has permitted a few Antarctic animal groups to reach, and to survive on, the very Equator. Thus a penguin and a southern fur seal inhabit the Galapagos.

To continue the account of floral subdivisions, about 400 miles west of Santiago, Chile, lie the islands of the Juan Fernandez group. These islands, unlike the Galapagos, are characterized by many distinctive plant endemics, and so are placed in a Province of their own, the Juan Fernandez Province. The tiny Desventuradas Islands, roughly 500 miles north of Juan Fernandez and 600 miles west of the Chilean coast, also belong to this Province. One plant species of Juan Fernandez is so distinctive as to merit a family of its own, the Lactoridaceae. The plant is a shrub with jointed branches, thickened leaves, and a flower without petals. The daisy family has produced endemic genera of woody species on the Juan Fernandez group and the Desventuradas. The pigweeds (*Chaenopodium*), small herbs in most parts of the world, include tree-sized species on Juan Fernandez.

Although the plants of this Province are, in the main, of American affinity, some find their nearest relatives in the Polynesian Subkingdom, and especially in Hawaii. The Juan Fernandez Province is not important in zoogeography, but the presence of a Polynesian element in its flora prompts us to ask whether there is a comparable element in the fauna of Pacific islands lying closer to South America than to Polynesia.

As already noted, some animals of New Guinean affinity range eastward for varying distances into Polynesia, certain small lizards—skinks and geckos—being especially widespread. The snake-eyed skink ranges all through Polynesia, to Hawaii and Easter Island, and has also crossed the Pacific to the west coast of South America, where it has been found both on the mainland and on nearby islands. The azure-tailed skink ranges through much of Polynesia, to Hawaii and the Tuamotus, and has reached Clipperton Island, which lies 1,600 miles west of Costa Rica. These two lizards may have reached the New World by natural means, and not as stowaways on vessels. Several Asian and Polynesian geckos, also present in the American tropics, are more readily identified as recent introductions; for in the New World they live mostly about seaport towns. At any rate, the trans-Pacific dispersal of lizards, by whatever means, has been from the Old World tropics to the New, and not the reverse.

The seventh and last Province of the Neotropical Kingdom is the Caribbean Province. It embraces the northern parts of Venezuela and Colombia; all of Central America; all of Mexico except the central highlands and the upper half of Baja California; the West Indies; southern Florida; and Bermuda. Little islands off the west coast of Mexico and Central America, from Guadalupe to Malpelo, also belong to this Province. The Caribbean Province corresponds roughly, although not precisely, to the Caribbean Transitions of zoogeography; and like the Transitions, the Province falls into two parts, one essentially West Indian and the other essentially Middle American.

Panama, eastern Costa Rica, and southeastern Nicaragua have a superlatively rich flora. This circumstance results in part from an auspicious combination of environmental factors in the general region: temperatures are tropical, rainfall is heavy, topography is varied. (Many localities report 200 to more than 300 inches of rain annually. Mountains rise to more than 5,000 feet in Panama, more than 12,000 in Costa Rica.) West of the mountains, on the Pacific slope, rainfall and floral richness are reduced.

Many plant genera are distributed over most of the Caribbean Province. Among them are *Hura*, which includes the sandbox tree; *Guajacum*, the trees called lignum vitae; *Bouvardia*, garden shrubs commonly called by the generic name; and *Swietenia*, whose best known species is the mahogany.

Mexican endemic genera include *Dahlia,* popular garden flowers; and *Polianthus,* the tuberoses. Among the West Indian endemic genera are *Catesbaea,* the lily thorns; and *Lagetta,* which includes the lacebark tree. Some genera centered in the Mexican portion of the Caribbean Province are also represented by a few outlying species in the southwestern United States, beyond the confines of the Province. A familiar example is *Agave,* to which genus belong the century plant and the sisal hemp. Some Agaves follow our southwestern arid lands northward into Utah. As a matter of fact, the Mexican portion of the Caribbean Province has many floral affinities with our southwest, a circumstance not surprising in view of the continuity of land and the similarity of some environmental factors.

A great many genera, while not confined to the Caribbean Province, are represented therein by endemic species. First we may list a few species endemic to the Middle American portion of the Province. Among them is *Castilla elastica,* which produces most of the rubber imported from Central America. Other endemic species include logwood, source of the dye haematoxylon; medicinal Smilax; and an orchid whose pod yields most of the world's supply of vanilla. Likewise endemic are the candle tree, whose long fruits resemble yellow wax candles; a Cordia; and the avocado.

The Middle American endemic species also include many ornamentals of the garden or greenhouse. Among these are the trumpet Achimenes, a pot plant; Antigonon or coralvine; the shrimp plant; a blue Ceanothus; cathedral bells, climbing members of the Phlox family; and the familiar Cosmos. Here too we could mention a Dalechampia with odd flowers, several Echeverias, a tree-sized Fuchsia, a Gesneria, a Lobelia, and a Lycaste orchid with huge flowers. Nor should we omit a red-flowered Rondeletia; a Russellia with drooping, rushlike stems; a Salvia; a Scutellaria or skullcap; a chalice vine; a tigerflower; and the frangipani. Some flowering species originally endemic to the Middle American portion of the Caribbean Province are so familiar as to warrant separate mention: the Poinsettia, conspicuous at Christmas time; Zebrina or wandering Jew; the Zinnia, ancestral to many cultivated varieties; and two kinds of marigold, the "African" and the "French."

Various plant genera not confined to the Caribbean Province have produced endemic species in the West Indian portion thereof. (Many of the West Indian species reach southern Florida, also.) Among the West Indian species are the pelican flower, the night jessamine, a Lantana, a Maurandia, and the

Pampas grass, at left, is native to the Pampas Province of South America, but is grown elsewhere as an ornamental. This clump was photographed in the Mesilla Valley near Las Cruces, New Mexico. (New Mexico Tourist Bureau) The genus Agave is centered in the Mexican portion of the Caribbean Province, but some of the species range farther north. Among them, above, is the century plant, here growing in the Gila National Forest of New Mexico. (New Mexico Department of Development)

Spanish bayonet. The soursop, custard apple, cascarilla bark, allspice, and soap-berry probably came from the West Indies; but they were spread widely by man, and their original homeland is hard to identify.

Obviously the Caribbean Province has provided many valuable plants, especially ornamentals. Many of these have been transported to other tropical lands, where they thrive in yards if not in the wild. Coralvine now grows in churchyards and cemeteries all over the Philippines; marigolds are left as offerings at shrines in India; frangipani grows wild in New Guinea.

Turning now to outlying parts of the Caribbean Province, Bermuda supports about 150 plant species. This is a sizable flora in proportion to the limited area available for plant growth (only 21 square miles) and to the monotony of the environment (mostly low volcanic rocks fringed by sandy beaches and coral reefs). There are no endemic plant genera, and only a dozen or so endemic species. Most of the Bermudan plants are to be found also in the Bahamas, and perhaps elsewhere in the West Indian portion of the Province. As noted previously, the Bermuda group lies about as far north as Montgomery, Alabama. However, the islands are climatically suitable for West Indian plants because local vagaries of temperature are modified by the warm Gulf Stream. This great oceanic current, of equatorial origin, sweeps northwestward through the Carib-bean, passes through the Yucatán Channel, swings around the Gulf of Mexico, escapes into the Atlantic through the Straits of Florida, and then turns north-eastward toward Bermuda.

The Revilla Gigedo Islands, lying nearly four hundred miles off the Mexican west coast, have only a small flora, but this includes some interesting endemic species. A few plants of the Revilla Gigedos have relatives on the Galapagos. A species of guava grows on both these archipelagos, but nowhere else. Two more outlying portions of the Caribbean Province remain to be considered: the southern half of the Baja California peninsula, and the southern tip of the Florida peninsula. These two areas are relegated to the next chapter, along with a discussion of faunal distribution within the Province.

16 *"On these islands is likewise a wood we call here the wood of many uses . . . also much fruits of many sorts, which I will not enumerate, as were I to attempt to do so, I should never finish."** **TWO PENINSULAS: BAJA CALIFORNIA AND FLORIDA**

The Caribbean Province includes the southern half of the Baja California peninsula. The northern limit of the Province, and hence of the Neotropical Kingdom in western North America, may be set at the northern edge of the Vizcaino Desert. A mountain chain extends the length of Baja California, reaching an altitude of about 6,500 feet east of the Vizcaino Desert, and 7,000 feet near the southern tip of the peninsula. Except at higher elevations, frosts are rare, and the lower coasts are frost free. Rainfall is everywhere low, although slightly higher toward the south.

From a standpoint of floral distribution, the tropical portion of Baja California may itself be divided into two components. The larger of these is a desert, which extends northward out of the Neotropical Kingdom; the smaller includes only the Cape San Lucas region, and the Sierra de Giganta which borders the Gulf of California. Naturally the Cape San Lucas region, at the extreme southern tip of the Baja California peninsula, harbors a greater percentage of tropical plants. Tropical species of the Cape San Lucas region often

* HERNANDO FONTANEDA, about 1575, on the Florida Keys.

range northward as much as 200 or even 250 miles up the eastern side of the peninsula, but seldom range more than 50 miles up the western side. This circumstance has come about mainly because the western side of the peninsula is chilled by the cold, southwardly flowing California Current, of subarctic origin. The eastern side escapes the effect of this oceanic current.

Tropical genera represented in southern Baja California include *Ficus*, the figs; *Jatropha*, the physic nuts; *Manihot*, the cassavas; *Agave*, the agaves; and *Tillandsia*, certain airplants. Also present are *Lantana*, the Lantanas; *Bignonia*, the trumpet creepers; *Fouquieria*, the candlewoods; *Pedilanthus*, cactus-like spurges; *Lysiloma*, the wild tamarinds; and *Bursera*, the torchwoods. These genera are tropical in the sense that most of their species are confined to the tropics; but some of them, such as *Tillandsia* and *Lantana*, have also invaded warm-temperate lands.

Along the southern and southeastern shores of Baja California grow the coconut, the red mangrove, the black mangrove, *Scaevola plumieri*, and a few other species widespread in the Neotropical Kingdom. A palm, the familiar Washingtonia, ranges into the Vizcaino Desert region from farther north. A *Glaucothea* or blue palm, present in the mountains to the east of the desert, is of Mexican derivation.

Endemism is marked in Baja California. Eight plant genera are nearly or quite confined to the tropical portion of the peninsula, but most of them are unfamiliar to the layman. *Triteleiopsis*, of the lily family, includes a blue-flowered species occasionally seen in rock gardens. *Burragea* is confined to the Magdalena and Santa Margarita islands, off the lower west coast of the peninsula; the genus, belonging to the evening-primrose family, is related to the better known *Zauschneria*. Another relative of the Zauschnerias, called *Xylonogra*, is restricted to Cedros Island and the shores of Sebastian Vizcaino Bay, at the northern edge of the tropics. *Machaerocereus*, a genus of the cactus family, is distributed over the southern and central portions of Baja California, and also follows the west coast of the peninsula northward to the Ensenada region.

The Gulf of California, 80 to 130 miles wide, has been no barrier to plants coming from the deserts of Sonora and Chihuahua; a very similar flora is found on opposite shores of this gulf. At least two plant genera (*Viscainoa* and *Idria*), otherwise endemic to Baja California, live also across the gulf in western Sonora. The flora of southern Baja California likewise includes species and genera of temperate rather than tropical affinity. Plants have invaded the peninsula not only from the mainland tropics but also from various parts of the southwestern United States and from the Mexican highlands. For example, the Cape San Lucas region has some oaks and an Arbutus that find their nearest relatives in the Sierra Madres.

Turning now to eastern North America, the Florida peninsula is a rough counterpart of Baja California: it extends from a warm-temperate land southward into the northern edge of the tropics, and its tropical portion is separated from other parts of the Neotropical Kingdom (that is, from Cuba and the larger Bahaman islands) by 65 to 135 miles of ocean. But Baja California and peninsular Florida are otherwise quite different, for the latter combines low elevation with high rainfall, rather than the reverse. The tropical portion of Florida seldom rises more than 15 feet above sea level. The northern limit of the Caribbean Province in Florida, and hence of the Neotropical Kingdom in eastern North America, may be set at Boca Raton on the east coast and Bonita Springs on the west.

The southern portion of peninsular Florida is really a meeting ground for temperate and tropical floras. Patches of tropical forest grow here and there in the limestone flatwoods, occupying detritus-filled hollows of the outcropping rock. Similar patches grow along water courses, on limestone elevations in the Everglades, and on huge shell heaps—the debris of ancient Indian villages—along the coasts and offshore islands.

Mahogany once was conspicuous in these patches of tropical forest, but has largely been cut out. Like the mahogany, the remaining trees and shrubs are mostly of West Indian affinity. We could compile a long list of them: lignum vitae, pigeon plum, mastic, poisonwood, marlberry, lancewood, white stopper, red stopper, boxleaf Eugenia, wild coffee, gumbo-limbo, black ironwood, strangler fig, shortleaf fig, bitterbush, Paradise tree. And there are many more: spicewood, bustic, fiddlewood, inkwood, crabwood, West Indian cherry, satinleaf, black bead, myrsine, wild tamarind, nakedwood, wild lime, whitewood, pond apple, coco plum, coral bean. Amongst the tropical plants may grow trees and shrubs of temperate affinity, especially live oak, laurel oak, cabbage palm, saw palmetto, red mulberry, French mulberry, red bay, dahoon holly, and hackberry.

Tropical plants range farther northward along the coasts and offshore islands than they do in the interior of Florida. Red mangrove, white mangrove, black mangrove, and a few other species form a coastal plant association known

A *water-retaining bromeliad of the genus* Tillandsia *grows on a pond cypress in southern Florida. (Charles Belden)*

as mangrove swamp; this association is essentially tropical, but follows the Florida coasts northward about to Cape Kennedy on the Atlantic and Tampa Bay on the Gulf of Mexico. Some of the mangroves turn up individually, or in scattered clumps, still farther north along the Gulf shores, but they do not predominate there.

From time to time, some of the more northerly populations of tropical plants are killed out by frosts. However, a few individual plants usually survive in protected places, and a series of mild winters will permit them to grow and reproduce. The effect of cold on the distribution of tropical plants was strikingly demonstrated on December 12, 1962, when a hard freeze swept over Florida. Tropical mangrove swamp is replaced farther north in coastal Florida by temperate saltmarsh, an association of reeds, grasses, and low herbs. A short distance north of Clearwater on the Gulf coast, the two associations meet; stands of mangrove trees are interspersed among patches of saltmarsh. But the freeze, bringing temperatures down to 14 degrees Fahrenheit in some spots, killed most of the mangroves while doing very little harm to the saltmarsh vegetation. Two years later the saltmarsh showed no sign of freeze damage, but almost all the living mangroves were seedlings, sprouting from seeds that had been buried in the coastal muck. Other tropical plants, growing not in mangrove swamps but on the open strand, were even more severely affected; the freeze wiped out some of the more northerly colonies of necklace pod, Florida privet, saffron plum, West Indian gutta percha, buttonwood, white stopper, West Indian leadwort, and gray nicker.

Florida plants of tropical origin vary considerably in their ability to tolerate cold, and so vary in the distance northward they are able to range. In this connection the family Bromeliaceae is of interest. The bromeliads are distributed throughout the Neotropical Kingdom. Most of them are epiphytes—plants that grow upon other plants, not parasitically but simply for support. About 15 epiphytic bromeliads, most of them widespread in South America, Middle America, and the West Indies, also reach Florida; but at least half of these species range northward out of the Neotropical portion of the state, and two of them reach Georgia. Another bromeliad, the "Spanish moss" (neither Spanish nor a moss), is distributed from South America to Virginia, Tennessee, and Texas.

Other plant groups, of tropical origin and largely of tropical distribution, have been even more successful in their invasion of temperate lands. For example, four species of passionflowers are confined in Florida to the Neotropical portion thereof. They are of West Indian derivation, and belong to a genus (*Passiflora*) whose 300-odd species are mostly Neotropical. However, two other Florida species of passionflower range northward, in one case to Virginia and in the other to Pennsylvania.

Florida harbors few really distinctive endemic plants of tropical affinity. This circumstance is thought to have come about as a result of certain past climatic fluctuations, the ones mentioned in Chapter 3 as having once brought northern plants and animals southward into Florida. Past periods of cooler climate apparently destroyed most of the state's tropical flora. The last cool period ended so recently that West Indian plants, reinvading southern Florida, have had insufficient time to evolve striking adaptations to the local environment.

Before leaving the topic of plant distribution at the northern edge of the Caribbean Province, it is instructive to look for floral similarities between Baja California and Florida; for the two peninsulas, however different in environment and history, should have floral elements in common if both are to be included in the same Province.

Many of Baja California's tropical genera are also present in southern Florida. The latter region has, for example, the gumbo-limbo tree, which is a *Bursera;* the strangler fig and shortleaf fig, belonging to the genus *Ficus;* false sisal and wild century plant, *Agave;* several Lantanas or shrub verbenas, *Lantana;* the trumpet creeper, *Bignonia;* the redbird flower, *Pedilanthus;* the wild tamarind, *Lysiloma;* three-seeded mercuries, *Acalypha;* Indian mallows, *Abutilon;* and a dozen air plants of the genus *Tillandsia.* One species of epiphytic air plant, *Tillandsia recurvata,* common on shrubs along the east coast of Baja California, is equally common on cabbage-palm trunks, and on telephone wires, in Florida.

Many tropical plants, requiring a moist environment, are present in Florida but not in arid Baja California. On the other hand a number of tropical plants, characteristic of a dry environment, do grow along the lower west coast of peninsular Florida, where the rainfall is comparatively scant and seasonal, and where the "soil" (limestone and dune sand) does not long retain moisture. At scattered localities from the Lower Keys to the mouth of the Manatee River, there develops a plant association called cactus thicket, dominated by stiff, crooked, spiny shrubs. The late John K. Small, eminent botanist, once called this association "the most remarkable natural cactus-garden east of the western American deserts." A majority of the cacti belong to the genus *Opuntia*, which is also well represented in Baja California. The species of this genus, variously known as prickly pear, tuna, cholla, and devil's pincushion, are concentrated in the Neotropical Kingdom, although cold-hardy outlying species reach Canada and Patagonia.

Red mangrove and black mangrove grow coastally in both Baja California and the Florida peninsula. *Scaevola plumieri*, whose stems creep beneath the sand of coastal dunes, is found in both regions. The coconut thrives at Cape San Lucas just as at Miami.

The discussion of the Caribbean Province may close with remarks about faunal distribution therein. The Province corresponds only roughly to the Caribbean Transitions of zoogeography. From a standpoint of animal distribution there is no special reason to bracket northern Colombia and northern Venezuela with Middle America or the West Indies, although many animals range from the forest of northwestern Colombia to that of Panama.

The fauna of lower Middle America, from Panama to southeastern Nicaragua, is but moderately rich. The floral richness of this area mostly reflects invasion by, and diversification of, numerous South American plant stocks; but South American animals were largely barred from Middle America by the Panama Portal, and with the closing of the Portal only a fraction of the South American fauna moved northward. Also, the southward movement of North American animals into Middle America was considerably restricted by the Isthmus of Tehuantepec, whose limited series of environments may have served especially to filter out animals adapted to rainforest existence. In short, Middle America perhaps offered ideal conditions for the development of a rich fauna, but did not receive all the animal stocks it could have supported.

As mentioned, the Caribbean Province falls into two portions, one corresponding roughly to the Middle American Transition of zoogeography, the other roughly to the West Indian Transition. However, Middle America and the West Indies are much more alike florally than faunally. This circumstance we might have expected, for a sea like the Caribbean largely prohibits the interchange of animals but not especially of plants. Faunal similarities between the West Indies and Middle America are most evident in the birds and the bats. The freshwater turtles (*Pseudemys*) of the West Indies are derived from Middle America. A

The American crocodile is one of the few tropical vertebrates to reach Florida. This one was caught in Biscayne Bay. (Ross Allen's)

Many West Indian animals show only minor differences from island to island. The Cuban boa, shown above, is marked with yellow and brown. A Bahaman relative is similar, but marked with gray and black. (Tod Swalm) The Cuban crocodile, below, is confined to Cuba. (Ross Allen's)

few West Indian lizards—certain geckos and anoles—have crossed the Caribbean to islands off the east coast of Middle America. The manatee, the West Indian wolf seal, the American crocodile, and various seabirds or shorebirds, are essentially circum-Caribbean in distribution; but they are really sea animals, and impart little zoogeographic unity to the lands bordering the Caribbean.

The small Bermudan fauna has been discussed previously. Its affinities are almost entirely with eastern North America. The Revilla Gigedan fauna, larger than the Bermudan, has its affinity mostly with the southwestern United States and northwestern Mexico; strong tropical relationships are evident chiefly among some of the birds. The Revilla Gigedos have one endemic whipsnake of the genus *Masticophis*, and two endemic iguanid lizards of the genus *Urosaurus*; neither genus is especially tropical in distribution.

The vertebrate fauna of Baja California was derived almost entirely from lands to the north. The Gulf of California has prevented much faunal interchange between the tropics of mainland Mexico and the southern part of the Baja California peninsula. (For that matter, the fauna of western Sonora and western Sinaloa, across the Gulf from Baja California, is not tropical; it is a desert fauna whose components often follow the arid environments far to the north.) In southern Baja California, tropical relationships are evident mostly among the birds and the bats, hardly at all among the terrestrial mammals, snakes, lizards, turtles, frogs, and salamanders. Fishes of Baja California rivers are, in the main, essentially saltwater species with some tolerance for fresh water; examples are snook, mullet, and snappers. The few true freshwater fishes find their relatives to the north. Endemism is marked among the reptiles, birds, and mammals of Baja California. The endemics, even when confined to the southern part of the peninsula, are mostly related to species of the western United States.

The vertebrate fauna of peninsular Florida also is derived almost entirely from lands to the north. The Straits of Florida have prevented much faunal interchange between the West Indian islands and the southern part of the Florida peninsula. A few West Indian birds nest also in extreme southern Florida, but a tropical element is weak even among the birds and the bats. A small, familiar iguanid lizard, miscalled "chameleon" because it may become either green or brown, reached Florida long ago from the West Indies, and has spread widely in eastern North America. There has also been but very little movement of Florida animals to the West Indies. A watersnake (*Natrix*), found on the north coast of Cuba, came from Florida; and a Bahaman raccoon is probably of Floridian ancestry.

Many vertebrate groups exhibit marked endemism in peninsular Florida. However, the endemics are concentrated in the higher parts of the peninsula, outside the Neotropical portion thereof; and they will be considered in a later chapter.

17 *"And so we came at last to a tundra vast and dark and grim and lone . . . ."** **THE ARCTIC-SUBARCTIC PROVINCE**

Last of the great floral subdivisions is the Boreal Kingdom. It stretches over all the lands that lie north of the Neotropical and Palaeotropical Kingdoms. Most of the Boreal Kingdom is temperate, but the more northerly portion is Arctic; and many higher mountain slopes and peaks provide arctic temperatures. The Boreal Kingdom corresponds fairly well to the Arctic-Temperate Region of zoogeography.

While there are plant families confined to the New World portion of the Boreal Kingdom, and others to the Old World portion, a great many families are represented in both areas. The Kingdom is not divided at the outset into Old World and New World components. Unless otherwise noted, plant families listed below as "temperate," or as "north-temperate," are represented in both the Old World and the New.

As regards plant distribution, temperate regions have two peculiarities that distinguish them from tropical ones, apart from climatic differences. First, they lie in two widely separated zones, one north of the tropics and the other south. Second, the temperate lands have no tropical areas within them, whereas tropical lands have temperate areas at higher elevations. As a result of these peculiarities, many families of plants, essentially temperate, have passed between the north-temperate and the south-temperate zones, chiefly by way of mountain ranges

* ROBERT W. SERVICE, *The Ballad of the Northern Lights.*
176

that trend north and south. Most plant families, today represented in both the north-temperate and the south-temperate zones, have at least a few additional members in the tropics, usually at high elevations.

In other words, high peaks form stepping stones whereby temperate families have often attained remarkably wide distributions. Take, for example, the blueberry family (Vacciniaceae). Blueberries are found all over the north-temperate zone, in scattered areas of Middle America and tropical South America, in the south-temperate zone of South America, in eastern Africa and Madagascar, and from southeastern Asia to northern Australia. The blueberries of tropical lands grow mostly in the mountains. Often a temperate family, having attained a foothold in some upland portion of the tropics, may eventually produce offshoots capable of invading the tropical lowlands also. For example, we have mentioned a blueberry on Rapa, far out in the South Pacific.

Only nine plant families, present in both Old World and New World portions of the north-temperate zone, do not range thence southward well into or across the tropics. These families are not all distributed in the same way, however. Thus the family Diapensiaceae, which includes the Lapland Diapensia seen in rock gardens, is widespread in the Arctic; and its more southerly members, such as Galax and Shortia, are largely confined to mountainous regions. The Trilliaceae, typified by the Trilliums or wake-robins, are concentrated south of the Arctic but north of the warm-temperate lands. The Podophyllaceae, of which the May apple is the most familiar member, and the Corylaceae or hazelnut family, are concentrated more toward the warm-temperate lands, although each has a few species well to the north.

The Boreal Kingdom is made up of eight Provinces. We may begin the discussion with the Arctic-Subarctic Province, which unites the Old World with the New. This Province includes the lands north of tree growth in Eurasia and North America, and also embraces all the islands of the Arctic Ocean. The Province corresponds closely to the Arctic Subregion of zoogeography, with the minor exception that Iceland is unclassifiable on the basis of its fauna. The flora of the Province is limited, and not solely on account of present environmental rigors. Much of the Arctic is ocean, not land; and the largest component of the Province, the great island of Greenland, is almost completely covered with glacial ice.

Furthermore, the Greenland glacier is but the remnant of a much greater one, a vast ice sheet that melted and retreated in what phytogeographers would call relatively recent times. There has been no great length of time for plants to invade the formerly glaciated parts of the Arctic-Subarctic Province, or for many invaders to evolve marked differences from their parent stocks. There are some Arctic endemic species, obviously derived from cold-temperate ancestors; but there are hardly any sharply distinct Arctic endemic genera. The flora of the Province is probably the youngest in the world.

Northern Alaska and parts of northern Siberia escaped wholesale glaciation, and perhaps this is why they have a richer flora than other portions of the Province. Northern Alaska harbors about 600 species of higher plants, the greatest concentration within the Arctic. But even Greenland, in spite of its glacier, has about 350 species of native plants, and another 50 introduced from Europe by early Norse settlers. A visitor to Greenland, during its brief summer, will be surprised by the variety of the local plants. Conspicuous are various grasses and sedges; a chickweed; saxifrages, some white-flowered and others pink; a poppy; buttercups; and Arctic rose (*Dryas*), with a delicate white blossom. Also in evidence are a Cochlearia, belonging to the mustard family; stonecrops; a Cassiope or mossflower; a yellow lousewort, more attractive than its name suggests; and eyebright, a Euphrasia with a yellow-centered white bloom. Still greater variety is lent by Arctic knotweed, which reproduces not by seeds but by fleshy leaf buds that fall off and sprout; a Pyrola or shinleaf; and a yellow-flowered Potentilla called silverweed.

The visitor to Greenland will also be impressed by the small size of the flowering plants; many bloom when but an inch high. In places the ground may be thickly carpeted with blossoming herbs, none above ankle height. When such carpets burn over, the wind-blown seeds of fireweed soon drift into the burned area, covering it with pink-flowered plants about four inches high. (In the more favorable environment of our gardens, the same fireweed may reach four feet.)

Some areas of Greenland support a low, scrubby vegetation of woody plants: willow, birch, a blueberry called bilberry, and an Empetrum or crowberry. But the plants all hug the ground, looking as though they had been trampled flat. On the very dry northern coasts of Greenland, two of the saxifrages bear clusters of fleshy, spine-edged leaves. The plants look much like little cacti.

The Greenland flora has been described here in some detail simply because

Furry anemone at Davos, Switzerland. Various anemones grow both in the Arctic and in the high mountains of Europe or Asia. (Swiss National Tourist Office)

it is so thoroughly Arctic in composition and character. Its species are from such families as the grasses, sedges, saxifrages, chickweeds, poppies, buttercups, roses and their kin, mustards, orpines, and heaths: also from the figworts, buckwheats, shinleafs, evening primroses, willows, birches, blueberries, and crowberries. The individual plants are either tiny herbs or prostrate shrubs, adapted to a harsh environment, rapidly sprouting and flowering when the upper soil begins to thaw.

By an interesting coincidence, the dwarfed Arctic plants grow thickly atop an ancient fossil bed on Disko Island, just off the west coast of Greenland. The bed is rich in the remains of large trees, especially elm, sassafras, fig, and sycamore. Obviously the Greenland climate has not always been as inhospitable as it is today.

Toward the southern edge of the Arctic-Subarctic Province, not far beyond the limit of tree growth, dwarfing of plants is not as pronounced as on Greenland. For example, on the northeastern coast of Labrador there are fair-sized shrubs such as the Lapland rhododendron; the Labrador willow; Ledum or Labrador tea; and the Arctic bunchberry, which is really a dogwood. Here, too, the herbaceous plants, such as alpine vetch and single-flowered bluebell, attain a respectable stature.

The flora of the Arctic-Subarctic Province has two components, as regards patterns of distribution. One of these components includes species that live only in the Arctic, and that do not range farther south in the mountains. In this category are the Arctic chrysanthemum, Pallas buttercup, some louseworts, and several dwarf willows. The other component of the Arctic-Subarctic flora exhibits a so-called arctic-alpine distribution. That is to say, the species of this category not only live in the Arctic but also at scattered localities farther south, on peaks and mountain slopes that provide an arctic climate. The Rocky Mountains extend into the Arctic; the Alps lie well south of the Arctic; the Himalayas are even farther south. It is therefore not surprising to find arctic-alpine species numerous in the Rockies, less numerous in the Alps, and rare in the Himalayas.

Arctic-alpine plants are not all distributed in the same way. For example, the alpine anemone, the glacier buttercup, the twin-flowered violet, and the opposite-leaved saxifrage all grow in the Arctic and in the Alps but nowhere else; the alpine forget-me-not in the Arctic, the Alps, and the peaks of Corsica. Scheuchzer's bellflower is found in the Arctic, the Alps, and the Altai of Mongolia; the snow gentian in the Arctic, the Alps, and the mountains of Asia

Minor; the pygmy buttercup in the Arctic, the Alps, and the Rockies. The herbaceous willow grows in the Arctic, the Alps, the Urals, and the Rockies.

What with the ability of plants to cross barriers of unsuitable habitat, it might be asked why most Arctic plants do not turn up also on high peaks farther south. But no peak south of the Arctic Circle can duplicate one important factor of the Arctic environment: the "midnight sun." The farther above the Circle, the more summer days with 24-hour illumination; and well into the Arctic-Subarctic Province the sun does not set at all during the plants' growing season. The "midnight sun" is never high in the sky, and its rays, striking the earth obliquely, provide only a feeble warmth; but plants need sunlight for more than warmth. A green plant carries on a remarkable chemical process, making starch and sugar from the water of the soil and the carbon dioxide of the air; and the process, vital to plant growth, is triggered only by sunlight. It is the constant illumination that permits many Arctic plants to thrive in spite of a brief growing season. Thus for many plants the alpine environment, with regular alternation of night and day, is more rigorous than the Arctic one. High peaks lying south of the Arctic Circle usually support only a small percentage of the Arctic-Subarctic flora.

Just as the Arctic is taken to be the region north of tree growth, so the alpine zone of a mountain is defined as the portion above timberline. But actually, various plants of the treeless Arctic may straggle out into the adjacent forested Provinces, and in like fashion the plants of the treeless alpine zone may find suitable habitats here and there a short distance below timberline. Arctic-alpine plants, as defined a little less than rigidly, are distributed southward to New Mexico in the Rockies, but only to New England and northern New York in the Appalachian uplift of eastern North America.

Toward the more northerly portion of the Arctic-Subarctic Province, the climate becomes more rigorous, not only colder but also drier. In the northernmost land areas, the temperature may rise above freezing only for a few days in July, and precipitation may not exceed two or three inches. The very cold, dry areas, known as fell-fields, are not covered with ice or snow; they are barren, windswept expanses of frozen ground and scattered rocks. Few higher plants can live in the fell-fields; the ones that do so never grow in clumps but are scattered individually, like plants in the driest parts of the warm deserts. Fell-fields are not restricted to the northernmost portions of the Arctic-Subarctic Province; they

Above, near the northern limit of tree growth in Canada's Northwest Territories. Between Fort Smith and Fort Reliance, coniferous forest gives way to treeless tundra. (Canadian Government Travel Bureau)

(Facing page) Above, near the upper limit of tree growth in the Sangre de Cristo Mountains of New Mexico. Above Little Blue Lake the coniferous forest gives way to a treeless expanse. (New Mexico Department of Development) Below, Arnica growing in the Swiss Alps. It ranges into sub-Arctic Europe. (Swiss National Tourist Office)

also develop farther south in the Province, at higher elevations or on rocky, wind-swept promontories. (And there are also alpine fell-fields in high mountains south of the Arctic.)

Some fell-fields offer such a rigorous environment that no higher plant can grow; yet they do support lower plants, notably lichens. Lower plants are not under discussion here, but lichens are mentioned because they permit certain herbivorous animals to survive even in the absence of higher plants. One lichen, the reindeer moss, is especially important in the diet of Arctic herbivores. The availability of lichens is not, however, the only factor that permits vertebrates to range into parts of the Arctic where higher plants are nearly or quite lacking. The ocean provides a rich source of food for Arctic animals such as the polar bear, various seals, and many birds.

In short, animals are not distributed precisely like plants in the far-northern lands, even though the Arctic Subregion does correspond closely in outline to the Arctic-Subarctic Province. The higher plants are concentrated in the more hospitable portions of the Province, but some of the most characteristic Arctic vertebrates can live where spermatophytes are scarce or absent. Accordingly, from a biogeographic standpoint it is desirable to distinguish between the "low Arctic," with its concentration of both shrubs and herbs; and the "high Arctic," with no shrubs, a few herbs, and some highly specialized animals.

Arctic animals are notably mobile. The polar bear swims for miles, lopes great distances across the ice, or drifts for days on floes and icebergs. The mouse-like lemmings emigrate in all directions when their local population becomes too large for the local food supply. (But contrary to legend, lemmings feel no urge to drown themselves in the nearest ocean.) Arctic birds fly between widely separated feeding grounds. The Arctic fox drifts on floes like the polar bear, follows the bear for days to feed upon its leavings, moves like the birds between widely separated feeding grounds, and emigrates like the lemmings when over-population threatens. This general mobility of Arctic animals, and the former existence of additional land bridges in the Arctic, have made for a remarkably homogeneous circumpolar fauna.

The Arctic lands are fragmented by seas, channels, sounds, bays, and straits. One would usually expect a much-fragmented region to be unified more by its plants than by its animals; but not so the Arctic. The fauna of the Arctic is more nearly homogeneous than the flora. Most Arctic animals are represented in both

The Old World reindeer is a close counterpart of the New World's Barren Ground caribou. These reindeer were photographed in Canada, whence they were imported and where they thrive. (Canadian Government Travel Bureau)

the Old World and the New. A notable exception is the musk ox, confined to northern Canada and Greenland; but in this case, as in some others, the animal has become extinct in parts of its formerly circumpolar range.

The Barren Ground caribou is a species of deer confined to the New World portion of the Arctic, ranging from central Alaska eastward to Ellesmere Island and Labrador; but in Greenland and Arctic Eurasia it is replaced by an extremely similar and closely related species, the reindeer. The existence of such "twin" species does not detract from the general impression of faunal homogeneity in the far-northern lands. As a matter of fact, almost any stock of plant or animal, in the Arctic or elsewhere, may be broken up into a series of slightly differentiated populations which replace each other geographically.

In connection with faunal or floral similarities throughout the Arctic, it is important to remember that homogeneity is not evident from north to south, but rather from east to west at any given distance from the tree line. The animals that are Arctic by our definition, the ones that live mostly to the north of tree growth, rarely inhabit treeless mountain peaks south of the Arctic. As might be expected, the closest approach to an alpine-arctic distribution is found among strong-flying birds. Thus the harlequin duck, which nests principally along the coasts of the Arctic Ocean, also rears its young in glacier-fed mountain streams as far south as California and Colorado.

In a comparison of faunal with floral distribution in the Arctic, attention must be given to migration, a seasonal adjustment of range possible to many animals but not to plants. Most Arctic animals migrate as winter approaches, and do not return to their old haunts until spring. The migration may be simply into a nearby and more sheltered habitat. Thus the willow ptarmigan, a quail-like bird of the treeless lands, moves into sheltered valleys, or into copses of dwarf willows, at the onset of winter. Usually, however, the migrating animal escapes the rigors of an Arctic winter by moving far southward, into a more favorable climate. For example the old-squaw, a beautiful but garrulous duck that nests mostly above the tree line, in winter flies down at least to the Aleutians and the St. Lawrence, sometimes even to Texas and Florida. The snow goose winters as far south as Mexico and Japan. The knot, a kind of sandpiper, may migrate from Ellesmere Island to southern Patagonia, or from northern Siberia to southern Africa and New Zealand.

Birds are not the only Arctic animals to migrate long distances. Each year

The common loon nests in and near the Arctic lands, but migrates to warm-temperate regions for the winter. (Canadian Government Travel Bureau)

the Barren Ground caribou treks southward to the forest for the winter, returning in spring to the coasts of the Arctic Ocean. The Old World reindeer similarly migrates; and it is followed in its movements by nomadic Laplanders who rely on it for subsistence. The little shrews and rodents, which overwinter beneath a thick (and relatively warm) blanket of snow, may be the only Arctic mammals that do not move at least a short distance overland to escape the brunt of the cold.

In our discussion of Arctic vertebrates we have made no mention of the ectotherms—the reptiles, amphibians, and fishes, which draw their body heat largely from the environment. One salamander and a few reptiles cross the Arctic Circle, but none of them reaches the Arctic-Subarctic Province. A few frogs range into the southern portion of the Province, all coming from temperate lands to the south. Especially interesting are the wood frogs. A single wood frog stock crossed Bering Bridge at the time of a milder climate than the present one and eventually diverged into two species. The Old World one today ranges northward to North Cape in Norway and Sakhalin Island in Siberia; the New World one reaches the Yukon River of Alaska and Ungava Bay in eastern Canada.

As for fishes, only one species is almost entirely confined to an Arctic habitat: the blackfish, which belongs to a family of its own (Dalliidae). This small fish inhabits ponds and shallow lakes of western Alaska, eastern Siberia, and St. Lawrence Island in Bering Strait. Strikingly adapted for life in the Arctic, the blackfish can recover after having been frozen for weeks. Other fishes of the Arctic generally inhabit rivers, or the large lakes along them. Few major river systems are confined to the Arctic; and most of the Arctic fishes live also in cold-temperate streams. However, it is convenient here to remark on certain fishes that do range into the Arctic, whatever their distribution in more southerly lands.

Of considerable zoogeographic interest are the longnose sucker and the northern pike. Neither species tolerates salt water or crosses a sea barrier. Both are found in more northerly streams of the New World and the Old, a distribution reflecting the former existence of rivers on Bering Bridge. The grayling, which ranges from the Yukon River to the upper Missouri, is replaced in northern Eurasia by a closely related species. Graylings are confined to fresh water, and may have followed a river system across Bering Bridge. However, their closest

A Canadian Eskimo eyes his catch of Arctic char, flame-colored relative of the brook trout. Widespread in the Arctic, the char migrates up rivers to spawn in lakes. (Canadian Government Travel Bureau)

relatives (the salmon and the whitefishes) are often very salt-tolerant, and graylings might have some ability to disperse through the sea.

Some of the best known fishes of Arctic streams are marine as adults, but periodically migrate upstream to a freshwater spawning ground. Salmon are especially famous for this habit; but certain whitefishes, sturgeon, and smelt may also invade Arctic streams, coming from the sea.

The strictly freshwater fishes do not range very far into the Arctic. In eastern North America, for example, they reach only as far as the George River, which harbors the longnose sucker, the northern pike, and a minnow. Such fishes are limited in their northward spread not so much by climate as by an inability to reinvade more northerly streams from which they were extirpated by glaciation. An ability to disperse through the sea probably explains why members of the whitefish family, and of the salmon family, have been able to range so much farther north into formerly glaciated territory. Certain trout, members of the salmon family, set a record for northernmost distribution of fishes in fresh water; one species reaches northern Ellesmere Island, another reaches Novaya Zemlya and the New Siberian Islands.

18 *"These woods stretch unbroken over a vast region . . . dense woods,*

and so still . . . in many places all the ground is hidden for miles under a

thick cushion of moss. . . . A rich cathedral gloom pervades the pillared

aisles . . . ."* **THE EURO-SIBERIAN**

PROVINCE

The Arctic-Subarctic Province is the only subdivision of the Boreal Kingdom to overlap the Old World and the New. South of it, in the Old World, lies the Euro-Siberian Province. This latter subdivision, as its name indicates, stretches from Europe to Siberia, more specifically from the Faeroes and the British Isles eastward through the Kamchatka Peninsula and the Komandorskie Islands. On the north it is bounded by the Arctic-Subarctic Province. To the south it extends into northern Spain, northern Italy (but not Corsica), northern Yugoslavia, and northern Bulgaria. From the region of Romania and northern Bulgaria it sends a tongue eastward to include the Caucasus. In Asia the Province extends south into northern Kazakhstan, extreme northern Mongolia, and northern Manchuria.

The Euro-Siberian Province is extremely long in its east-west extent; it reaches half way around the globe. To some degree the opposite ends of the subdivision differ from each other in their respective floras. To reflect this difference, the Province can be divided into eastern and western components, separated by the Ural Mountains. Not that these low mountains are a barrier to plant distribution; but they are generally taken to separate Europe from Asia,

* MARK TWAIN, *A Tramp Abroad*, 1880, on the Black Forest of Germany.

and so they provide the phytogeographer with a convenient, if somewhat arbitrary, line along which to split this elongated Province. The Urals, which trend in a north-south direction, roughly follow the parallel of 60° E. longitude, and this parallel is sometimes used to divide the Province. Within the Euro-Siberian subdivision, the main pattern of plant distribution is not one of different and comparable floral assemblages at opposite ends of the Province; rather, it is one of floral richness to the west, with increasing impoverishment toward bleak Siberia.

The Asian portion of the Province may experience, in its northern sector, winter temperatures colder than any in the Arctic. The Siberian city of Verkhoyansk, lying 1,500 miles south of the North Pole, and just south of the Arctic, has the dubious honor of reporting the lowest winter temperatures of the northern hemisphere. At Verkhoyansk the Fahrenheit temperatures in January average 59° below zero, and a cold wave one February twice brought the local temperature down to 90° below zero. It may seem strange that the lowest temperatures should be recorded so far from the Pole. The explanation lies in the effect of the Arctic Ocean on climate. This ocean is a cold one, but it is yet a great mass of water; and water does not change temperature as rapidly as land. The Arctic Ocean ameliorates the winter temperatures of all the bordering regions. Localities such as Verkhoyansk, although farther south than the Arctic, are also farther inland, and so farther away from the climatic influence of the sea.

One might also be puzzled by the existence of cold-temperate forests at localities that become colder than the treeless Arctic. Actually, the poleward distribution of trees is determined not by the coldness of the local winter but rather by the warmth of the local summer. Verkhoyansk may have a cold winter; but in July (the warmest month) its average temperature is 60° above zero, and its maximum recorded temperature is 94° above. It is interesting, by way of contrast, to note the temperatures at Bulun, in the Siberian Arctic about 300 miles farther north than Verkhoyansk. At Bulun the average January temperature is 40° below zero, and the minimum recorded low is 75° below. Obviously Bulun, on the Lena River about 50 miles from the Arctic Ocean, does not become as cold in winter as the more southerly Verkhoyansk. But on the other hand, Bulun does not warm up in summer as strikingly as does Verkhoyansk. The average July temperature at Bulun is 52° above zero, and the maximum recorded high 85° above.

Tree growth is not possible unless the average temperature of the warmest month exceeds 50° above zero, even if all other environmental conditions are ideal. At most localities these other conditions are not likely to be ideal; soil may be poor, topography unfavorable, moisture too scant, the winds chilling or drying, the nights occasionally very cold in spite of mild daytime temperatures, unseasonable cold snaps so extreme as to nip new plant growth. Some weather stations in the Siberian Arctic report an average July temperature as high as 55° or 56° above zero; yet the local woody plants do not reach tree size. But Verkhoyansk's July average of 60° suffices for the growth of some trees, notably conifers such as fir and spruce.

The western or European end of the Euro-Siberian Province is surprisingly warm for its latitude. (The sunny beaches of the French Riviera lie as far north as Portland, Maine; Aberdeen in Scotland is as far north as Sitka, Alaska.) The comparative warmth of western Europe largely reflects the climatic influence of the Gulf Stream. This great oceanic current, originating in the tropics, sweeps northwestward through the Caribbean, swings around the Gulf of Mexico, passes into the Atlantic through the Straits of Florida, and then turns northeastward. In the North Atlantic the current splits, a part turning southward; but the other part continues northeastward, to pass off the coasts of the British Isles and of Norway.

Because of the Gulf Stream's warming influence, the Province widens at its western end, providing more room and a greater variety of environments in which plants might diversify. Thus the flora of Europe is relatively rich, and includes many genera that are lacking from the eastern end of the subdivision. Before man cleared the land, forests of broad-leaved trees—oak, elm, basswood, birch, sycamore, holly, aspen, beech—grew over much of Europe; but in the Asian portion of the Province, forests are largely made up of the cold-hardy conifers. Also, in western Europe some plants of the Mediterranean region have been able to push northward into the relatively warm southern part of the Province; and the presence of a Mediterranean floral element enhances the phytogeographic distinctiveness of Europe.

Scarcely any plant families are confined to the Euro-Siberian Province. The Euro-Siberian families are often represented also in the Arctic, and in cold-temperature parts of North America. They may also be represented in warm-temperature lands, especially of the Mediterranean region, less frequently

At the left, Nigritella growing in the Bernese Oberland of Switzerland. Its genus is distributed from the Alps to northern Europe. Above, silver thistle, common in Alpine pasturelands. (Swiss National Tourist Office)

of eastern Asia and North America. About one hundred genera of plants are distributed from the eastern to the western end of the Euro-Siberian Province, yet are nearly or quite lacking from adjoining Provinces. Many of the Euro-Siberian endemic genera belong to some one of four very widespread families: the Umbelliferae or parsley family, the Gramineae or grasses, the Cruciferae or mustard family, and the Compositae which include asters and daisies. However, other well-known families have produced genera endemic to, and widespread in, the Euro-Siberian Province; among them are the buttercups, the ginsengs, the mints, the lilies, the olives, and the orchids.

Some of the Euro-Siberian endemic genera include species that are familiar to us as garden flowers. Thus *Doronicum*, of the Compositae, includes the leopard's-bane, whose daisy-like flowers open very early in the spring. The winter aconite, grown in northern rock gardens, belongs to the Euro-Siberian genus *Eranthis*, of the buttercup family. *Hedera*, of the ginseng family (Araliaceae), includes the popular English ivy. This genus is native to the Euro-Siberian Province and a few little islands off northwestern Africa, although several of its species have been carried by man to many countries. The dead nettles, *Lamium*, belong to the mint family; they get their name because they resemble nettles but do not sting. Several species of this genus are grown as ornamentals; but one *Lamium*, small and unattractive, has become a lawn weed in the New World. The unwelcome *Lamium* is the well-known henbit, widespread from Florida to New Brunswick, and from California to British Columbia. Herb paris, occasionally planted in rock gardens, is a member of the lily family; its genus, *Paris*, resembles our *Trillium* but has the floral parts in fours rather than in threes.

One of the best-loved flowers of the United States belongs to a genus that is essentially Euro-Siberian. This is the common lilac, *Syringa*, a member of the olive family. The genus is most diverse in the Asian portion of the Province, and some of the species extend also into warm-temperate Asia. A dozen or more species of Asian lilacs have been brought under cultivation, and variously hybridized or selected for characteristics uncommon in wild stocks. About as well known as the lilacs are the tulips. Their genus, *Tulipa*, of the lily family, is concentrated in the Euro-Siberian Province, although a few species range a little south thereof in both Europe and Asia. Most of our common garden tulips, such as the Darwins and the Parrots, are descended from *Tulipa gesneriana*, a species brought under cultivation so long ago that its original homeland is uncertain; it may have come from northern Iran.

The orchid family, although concentrated in the tropics, is represented by a few genera in cold-temperate lands. *Neottia* is an interesting orchid genus endemic to the Euro-Siberian Province. Like all the more northerly orchids, the Neottias do not grow on the trunk or branch of a tree, but take root in the ground like most plants. Best known of the genus is the brown-stemmed, leafless bird's-nest orchid, a favorite of nature lovers in the British Isles.

Just a few plant genera, perhaps not over a dozen, are restricted to the Asian portion of the Euro-Siberian Province. Several of these genera belong to the mustard family. None is especially familiar to us. In contrast, roughly a hundred genera are confined to Europe, or else extend from this region into western but not eastern Asia. Many of these European genera include species that are familiar in the United States. Among such genera are *Bulbocodium*, similar to a crocus and about as early to bloom; *Lunaria*, the moonwort and satinflower, members of the mustard family, with papery seed pods that are dried for winter bouquets; and *Melittis*, pink-flowered mints. Also familiar are *Stratiotes*, the water soldier, related to eelgrass and grown in aquariums; and *Pulmonaria*, the lungworts, very similar to their relative the forget-me-not.

Among the genera that range from Europe into western Asia are *Astrantia*, the masterworts, of the parsley family; *Laburnum*, the golden chains, of the pea family; and *Vinca*, the periwinkles, of the dogbane family. A somewhat different pattern of distribution is exemplified by *Eremurus*, of the lily family. Its members, called desert candles or foxtail lilies, are adapted for life in arid lands, and are concentrated toward southwestern and south-central Asia, and the Mediterranean region. About 20 Asian species of the genus, and many showy hybrids thereof, are planted in the southwestern United States, where they are frequently—but mistakenly—pointed out as spectacular plants native to the American desert.

Quite a few plant genera are localized in one or more European mountain ranges. One such is *Erinus*, of the figwort family. *Erinus alpinus*, which may be white-flowered or red, is grown in the northern United States, usually in the crevices of rock walls; it is native to the Alps and the Pyrenees. Another localized, montane genus is *Ramonda*, of the Gesneria family. Two of the Ramondas, one

The Euro-Siberian landscape often suggests northern and upland portions of North America. Above is Albula Valley in the Grisons, Switzerland. (Swiss National Tourist Office)

Rocky Mountain National Park in Colorado, U.S.A. (Colorado Department of Public Relations)

from the Pyrenees and the other from the Balkan Mountains, are beautiful plants of the rock garden, difficult to grow but worth the care they must have. *Soldanella* is a genus of the primrose family, endemic to the Alps; its species, with blue, violet, or white flowers, thrive in moist, shady spots of gardens in the northern United States.

Some genera, largely confined to the western part of the Euro-Siberian Province, are also represented outside the Province on some of the Atlantic islands from the Azores to the Cape Verdes. Included here are *Bryonia,* the bryony, of the gourd family; *Carlina,* of the daisy family; and *Crithmum,* seashore plants of the parsley family, grown in rock gardens or, under the name "samphire," eaten in salads.

One could compile a lengthy list of plant species endemic to the Euro-Siberian Province, although belonging to genera with members in other parts of the world. The Euro-Siberian species that reach Europe have generally been taken into cultivation if at all edible or attractive, and most of them have subsequently been exported into North America. Examples include asparagus, an important vegetable; the poison hemlock, a toxic member of the parsley family; a lady-slipper orchid; and a crane's-bill with purplish flowers. We might also mention a cow parsnip with tiny pink or white flowers aggregated into large clusters; a buttercup; and the bittersweet nightshade, a climber with poisonous berries. All these species range from one end of the Province to the other.

Other species, although very widespread, are lacking at least from the British Isles, if not from other westerly areas of the Province. In this category are an aconite or monkshood, with poisonous roots; a pink; the curious gas plant, whose flowers at times emit an inflammable vapor; a blue-flowered *Eritrichium* whose stems and leaves are covered with white hairs; a blackberry; and a clover. Still other species are concentrated toward the western part of the Province, reaching the British Isles on the west but not the Pacific on the east. Here one might list a wormwood; a Butomus or flowering rush, raised in ponds; a sweet pea; and a little, ground-growing orchid called twayblade. Also in this last category are a forget-me-not whose blue flowers have a white, pink, or yellow center; the tansy ragwort, of the daisy family; and a betony, of the mint family.

From the Siberian end of the Province come an onion grown for its flowers; two Bergenias, related to the saxifrages; a larkspur, ancestral to several cultivated varieties; the Siberian iris, similarly ancestral to garden varieties; a Lychnis, akin to the campions; and the Siberian crab apple, valued for its white blossoms. Some species are widespread in the Province, but are very discontinuously distributed and lacking from many areas. Among these are a cotton-grass, actually a sedge, planted around pond borders; the yellow star-of-Bethlehem, a member of the lily family; and an Easter-bell, related to the pinks.

The Euro-Siberian Province has given us many desirable plants; it has also given us many that are far less desirable. Much of Europe has been under cultivation for several thousand years. Various plants, mostly small and fast-growing, were able to invade fields, yards, and roadsides—the areas cleared by man. When Europeans colonized other countries, inadvertently they took the seeds of these plants with them, especially in sacks of grain. Often the seeds would sprout in the new land, and the introduced plants might spread widely.

If a plant is adapted for rapid growth in areas from which the natural cover has been removed, it is very likely to appear, all unwanted, in lawns, gardens, or cultivated fields. Such plants, the ones that grow where we do not welcome them, are generally called weeds. Even if scrupulously eradicated from yards and fields, weeds generally find a home in waste places and especially along roadsides.

Not all weeds are imported, although the really troublesome ones usually are. Nor has the transport of weed seeds been entirely from Europe to other countries. Weeds are still being spread around the world in complex fashion. But of the so-called "widespread roadside weeds," found on most continents, a majority came from Europe. Almost any roadside, from New York to New Guinea, from British Columbia to Brazil, from South Carolina to South Africa, will probably yield shepherd's purse, pigweed, fleabane, one or two little spurges, plantain, six-week's grass, a knotweed, common nightshade, sow thistle, chickweed, a dandelion, stinging nettle, and perhaps two dozen other weeds, most of them originally from Europe.

The Euro-Siberian flora reminds us of the North American one. Especially is it similar to the flora of the northern United States, and of Canada below the Arctic. The coniferous forest, stretching from Scandinavia eastward to the Pacific coast of Siberia, continues also across Alaska and Canada to the Atlantic coast of Labrador. Like Europe, northern North America has oaks, elms, basswoods, birches, hornbeam, crab apples, aspen, elder, and beech, among the trees. It also has crane's-bills, buttercups, cow parsnips, nightshades, monkshoods, blackberries, clovers, twayblade orchids, forget-me-nots, cotton-grasses, larkspurs, and irises, among the herbaceous plants. Often the American species is closely related to, if not identical with, some Old World one.

The animals of the Euro-Siberian Province often have a counterpart in North America, also. Thus the European bison, a forest dweller, belongs to the same species as the plains bison and the woodland bison of North America. The European red deer is closely related to the Siberian wapiti and to the American wapiti which we often miscall "elk." The true elk of Europe is a close ally of the American moose. Among carnivores, the Eurasian brown bear belongs to the same species as our grizzly; and the wolf, which figures so prominently in European legend, is of the same species as the North American timber wolf. The red fox, which ranges from the British Isles to eastern Siberia, also inhabits much of North America. At one time it was suspected that the red fox of New England might be a descendant of British stock imported for fox-hunting; but eventually red fox bones were found in the debris of American Indian encampments dating back to prehistoric times.

Both North America and Eurasia have a lynx, a badger, a wolverine, an otter, a marten, weasels, a pika, marmots, chipmunks, a beaver, jackrabbits, and wild sheep. The European hazel grouse finds counterparts in the American blue grouse and spruce grouse. The Eurasian sport of falconry, in which birds of prey were flown to capture small game, was based in large part upon the peregrine falcon, the gyrfalcon, the goshawk, and the golden eagle—species found also in North America. The little European falcon called kestrel is a close relative of the American sparrow hawk. The ern, or erne (an obsolete name revived in cross-word puzzles), is a Siberian eagle closely related to the bald eagle. It will be seen that faunal similarities between the Euro-Siberian Province and North America mostly involve some cold-hardy mammals and birds.

The fauna of the Euro-Siberian Province differs in some ways from that of temperate North America, for many animal groups never crossed Bering Bridge, or Bering Strait, between the Old World and the New. Thus North America never had a counterpart of the European aurochs, a gigantic wild bull with "handlebar" horns and a shaggy forelock; goats and swine reached the New World only as domesticated stock from Europe. On the other hand, Eurasia never had pronghorns, raccoons, or gray foxes.

Just as some of the commonest weeds came from Europe, thence also came the most widespread of vertebrate pests: the house mouse and the brown rat.

Other European animals, often a pest at least in parts of the United States, include the house sparrow, also called English sparrow; the European starling; the "razorback," a wild hog descended from domestic stock; and the carp, a fish that roots hoglike on lake bottoms.

This account of the Euro-Siberian Province may be rounded out with comments on various faunal and floral diversities from place to place within the great land mass that is Eurasia. These diversities reflect several factors. Today the Mediterranean separates Europe from Africa. However, the Mediterranean is but a remnant of a much greater sea that once stretched all the way across southern Asia to the Pacific. The Eurasia of that day was well separated from both Africa and India. The ancient seaway, dubbed the Tethys Sea, was finally split by the rising of land in its middle portion. Even today the fish fauna of the Mediterranean is much like that of the seas around Japan, but markedly unlike that of the Red Sea just a hundred miles away. (Of course the Suez Canal has encouraged interchange of Mediterranean and Red Sea fish faunas in modern times.)

Another ancient sea existed, off and on, in the area where the Urals now rise. Thus at times the east-west movement of animals across Eurasia was inhibited. Even with Bering Bridge permitting faunal interchange between Asia and North America, some European animal groups could not reach Asia because this sea barred the way. At times, European organisms might have been able to reach North America more easily across the North Atlantic than by any other route. As parts of the Tethys Sea vanished, the Himalayas rose, their rocks rich with marine fossils. Thus a new barrier separated Eurasia proper from India. Ages later, with the coming of glacial ice and a colder climate, much animal life fell back to the south, but many species found their way hopelessly blocked by the Himalayas and the Mediterranean. The species that crossed these barriers had the problem of recrossing when the glaciers melted and the climate warmed again. A great many animals and plants vanished from Eurasia as a result of climatic fluctuations.

In some places the advancing glaciers scraped away the fertile topsoil, which the retreating glaciers replaced with rocks, gravel, and sand. Such places might today be climatically suitable for the growth of forests, but the soil permits the development of moorland only. We hear of moors chiefly in connection with the British Isles, but they are widely scattered over Eurasia. A good many moors

At left, red deer in a zoological garden at Berne, Switzerland. The species is a
European counterpart of the American wapiti. Above, ibexes in Switzerland.
They are true goats; their genus, Capra, was restricted to Eurasia and northern
Africa, until a domestic species was carried by man to other lands. (Swiss Na-
tional Tourist Office)

have been drained or otherwise altered by modern man, and then planted with hardy pines brought from other lands.

When the glaciers retreated for the last time, man began to spread, and soon to modify the landscape. In Europe, the hardwood forests suffered egregiously at the hands of man. Today it is difficult to visualize the hardwood belt that once stretched from parts of the British Isles, northern Spain, western France, Denmark, and southern Sweden eastward through northern Italy, Germany, Poland, and western Russia. Much of this belt has been replaced by wheat fields, vineyards, potato fields, and cities.

Where bits of hardwood forest remain, trees from other countries have often been introduced for their commercial value. Germany's famous Black Forest, for example, had oaks and beeches at lower elevations, and native firs at higher; but American pines and Douglas firs are now extensively planted and timbered there.

Few indeed are the undisturbed remnants of the European hardwood forest. One such relic, the preserve of Bialowieza, exists in Poland, near the Russian border. Here are 11,000 acres of oaks, elder, ash, and hornbeam, along with some towering conifers and a scattering of peat bogs. Bialowieza is especially noted for its population of the European lynx, and for a small herd of European bison. But the bison once ranged all over temperate Europe, and the present Bialowieza herd was established from captive stock. Gone even from Bialowieza is the aurochs, which once lived throughout the hardwood belt. Wild boars and roe deer are still widely but spottily distributed over Eurasia. The red deer persists in many parts of Europe, but is often restocked for hunting purposes.

The wolf was formerly distributed from the British Isles eastward across Eurasia and much of North America; but today in Europe it is rare outside Russia and Yugoslavia. A few wolf packs persist in Spain, Italy, Czechoslovakia, Poland, and Scandinavia. The European brown bear, called the grizzly in America, was once distributed much like the wolf. European populations of this animal still exist, chiefly in mountainous country, in Spain, France, Italy, Yugoslavia, Albania, Bulgaria, Scandinavia, and especially Russia. The lynx, once common over Europe, survives in Spain, northern Scandinavia, Poland, Yugoslavia, Hungary, and especially Russia. A smaller feline, the European wildcat, formerly distributed from the British Isles to the Caucasus, holds out in Scotland, Spain, France, Belgium, Germany, Italy, Hungary, the Balkan Mountains, and the Caucasus. Smaller carnivores—red fox, badger, weasel, and polecat (the last a ferret, not a skunk)—are better able to coexist with man and are widespread over the Euro-Siberian Province.

To summarize: The diversity from place to place, within the Euro-Siberian Province, in part represents climatic differences, especially from east to west; but a vanished sea further helped to split Eurasian wildlife into eastern and western assemblages. Differential extinction, resulting from glacial fluctuations and later from man's activities, also accounts for much faunal diversity within the Province. Finally, the mountains of Eurasia do not form a continuous series of uplifts, but instead exist as a number of well-separated blocks; and the circumstance has militated against some interchange among montane animals and (to a lesser degree) plants.

19 *"Blessed be agriculture! if one does not have too much of it."** **THE MEDITERRANEAN PROVINCE, A RAVAGED LAND**

Continuing the account of the Boreal Kingdom, the next subdivision to be considered is the Mediterranean Province. Lying south of the Euro-Siberian Province, it includes southern Portugal, southern Spain, a bit of the French coast around Toulon, central and southern Italy, southern Yugoslavia and southern Bulgaria, all of Greece and Albania, and western and central Turkey. It also embraces a narrow strip along the eastern edge of the Mediterranean from Syria into Sinai. In Africa the Province encompasses northern Egypt and northeastern Libya, as well as northern Tunisia, northern Algeria, and most of Morocco. All the islands of the Mediterranean, from the Balearics on the west to Cyprus on the east, likewise belong to the Province.

Numerous distinctive plant groups are centered in the Mediterranean Province; but often they transgress the provincial boundaries, which are therefore but approximations. As noted in the preceding chapter, many essentially Mediterranean plants range also into the western end of the Euro-Siberian Province; quite a few of them reach the British Isles. Other Mediterranean groups have

* CHARLES DUDLEY WARREN, *My Summer in a Garden.*

spread to Atlantic islands such as the Azores, Canaries, and Cape Verdes. Along the northern edge of the Mediterranean Province, the plants of that subdivision intermingle, especially at the higher elevations, with species that are primarily Euro-Siberian. To the east, some Mediterranean plants may invade the desert lands from Syria to Iran or even beyond.

The Mediterranean Province is sharply delimited only on the south, where the Atlas Mountains separate it from the Sahara Desert. The mountains introduce phytogeographic complications, however. During glacial periods, when vast sheets of ice spread over Europe, an old European forest flora moved into the Atlas uplift from the north. With the final retreat of the ice, many plants spread northward again into the newly deglaciated lands; but others found suitable habitats on the upper slopes of the mountains, where they have persisted. These latter plants form a decidedly atypical component of the Mediterranean flora. They may find their nearest relatives in various parts of China and Japan, or even of the United States; for in Asia and North America, just as in Europe, an old forest flora moved southward as glaciers advanced, and persisted (at least in part) in the mountains after the glaciers retreated.

When the montane relics are disregarded, the distinctiveness of the Mediterranean Province results mostly from the presence of plants specialized for existence under the rather unusual climatic conditions that prevail in the lands bordering the Mediterranean. In winter the prevailing winds, blowing from the north and west, are warmed as they pass over the western Mediterranean, from which they also pick up moisture. As these winds reach the mountainous uplifts that ring the Mediterranean basin, they rise, become chilled, and drop their moisture as rain or snow. Only at the eastern end of the Mediterranean, where the mountains are lower, do they fail to provide considerable precipitation in winter. But in summer the winds die down and the mountains warm up, so that this season is very dry; days are usually bright, and the summer sun further evaporates the land. In other words the life-giving rains arrive in winter, normally a dormant season for most plants in temperate climes.

Rivers bring the Mediterranean less water than this sea loses by evaporation. If the Mediterranean were a completely closed basin, it would rapidly dwindle, becoming saltier and saltier; eventually it would become a counterpart of Utah's Great Salt Lake. But the Mediterranean is open to the Atlantic by the Strait of Gibraltar. Through this relatively small channel (in places less than eight miles wide, and not much over 1,000 feet deep) the Atlantic constantly pours into the Mediterranean, making up for the water that the latter loses by evaporation. Thus the Mediterranean maintains a remarkably constant water temperature, which in turn moderates the vagaries of land temperature in bordering regions. (However, this beneficent climatic effect is not felt at any distance from the coast, and even northern Africa may experience severe cold snaps. Twice within historic times the Nile River at Cairo has frozen over!)

The Mediterranean Province has been profoundly altered by man. Oaks and pines once grew on the barren hills where Athens now stands; pines and other conifers, interspersed with laurels, once grew thickly at Cannes. The best soils formerly supported forests of holm oak, green pistachio, laurel, and strawberry tree (an *Arbutus*). But more than 2,300 years ago the Greek philosopher Plato deplored the disappearance of forests from his native land. In his *Critias* he told how the mountains of Attica had once been forested; but after timbering, the unprotected topsoil washed into the sea, and soon trees would no longer grow. He said that some mountains, producing nothing but blossoms for the bees, had once been covered with forest; and that many barren and stony fields of the lowlands had once been fertile plains.

Some elements of the Mediterranean flora may have vanished completely; others now occupy only a fraction of their original range. There are still a good many stands of cork oak in the western part of the Province, and especially in the African portion thereof. In many parts of northern Africa, Aleppo pines rise above an understory of juniper, green pistachio, and thyme. Stone pine is common in the European section of the Province. The cedars of Lebanon have nearly vanished from the land with which they are traditionally associated, but persist in the Atlas Mountains. Olives line the Mediterranean hillsides and mountain slopes, but they are under cultivation, as are the grapes which thrive in a soil that is scarcely more than crumbled rock.

The former oak forests of the Mediterranean Province have largely been replaced by a scrubland of stunted oaks, myrtles, strawberry trees, and other woody shrubs, bound together with thorny vines. This plant association is the famous "maquis," and it develops as a result of improvident timbering, burning, and grazing. The maquis may be still further degraded into the so-called garrigue, which will not support even brushy thickets but only a low and scattered growth of rockroses, thyme, rosemary, lavender, the dwarfed kermes oak, a palmetto, and a few other hardy species.

Five plant families are concentrated in the Mediterranean Province, to which they are nearly if not strictly endemic. Only three of these are at all familiar. The Globulariaceae include the Globularias or globe "daisies," plants of the rock garden. The Punicaceae include only two species. One is the pomegranate, now grown for its fruit in the warmer parts of the United States; the other lives outside the Province, on Socotra. The Ruscaceae include the butcher's-broom, well known in Europe. Brooms also grow on Atlantic islands outside the Province.

Perhaps 250 genera are concentrated in the Mediterranean lands. However, some of these have representatives in western Europe also. Many of the genera include some familiar species. *Atropa,* of the nightshade family, includes the belladonna, source of atropine. The species of *Bellis,* in the daisy family, are called English daisies. *Jasione,* of the bellflower family, provides gardens with the blue-flowered sheep's-bit. Species of *Medicago* are sometimes grown for ornament, but *Medicago sativa* is a valuable forage and cover crop; it is the well-known alfalfa, a legume (that is, a member of the bean and pea family). The root system of the alfalfa plant may probe the soil to a depth of fifteen feet in search of moisture; and this characteristic, an adaptation that permits growth during the warm, dry Mediterranean summer, also renders the species especially useful in other arid regions such as Chile and the west-central United States. Another leguminous genus, *Ulex,* includes the European plant known as furze, or gorse. *Origanum,* of the mint family, provides a spice: not oregano, as one might guess, but sweet marjoram. Another *Origanum* is the wild marjoram, naturalized in the United States. *Verbascum,* of the figwort family, includes the mulleins, one of which is a familiar weed in the United States.

All the above genera, and many more, are essentially Mediterranean, but have at least one species that ranges as far north as the British Isles. Such genera have often become known in North America only through their British species. Thus it is surprising to learn that English daisies, mullein, and gorse are but British outliers of Mediterranean stocks.

A good many Mediterranean genera do not reach the British Isles, yet include garden species familiar there and in the United States. Examples are *Centranthus,* of the valerian family, with flowers of white, pink, or red; *Crocus,* of the Iris family, one of its species yielding saffron and others providing beautiful blooms in earliest spring; *Galanthus,* the snowdrops, of the Amaryllis family; and *Gypsophila,* of the pink family, the best known species being baby's breath.

Helleborus, a genus of the buttercup family, includes the Christmas rose; and *Nigella,* likewise of the buttercup family, includes love-in-a-mist and the fennel-flower or black cumin. Other examples are *Lavandula,* the lavenders, of the mint family; and *Muscari,* the grape hyacinths, members of the lily family, one of them well established as a wild plant in the eastern United States. *Narcissus,* of the Amaryllis family, embraces the daffodils and the jonquil, as well as the sweet-scented flower popularly called by the generic name.

An important genus, widespread throughout the Mediterranean Province, is *Cistus,* of the Cistaceae. Its numerous members, mostly evergreen shrubs, are known as rockroses. Often they will grow in the poor, stony soil of the maquis or of the garrigue, and in some Mediterranean wastelands they are almost the only plants to be seen. Rockroses are popular flowers of the rock garden, and several of the species also yield the gum called ladanum (not laudanum). Among other Mediterranean genera are *Aubrietia,* the false rock cresses, of the mustard family; and *Rosmarinus,* which belongs to the mint family and which includes the rosemary, source of a medicinal oil and of a seasoning. *Santolina,* of the daisy family, includes the lavender cotton, one of the first plants ever to be raised in American gardens although today rarely seen.

Several genera are restricted to some small part of the Mediterranean Province, perhaps Spain, or northern Africa, or these two together, or some one or two of the Mediterranean islands. The more localized endemic genera are not well known, and need not be listed; several of them belong either to the mustard family or to the parsley family. Various genera are represented more or less continuously from central and western Asia westward into the eastern end of the Mediterranean Province. In this category are *Chionodoxa,* of the lily family, a genus including glory-of-the-snow; *Danae,* also of the lily family, its best known species the Alexandrian "laurel," grown for its evergreen leaves and red berries; and *Cicer,* a legume, one of its species the garbanzo bean or chick-pea.

Numerous genera, not confined to the Mediterranean Province, have produced species that are nearly if not strictly endemic thereto. Many of these species have long been under cultivation, and are very familiar to us today. From the Mediterranean came the leek, a mild-flavored onion; caperbush, whose flower buds are called capers; the "French" artichoke; the edible fig; curl-leaved parsley; and the "Spanish oyster" with parsnip-like roots. The olive, source of oil, a fine wood, and a fruit of several uses, is the cultivated descendant of a small tree that grows wild in the coastal maquis of the Mediterranean. The cork of commerce is taken from the outer bark of the cork oak; in nature it protects the tree trunk from evaporation, fires, and frost. A red dye was formerly derived from a scale insect that feeds on the kermes oak. The Judas tree, a close relative of the American redbud, received its name because Judas is said to have hung himself from its branches. The love apple or mandrake, with its vaguely man-shaped root, gave rise to fantastic superstitions in medieval times. The hairy-leaved *Acanthus mollis*, which we call bear's-breech, served as the model for the graceful leaves carved on the capitals of Greek Corinthian pillars.

Garden flowers from the Mediterranean include the poppy-flowered anemone; the garland chrysanthemum; the Persian Cyclamen, a wild ancestor of most cultivated Cyclamens; and the Oriental hyacinth, ancestral to most cultivated hyacinths. Also familiar are the Madonna lily and the scarlet Turk's-cap lily; a peony, the old-time garden variety; mignonette, whose sweet-scented flowers are laden with honey; and the dusty miller, whose woolly leaves look as though they had been powdered with flour. Among ornamental trees and shrubs from the Mediterranean are the oleander, a bush with beautiful flowers and a highly poisonous juice; the "English" laurel, really an evergreen cherry with a close relative in the United States; a red-berried Pyracantha or fire thorn; and the holm oak, also called holly oak on account of its stiff, holly-like leaves. Others include the Pontic Rhododendron, a purple-flowered species often used as a hardy rootstock upon which to graft hybrid Rhododendrons; an evergreen Viburnum; and the laurel of Classical fame.

A Mediterranean type of climate—summers warm and dry, winters cool and wet—often characterizes areas along the western coasts of continents, in the mid-latitudes. For example, much of California, from the Pacific coast inland to the foothills of the mountains, has a Mediterranean climate. However, the native flora of coastal California shows no special affinity to that of the Mediterranean Province; and in general, each area of Mediterranean climate has its own distinctive assemblage of plants. But it is true that vegetation has much the same general appearance in all the areas with a Mediterranean climate. Trees are slow-growing, and few attain great stature; shrubs are often thorny, stiff, and angular of branch. Trees and shrubs alike have relatively small leaves that are crisp, coarse, glossy, and toothed or spined. Woody plants usually have evergreen leaves, and the herbaceous or nonwoody plants often sprout from a bulb

Vineyards at Fondi, Italy. (Italian State Tourist Office)

or tuber. Thus the coastal chaparral of California looks much like parts of the Mediterranean maquis, although the two plant associations differ in the species of which they are composed.

The similarity of California to the Mediterranean is enhanced by the success in the former area of plants from the latter. Thus the Mediterranean grape was introduced into California by Franciscan fathers a little more than two hundred years ago, and today California's wine production is exceeded only by that of France and of Italy. Olives, grapes, and French artichokes do especially well in California, also; and garden flowers from the Mediterranean are more conspicuous in this state than in any other.

Most vertebrates of the Mediterranean Province are not endemic thereto, but are widespread in Asia or Africa. The Mediterranean fauna, like the flora, has suffered at the hands of man. Lions roamed the Middle East and southeastern Europe in Classical times, but no longer. Probably they fed upon wild sheep, gazelles, and wild goats—herbivores replaced in most areas by man's domestic flocks. The mouflon, a wild sheep, survives only on Corsica and Sardinia; the Barbary sheep is restricted to the mountains of northern Africa; wild goats, unmixed with domestic stock, persist mostly on the islands, and in the mountains that rim the Province. In Roman times the African elephant ranged northward into Tunisia, the Indian elephant westward into Syria; but neither lives anywhere near the Province today. The Persian fallow deer has fallen back to one or two wooded river valleys in Iran.

The Asiatic jackal still ranges westward into northern Africa and the Balkan Peninsula. The striped hyena is distributed from Pakistan westward into Morocco. The European fallow deer has fared better than most other large mammals, having been restocked in the Mediterranean region and introduced into other countries. The African wildcat survives on Majorca, Corsica, and Sardinia, as well as in northern Africa and elsewhere. Although a fierce little animal when cornered, it is the ancestor of the house cat. In fact it can be crossbred to its domestic descendant.

A famous endemic animal of the Mediterranean Province is the Barbary "ape," confined to Morocco, northern Algeria, and the Rock of Gibraltar at the southern tip of Spain. Actually the animal is a tailless monkey, a kind of macaque, and its genus (*Macaca*) is widespread from Pakistan to Japan and Indonesia. It is suspected that the monkey was brought to Gibraltar from northern

Africa by Arabs during the Moorish conquest of Spain. At any rate the animal would have died out from Gibraltar had it not been for the British, who came to occupy the Rock in the eighteenth century. According to local superstition, the disappearance of this monkey from Gibraltar would herald the end of British control there; and in 1863, when the Gibraltar monkey population had dwindled to three lonely individuals, fresh stock was imported from Morocco. And after World War II, Winston Churchill again restocked the Rock with Moroccan specimens.

Birds have had a difficult time in the Mediterranean region, where their edibility is their only characteristic to elicit much enthusiasm from the general populace. In Italy alone, nearly 100 million birds, many of them diminutive songsters, are shot, snared, or bird-limed annually. A large percentage of these are migrants; destruction of the habitat has decimated the resident bird life.

European birdwatchers visit the Mediterranean with the hope of seeing the greater flamingo, the European bee eater, and the common roller. These colorful species are characteristic of the Mediterranean area, but are not endemic thereto. Audouin's gull, a vanishing species which nests on a few small islands, is the only sea bird endemic to the Mediterranean; but there are quite a few endemic land birds, including a variety of Old World warblers, and a little partridge of northern Africa.

Several birds of the Province exhibit highly interesting patterns of distribution. Thus the white-headed duck nests in Morocco, but also ranges eastward into central Asia. Its distribution therefore parallels that of such plant genera as *Chionodoxa, Cicer, Danae,* and *Michauxia.* The azure-winged magpie lives in two widely separated areas. One includes southern Spain and Portugal, the other northern China, Manchuria, and Japan. Widely disjunct distributions are often encountered in the bird group that includes the magpies and jays. These birds are often quite specialized in their feeding and nesting behavior; and a species of this group might easily be wiped out from a large part of its range by a change of climate and accompanying change of vegetation. Of course one suspects the range of the azure-winged magpie to have been fragmented during a time of glacial advance.

The Corsican nuthatch is confined in the Mediterranean region to the pine forests of Corsica, but the same species reappears in northern China and Korea, and again in Canada. The distribution calls to mind that of the old north-temperate flora, some of whose components have persisted in China, Japan, North America, and the mountains of the Mediterranean Province. Also called to mind are the web-toed salamanders mentioned in Chapter 1; their genus, *Hydromantes*, is represented in the mountains of Sardinia, the Apennines, and the Maritime Alps of France and Italy, as well as in the mountains of California.

The vertebrate fauna of the Mediterranean Province shows no strong affinity to that of any other region having a Mediterranean type of climate. This is not surprising, for such areas have always been separated from each other by wide expanses having some other type of climate. As noted in the opening chapter of the book, there are some faunal resemblances between Europe and western North America, but these resemblances are not confined to animal groups that are exposed to the rigors of a Mediterranean climatic regimen.

20 *"Upon leaving Kieriman and traveling three days, you reach the border of a desert. . . . The first three days but little water is to be met with, and that little is impregnated with salt, green as grass, and so nauseous that none can use it as drink."** **ATLANTIC ISLANDS; ASIAN DESERTS AND GRASSLANDS**

The Azores are a group of islands in the Atlantic, about 800 miles west of Portugal. They are volcanic peaks, and the highest of them rises an incredible 24,000 feet above the ocean floor, although only 7,600 feet are above water. Farther south, 350 miles west of Morocco, lie the mountainous Madeira Islands, with Madeira itself towering 6,890 feet above sea level. About 300 miles south of Madeira are the Canary Islands, with peaks from 2,000 to more than 12,000 feet high. Much farther south, almost 400 miles west of Senegal, lie the Cape Verde Islands, which rise from 1,300 to more than 9,000 feet. No familiar geographic name embraces all four of these archipelagos; and phytogeographers, including all four in a single subdivision, coined the term Macaronesian Province. ("Macaronesia" is merely a formal way of saying "miscellaneous islands.") The Province is not an especially important one. It harbors perhaps two dozen endemic genera, about half of them confined to the Canaries.

* MARCO POLO, thirteenth-century traveler, on Iran.

Madeira has a few endemic genera, only one particularly interesting. This one is *Melanoselinum*, of the parsley family. In other parts of the world, the members of this family are usually herbaceous plants attaining no great stature; familiar examples include the snakeroots, fennel, angelica, sweet cicely, carrot, water hemlock, celery, and dill. On Madeira, however, *Melanoselinum* reaches tree size.

Two genera are confined to the Cape Verdes, but no genus is limited to the Azores. Three genera are restricted to the Canaries and Madeira; one genus to the Canaries and the Azores; and one to the Canaries. Madeira, and the Azores. In this fashion the Province is unified.

Several genera are shared between the Macaronesian and the Mediterranean Province. Thus *Ecballium*, which includes the oddity known as squirting cucumber, is represented in both subdivisions. *Echium*, whose members are called viper's bugloss, is similarly distributed. The genus *Ruscus*, whose best known species is butcher's-broom, includes another species that is distributed from Madeira to the Caucasus. *Centranthus*, mentioned in connection with the Mediterranean Province, has an endemic species on Madeira and the Canaries.

Each Macaronesian archipelago has peculiarities. The Azores have the poorest flora and the smallest percentage of endemic species. This circumstance is not attributable solely to the remoteness of the archipelago; surface currents also militate against the drifting of seeds to these islands. As noted previously, the Gulf Stream splits in the North Atlantic, one branch continuing northeastward toward the British Isles and Norway. The other branch sweeps eastward past the Azores on its way toward the coast of Portugal. Thus the archipelago is in a position to receive ocean-borne seeds only from the New World; and most seeds would be unable to withstand such a prolonged immersion in salt water. Furthermore, comparatively recent volcanic activity may have wiped out a portion of the local flora; for several of these islands, their craters still bubbling with hot springs, have known eruptions and earthquakes in historic times. The Azores have received their small flora chiefly from western Europe, probably through the agency of winds and seabirds.

The aforesaid branch of the Gulf Stream, flowing eastward through the Azores, eventually turns southward; it passes the shores of Portugal and Morocco, and thence sweeps through the Madeiras and the Canaries. These two archipelagos are therefore in a good position to receive seeds from the western end of the Mediterranean Province; and the distances involved are not great. The Madeiran flora is richer than that of the Azores, and includes a higher percentage of endemic species. The affinities of the flora are clearly toward the Mediterranean lands.

The Canaries do not arise from the ocean depths but rather from the continental shelf, the shallowly submerged edge of the African mainland. Islands in such a position usually have had continental connections; and it may formerly have been possible for African plants to spread overland into the Canaries. Of the four Macaronesian archipelagos, the Canaries have by far the richest flora and the highest percentage of endemics. The floral relationships are largely toward northern Africa.

After passing through the Canaries, the oceanic current again splits, a part swinging back westward to recross the Atlantic; but another part continues southward past the Cape Verdes. These islands have a flora a little larger than that of the Azores, and with a slightly higher percentage of endemics. The endemic species usually have Mediterranean kin. Unlike the other three archipelagos of the Province, the Cape Verdes are fully tropical, and harbor a variety of plants that grow also in the African tropics.

In Macaronesia, as elsewhere, some conspicuous plants may be of very widespread distribution, transgressing the boundaries of several Provinces. Thus a silk vine of the islands ranges also to Somaliland; the tree heath ranges from Macaronesia northeastward to Greece and southeastward to Tanzania. Numerous useful or ornamental plants have been introduced into Macaronesia from the Mediterranean Province and elsewhere. On Madeira, especially, a good bit of the natural vegetation has been replaced by vineyards.

The faunal relationships of Macaronesia are toward the Mediterranean lands and northern Africa, except in the case of the Cape Verdes where they are more toward the African tropics. The Province harbors no native land mammals, snakes, amphibians, or freshwater fishes. Lizards are the only Macaronesian land vertebrates of much geographic interest. Skinks and geckos predominate. There are no endemic genera of lizards in the Province, but there are several species endemic to the Canaries, and others to the Cape Verdes. Madeira has one endemic lizard species, but the Azores have none.

One lizard genus merits special attention: *Tarentola*, of the gecko family. These big geckos are characteristic of the Mediterranean lands, but there is also a species on the Canaries and another across the Atlantic on Cuba and the

Bahamas. One recalls that a strong ocean current sweeps past Portugal and Morocco to the Canaries, where it turns westward and makes directly for the West Indies. It is tempting to suggest that a floating log, with clutches of gecko eggs beneath its bark, first brought *Tarentola* to the New World. The West Indian *Tarentola* is a distinct species, and its bones have been discovered in fossil deposits on Cuba; so we cannot include *Tarentola* among the Old World geckos that arrived in the New World as stowaways on ships.

Previously it was noted that the Mediterranean Province had floral relationships on the one hand with certain Atlantic islands, and on the other hand with western and central Asia. Having discussed these islands, which make up the Macaronesian Province, we may turn to a larger subdivision bordering the Mediterranean Province on the east: the Western and Central Asiatic Province. This includes southern Turkey, northern Iraq, and most of Iran; southern Russia, most of Kazakhstan, and Trans-Caspia; most of Sinkiang and Mongolia; and the Tibetan Plateau.

At the heart of this Province lies a series of deserts: the Iranian, the Turkestan, the Takla Makan, and the Gobi. In these deserts the summers are scorching, the winters bitterly cold, the annual rainfall usually less than six inches. In the poorest, sandiest soils the only plant may be white saxaul, a curious, scraggly bush whose branches, nearly leafless, cast but little shade. In places the soil is salty, and here grows the leafless black saxaul. Sagebrush and saltbush appear where the soil is a little better.

Toward the edges of the Province, soils become richer and rainfall a bit more plentiful; the desert passes into a sparse grassland, characterized by scattered clumps of fescue and feather grasses, sagebrush, and even some short-lived annual tulips and buttercups. Along the northern edge of the Province, in southern Russia and southwestern Siberia, the soil becomes rich and black; the rainfall is adequate to support a lusher grassland, with the grasses tall and thickly clustered. Here tulips, Irises, peonies, hyacinths, sainfoin, broomrape, white Draba, and sages burst into bloom early each year, before the grasses attain full height. This Province harbors roughly 150 endemic genera, most of them distinct by virtue of adaptation to a very dry environment, or to saline soils. A majority of the endemic genera belong to the mustard family, the goosefoot family, or the parsley family.

Two of the endemic genera are especially familiar. *Cannabis*, of the mulberry family, includes but one species, although many varieties of this species have been developed during thousands of years of cultivation. *Cannabis sativa* is grown for the fiber called hemp, and for the drug called hashish. The plant has been introduced into the New World, where it and its drug extract are called marijuana. *Spinacia*, of the goosefoot family, includes the popular vegetable spinach. Two other endemic genera of the Province are known at least to the gardening enthusiast. *Exochorda*, of the rose family, is the genus of the pearl-bushes, ornamental shrubs resembling a Spiraea. *Ostrowskia*, of the bellflower family, is restricted to the eastern part of the Province, and includes the giant bellflower which reaches a height of five feet.

The western end of the Province is less severe of climate than the eastern, and harbors a larger flora. It has a considerable number of endemic species belonging to widespread genera; and many of these species were taken into cultivation, especially in Iran and Turkey, whence they were carried to other lands. From the western part of the Province came the fritillary called crown imperial; the hyssop, an ornamental, culinary, and medicinal herb; a jasmine; and Molucella, once known as shellflower but now generally called bells of Ireland. To the list we could also add a mint; the Oriental poppy; the European mock orange, a Philadelphus; the Oriental plane tree, a congener of the sycamore; and the almond.

In and near the southwestern end of this Province, man first began to cultivate wheat and barley. Large-grained wild wheat, small-grained wild wheat, and wild barley still exist in the area today. But a good 8,000 years ago, people at Jarmo in northern Iraq had already begun to take the cereal grains into cultivation. It was a step of enormous consequence in man's rise to civilized status.

A few well-known plant species are native to the high, barren lands that lie in the eastern portion of the Province. Among them are a tragacanth, an Iris, a honeysuckle, a fleecevine, a pea tree, a false tamarisk, and a primrose. Better known than any of these is the rhubarb, which on the Tibetan Plateau will grow at an elevation of more than 14,000 feet. Some plants live at even higher altitudes on the Plateau. *Saussurea leucocoma*, a jug-shaped plant of the daisy family, reaches 19,000 feet. It wears a cottony coat for retention of the sun's heat and for protection against evaporation; it is pollinated by shaggy, high-altitude bumblebees. A little grass of the genus *Poa* reaches 19,000 feet, too. However, plants that grow at very high altitudes on the Tibetan Plateau seldom are endemic thereto; usually they range also into the Himalayas and sometimes into other mountain chains.

On the southern part of the Tibetan Plateau lives a small plant whose yellow flower heads are each surrounded by a starlike cluster of woolly, white leaves. Any European mountain-climber would recognize the plant, and call it by its German name of edelweiss—"noble-white." Actually, edelweiss is not a single species of plant but a group of closely related species, distributed in the mountains all the way from Europe to Japan.

In parts of the Western and Central Asiatic Province, rainfall is insufficient to leach out the salts formed in the ground by the chemical breakdown of rocks. Streams, while few and perhaps intermittent, occasionally overflow their banks; floodwaters dissolve the salts, which are then deposited in the upper soil when the waters evaporate. In places the ground has become very salty through the ages. It is interesting to note that some plant species, characteristic of saline soils in central Asia, have relatives that live on seashores in other Provinces. For example, a Statice of the Tibetan Plateau is closely related to the sea lavenders (*Limonium*), plants of the American saltmarsh. The Russian thistle, an Asian plant that has become a pest in the western United States, belongs to the same genus as the saltworts (*Salsola*), Eurasian saltmarsh plants introduced into America.

Birds are the only conspicuous vertebrates in the deserts of this Province, but a remarkable lot of herbivorous mammals live in the bordering steppes and prairies. Przewalski's wild horse once roamed the lusher grasslands, but constant hunting pushed it back into the semidesert. In the last few decades it has vanished from the wild, although a few herds are kept in captivity. The kulan, or Mongolian wild ass, was once widespread in Eurasia, but today persists only in central Mongolia; its more southerly race, the onager, survives only in a game preserve on the Iran-Turkmenistan border. The Bactrian camel was formerly common in the grasslands of central Asia, but only two small colonies of this animal now exist in the wild state, one in the Mongolian Altai and the other in Sinkiang. Strangest animal of the steppes is the saiga, an antelope with a moose-like snout and downwardly directed nostrils. It was hunted almost to extinction for its horns, valued as medicine in China; but when given protection in southern Russia, the saiga made a spectacular recovery.

The Tibetan Plateau harbors several distinctive herbivores. On the southern part of the Plateau, where the grass is comparatively lush, the Tibetan gazelle lives at an altitude of 12,000 feet. Farther north the Plateau becomes higher and drier, and here roams the kiang or Tibetan wild ass. Still farther north, on cold

Edelweiss, common on the Tibetan Plateau, ranges from the Alps to the mountains of Japan. (Swiss National Tourist Office)

and desolate uplands that rise to 20,000 feet, lives the Tibetan yak, a shaggy member of the cattle family. This barren land is also the home of the chiru, a graceful antelope covered with wool; and of the argali, a great wild sheep that follows mountain chains to localities off the Plateau. Farther west, toward the Pamirs and the Hindu Kush, the argali is replaced by a near relative, the Marco Polo sheep. Here, too, is found the markhor, a spectacular wild goat with corkscrew horns. The large predators of the Province, in addition to man, are the Eurasian brown bear and the wolf, both species widespread in north-temperate lands. Burrowing rodents—ground squirrels, marmots, and meadow mice—are common in the steppes.

Some geographically interesting birds inhabit the Province. The family Prunellidae includes a dozen species of small birds known as accentors. All but one of these species are confined to high plateaus and mountains, above timberline; or else to the treeless Arctic. Accentors are characteristic of the Province, although certain of the species range variously to the Alps, the Atlas Mountains, the Himalayas, or the mountains of Japan. The one lowland accentor, called dunnock or hedge sparrow, is widespread in Europe; and a Siberian species has turned up occasionally on islands off the coast of Alaska.

The bustard family (Otidae) is represented in Eurasia, Africa, and Australia; but its most impressive species, the greater bustard, is characteristic of western and central Asia (although ranging both to the east and the west thereof). Heaviest of flying birds, it has been known to reach a weight of 37 pounds; but on the plains and steppes it runs more often than it flies.

The rock ptarmigan, a circumpolar bird of the Arctic, reappears in the high mountains of central Asia, as well as of Europe and Japan. It affords another example of an arctic-alpine distribution. Other birds, characteristic of the Province although not necessarily restricted thereto, include Hodgson's partridge, allied to the European gray partridge; a snow partridge; the snowcock, a grouse-like bird of the high altitudes; a variety of larks; a rosy finch and the snow finch; the grandala; a ground jay; and a small crow.

There are faunal and floral similarities between this Province and western North America, but they are likely to involve other regions as well. Sagebrushes, sages, and fescue grasses are conspicuous in parts of western North America, just as in parts of western and central Asia; but they belong to genera (respectively *Artemisia, Salvia,* and *Festuca*) with representatives in many lands. Various wolves, bears, ground squirrels, marmots, and meadow mice help unify the Arctic-Temperate Region, but do not specifically link western and central Asia with some particular portion of the New World.

The argali and Marco Polo sheep are related to the American bighorn and Dall sheep; but otherwise, the big herbivores of central Asia lack New World kin. North America once had native camels and wild horses, but they have been extinct for millennia. The American "wild" horses of historic times are but the descendants of domestic stock.

A final comment on the Western and Central Asiatic Province: In 1922 the American naturalist and explorer Roy Chapman Andrews led an expedition deep into the Gobi Desert. In an area of sandstone outcrops he found clutches of fossilized eggs, laid by a small, herbivorous dinosaur some 95 million years before. Search revealed many eggs, two of them with fossilized embryos; and there were also the skeletal remains of numerous hatchlings and adults, some of the latter up to nine feet long. Today the locality is one of barren sand, gravel, and exposed bedrock, with naught but a few scattered clumps of coarse grass and stunted sagebrush. Fierce winds blow, and in winter the temperature drops to 40 degrees below zero. Surely central Asia must have had a different climate and vegetation in the days when plant-eating dinosaurs nested in the heart of the Gobi. Fossil beds of Mongolia and Kazakhstan reveal that the Province once supported oaks, maples, walnuts, poplars, willows, and conifers—a forest vegetation, fed upon by herbivorous dinosaurs which in turn were eaten by great reptilian carnivores. Thus the Western and Central Asiatic Province, like the Arctic, is today inhospitable of climate and sparse of vegetation, yet has not always been so.

21 *"In 1868 I had the pleasure of announcing the discovery of this genus, not indeed where we were looking for it, but where experience had led me to expect that any or every peculiarly Atlantic states type might recur, namely in Japan."** **THE SINO-JAPANESE PROVINCE**

One more Province of Asia, the Sino-Japanese, belongs to the Boreal Kingdom. Its description will complete the account of the Old World flora, although two North American Provinces remain to be considered. This Province includes Korea, most of Manchuria, the extreme southeastern corner of Siberia, the southern part of Sakhalin Island, the Japanese archipelago, the Kurils, and most of northern and central China. Appended to the Province are the world's highest mountains, the Himalayas. In its north-south extent this interesting Province fills the gap between the cold-temperate coniferous forest of Siberia and the tropical rainforest of southeastern Asia; in its east-west extent the same Province accomplishes a transition from the humid Pacific lowlands to the arid uplands of central Asia.

The flora of this subdivision is rich and noteworthy for a great variety of trees. There are more species of trees in this subdivision than in all other north-temperate Provinces together. The floral richness of Japan and eastern China led botanical explorers to wonder what useful or ornamental species might be found in remoter parts of the Province. Today the gardens of Europe and North

* ASA GRAY, nineteenth-century botanist, on the "lost" Shortia of South Carolina.

227

America abound with showy plants brought from the mountains and valleys of western China, Tibet, Bhutan, and Nepal.

Roughly three hundred plant genera are restricted to the Sino-Japanese Province, but very few of these have representatives in all parts of this diverse subdivision. Among the most widely ranging genera of the Province are *Caryopteris*, the bluebeards; *Euptelea*, shrubs whose blossoms have bright red anthers but no petals or sepals; and *Helwingia*, of the dogwood family, remarkable because its flowers are borne on the upper surfaces of the leaves. Also widely ranging are *Stachyurus*, shrubs whose bell-shaped flowers appear before the leaves; and *Tricyrtis*, the toad lilies.

Almost as widely distributed, represented at least throughout most of China and Japan, are *Callistephus*, the China asters; *Hosta*, the plantain lilies; and *Liriope*, the lilyturfs, popular as ground cover or for the edging of flower beds. Here we might also mention *Lycoris*, relatives of the Amaryllis; *Platycodon*, the balloonflowers; *Rhodotypus*, the jetbeads; and *Rohdea*, lilies grown for their foliage rather than their blossoms. Larger ornamentals include *Akebia*, climbing shrubs with an edible fruit, and with a pliable stem used in wickerwork; *Nandina*, a hardy shrub with evergreen leaves and bright red berries; and *Paulownia*, the empress trees.

Many genera are widespread in China but lacking from Japan. Among them are *Hovenia*, the "Japanese" raisin tree; *Kerria*, whose single species is called by the generic name; and *Kolkwitzia*, its best known member the beauty bush. More familiar, perhaps, are *Litchi*, the Chinese litchi nut; and *Poncirus*, the trifoliate orange, which makes a thick, spiny hedge. At least seventy genera are confined to Japan, but few of them are familiar. Best known of the Japanese genera may be *Anemonopsis*, attractive relatives of the Anemones. The genus *Fatsia* was probably endemic to Japan, but one of its species, the source of "rice" paper, has been spread widely in eastern Asia. (The odd generic name is from the native name of *fatsi*.)

Many genera, not restricted to the Province, have produced endemic species therein. Sino-Japanese species, well known in both Europe and North America, include a climbing Clematis; the popular shrub called Weigela; an Iris; the Asiatic ginseng, highly regarded in China as a medicine; and the hardy rose called "*rugosa*." Garden vegetables from the Province include the skirret, related to parsley but with an edible root; and the Japanese artichoke. All of these species are widespread in the Province, and there is no certainty about where they were first brought under cultivation.

Absent from outlying parts of the Province, and originally restricted to China and Japan, were the Japanese Anemone; the gold-dust tree; a large Hydrangea; and the tiger lily. Similarly distributed were the "Boston" ivy, also called Japanese ivy; the memorial rose, ancestor of many hardy climbing roses; and the strawberry geranium, actually a saxifrage. Better known than any of these is *Camellia japonica*, called either Camellia or japonica.

Originally endemic to China alone were the Chinese silvervine; the familiar Aspidistra; a jasmine; the Chinese fan palm; and a bush honeysuckle with evergreen leaves. Other such endemics include the fairy primrose, the Chinese primrose, and another primrose with fine hairs that may irritate the skin; also the thornless "*banksia*" rose and the shrubby Omei rose; and the Chinese Wisteria (properly, Wistaria). Large, shrubby species from China include the creeping fig, the common Forsythia or golden bell, and an Osmanthus with extremely fragrant blooms. Trees include the "Japanese" persimmon, the Chinese angelica tree, and the white mulberry.

Native to Japan alone were such species as *Astilbe japonica*, often miscalled Spiraea; a Dicentra, the bleeding heart of old-time gardens; the Japanese witch hazel; the golden-banded lily; and a Magnolia. To this list could also be added the Japanese primrose; the "*multiflora*" rose which, hybridized with the Chinese rose, gives us the "*polyantha*" roses; a climbing hydrangea; and a Stephanandra, prized for the bright color of its leaves in autumn. Best known of the Japanese endemic species is the Easter lily.

Some familiar species of the Sino-Japanese Province were originally very restricted in range. From Manchuria came a pea tree and the soybean, the latter spread widely by man. Siebold's beech, and an elmlike tree called Zelkova, were endemic to Korea and Japan. From the Himalayas came a large Cotoneaster with exceptionally attractive flowers and fruits; a fleabane; two gentians of the rock garden; and an Incarvillea, miscalled "hardy Gloxinia." Likewise Himalayan were Campbell's Magnolia; Bulley's primrose; and a Meconopsis or blue poppy, one of the finest blue flowers but difficult to grow.

The Himalayas form an aberrant part of the Province, harboring high-altitude species that follow mountain chains into other subdivisions. For example, *Saussurea* and rhubarb, mentioned in connection with the Tibetan Plateau,

China is called "the mother of gardens" because so many of its flowers have been taken into cultivation. Hollyhock was under cultivation in China when first encountered there by visiting Europeans. Here it lines the wall of an old-fashioned garden in Berkshire, England. (British Travel Association)

live also in the Himalayas. Some plants of these mountains merit comment because of the altitudes at which they grow. The altitude record for flowering plants is held by *Stellaria decumbens*, a member of the pink family and a congener of the familiar Easter bells. The Himalayan Stellaria has been found growing in rock crevices at 20,130 feet. The Himalayan Anaphalis, a congener of the pearly everlasting, almost equals the record of Stellaria, and hardly ever grows below 17,000 feet. Gentians, buttercups, Allardias, and cinquefoils reach impressive altitudes, also.

Plants from China and Japan have been especially successful, and popular, in the southeastern United States. Stroll through the suburbs of any large town in, say, Georgia or South Carolina, and you will be sure to see Nandina bushes beside the porch steps, impassable hedges of thorny Poncirus, rows of graceful Forsythias, lilyturf bordering the walks and the tiger-lily beds. The native oaks are draped with Chinese Wisteria. House foundations are concealed by Hydrangeas, Weigelas, Camellias, Japanese Azaleas, and Osmanthus bushes. In the yards stand Japanese persimmons, pink-flowered Asiatic Magnolias, and Japanese maples; along the roadsides are Chinese elms and Paulownia trees. White mulberries are so common as to seem native.

Floral similarities between eastern North America and the Sino-Japanese Province involve more than the introduced ornamentals of the Deep South. Many native plants of eastern North America have their closest relatives in China and Japan. For example, the genus *Liriodendron* includes only two species. One is the tulip tree, often miscalled tulip "poplar," widespread in eastern North America; the other is confined to eastern China. The native trumpet vine of eastern North America belongs to a genus (*Campsis*) that has but one other representative, and this in China. The wood-vamp vine (*Decumaria*), native to the southeastern United States, likewise has its only congener in China. The genus *Menispermum* includes but two species. One is the Canada moonseed, a vine found in cooler parts of eastern North America, the other a similar vine found chiefly in the northeastern part of the Sino-Japanese Province. The genus *Chiogenes* is often said to embrace but two species, one in eastern North America and the other in Japan; but actually the two are scarcely distinguishable, and might be thought of as disjunct populations of a single species. This species, called creeping snowberry or moxieplum, is often transplanted to the rock garden.

The Himalayas, world's loftiest uplift. (Government of India Tourist Office)

The milkweed genus *Amsonia*, the orchid genus *Arethusa*, the Croomias, the Elliottias, the trailing arbutuses, the orangeroots, and the Glaucidiums characterize eastern North America and Japan. Other groups, such as the Abelias, the Aralias, the supplejacks, the catalpas, the persimmons, the yellow jessamines, the loblolly bays, the Hydrangeas, the Illiciums or star anises, the Magnolias, the Photinias, and the Schisandras, are concentrated in eastern Asia and eastern North America, but also extend southward in one or both hemispheres. The partridgeberries, *Mitchella*, include only two species, one of eastern Asia and the other essentially of eastern North America; but the latter species reappears on certain high mountains of Middle America. Several other plant groups are distributed like the partridgeberries.

Chapter 1 suggested other floral relationships between eastern North America and eastern Asia. Remarks in several chapters may have already afforded clues to why these relationships exist. However, a further discussion of the topic is now in order. Our present type of world climate—warm and equable at the Equator, frigid at the poles, seasonally variable in the mid-latitudes—has not prevailed through any great percentage of geological time. Indeed, this type of climate has characterized only some relatively brief periods of mountain-making and continental uplift, during which time the patterns of temperature and rainfall would vary widely from place to place. There have been much longer periods when the earth's warmth and precipitation were more evenly distributed. During one such long period of generally mild climate, a temperate hardwood forest extended all the way across Canada, and southward into the north-central and northeastern United States, as well as across northern Eurasia. At many localities the plants of this forest chanced to be fossilized, and so today we can tabulate numerous genera and species that made up the old north-temperate flora.

This flora is sometimes called the Arcto-Tertiary forest. "Arcto-" means northern. "Tertiary" is the name of the aforesaid time period, which began when the last dinosaurs vanished and which continued until vast ice sheets began to form in the high latitudes. The composition of this forest varied from place to place and through the ages; but the variation does not seem great when one considers either the large area occupied by the forest, or the great length of time that it persisted. During the Tertiary, a few plants of the old northern flora managed to reach various mountains of Mexico and Central America; and these isolated, Middle American plant populations escaped the disaster that otherwise overtook the Arcto-Tertiary forest. Toward the latter Tertiary the world climate began to diversify, the north becoming colder and some regions becoming much drier. Then, about a million years ago, glaciers began to develop in the far north; and as they advanced, the two great segments of the Arcto-Tertiary forest—the Eurasian and the North American—both fell back to the south.

But in many lands the southward retreat of this flora was blocked by seas, or by high mountain ranges, or by areas with insufficient rainfall. In Europe a large part of the Arcto-Tertiary flora was wiped out. A few Arcto-Tertiary relics persist here and there in Europe, especially toward the Mediterranean lands. Others managed to cross the Mediterranean Sea, and today survive in northern Africa. There were small refuges available at scattered localities in southwestern and south-central Asia. However, in all Eurasia only the eastern portion of the continent afforded a large, easily accessible area with a climate suitable for the Arcto-Tertiary flora. And in North America, just as in Eurasia, the well-watered east offered the only sizable refuge area. With the retreat of the glaciers, and a shift toward a warmer climate, the surviving remnants of the old north-temperate flora pushed northward again, or moved up into the mountains. But never again did this flora recapture its former territory, most parts of which remained too cold or to dry to support a hardwood forest.

The movements of the Arcto-Tertiary forest were more complex than we have indicated, for Tertiary climate, while generally mild, was not absolutely stable; and the glaciers, whose formation brought the Tertiary to an end, came down not once but four times, the four advances of ice being separated by relatively warm, dry periods. However, the general effect of climatic change was to leave us today with no more than scattered remnants of the Arcto-Tertiary flora. The largest remnant now occupies the Sino-Japanese Province, where it is concentrated in Japan and China. The next largest occupies eastern North America, concentrated toward the Southeast and the Appalachians. The Middle American remnant is small but interesting; it includes sweet gum, black gum, beech, dogwood, basswood, sugar maple, witch hazel, star anise, redbud, partridgeberry, and white pine, among others. In some but not all cases the Middle American relic belongs to a species that survived also in eastern North America. Of course the Sino-Japanese Province has also been invaded by plants that were not part of the old north-temperate flora, and it should not be assumed that all the plants of the Province have had the same history.

The faunal similarities between eastern Asia and eastern North America are not as numerous as the floral ones. Many animal groups never passed between Eurasia and North America, and the old north-temperate fauna was not especially homogeneous as compared with the flora. Some widespread Eurasian-American animal groups were beginning to lose ground well before the coming of the glaciers. For example, the cryptobranchid salamanders are today represented only by the hellbenders of eastern North America and the giant salamanders of China and Japan; but fossil cryptobranchids are known from the Great Plains of the United States. When the Rocky Mountains first rose, their rain shadow brought dry conditions to the Plains, from which the cryptobranchids vanished; and all this took place in the latter, but not latest, Tertiary. Alligators, today restricted to China and the southeastern United States, vanished from the Great Plains at the same time the cryptobranchids did, and for the same reason. Of course plants as well as animals were extirpated as conditions grew drier in that part of North America to be affected by the Rocky Mountain rain shadow. The environmental change favored the evolution, spread, and rise to local predominance of drought-adapted plants.

Now to list a few faunal similarities between the Sino-Japanese Province and eastern North America. As noted in Chapter 1, the paddlefishes are today localized in east-central North America and in China. The paddlefish family had attained a wide Eurasian-American distribution even before the dinosaurs vanished; paddlefish fossils have been found in both western North America and Europe, in deposits older than the Tertiary. The fish family Catostomidae, whose members are called suckers, is concentrated in eastern North America, although a few outlying species carry the distribution to Guatemala and to the Pacific coast of North America. One sucker of northern North America has extended its range into Siberia; but otherwise, the only Old World sucker is a single, very distinct species of east-central China.

The jumping-mice, genus *Zapus*, include one species in northern and eastern North America, and another in western North America. The only other living species of the genus inhabits western China. Small lizards called blue-tailed skinks (*Eumeces*) are widespread on several continents, but the three species of eastern North America are most nearly related to species of the Sino-Japanese Province. Ground skinks (*Lygosoma*) are also widespread, but the single species of eastern North America is most closely related to those of eastern Asia, and especially to one of China.

Faunal similarities between eastern Asia and eastern North America are interesting, but attention must also be given to many Sino-Japanese animal groups that do not have counterparts in the New World. Hoofed animals, especially, are diverse and distinctive in the mountains of the Sino-Japanese Province. The takin ranges from the eastern Himalayas into western China. Looking like a huge ox with a sheeplike head and the horns of a buffalo, this animal inhabits mountain forests, and often descends into the valleys to browse. It is somewhat akin to the muskox, but has no really close relative anywhere in the world today. The bharal, or blue sheep, occupies roughly the same geographic range as the takin, but lives at much higher elevations. Although called a sheep, the bharal is somewhat intermediate between sheep and goats. The goral, distributed from the Himalayas to southeastern Siberia, is a rock goat that lives on precipitous slopes. Other rock goats, called serows, are widespread in the mountains of eastern Asia, including those of the Sino-Japanese Province.

A group of small deer, known as musk deer, were originally distributed about like the goral; but in many areas they were killed out for the musk gland, the secretion of which is used as a perfume fixative. Another group of deer, the sikas, are distributed from southeastern Siberia and Japan southward to northern China and Formosa. Once held sacred by the Japanese, they have become favorite park animals in Europe and North America. A rare beast is Thorold's deer, restricted to western China and eastern Tibet. It is related to the European red deer and the American wapiti, as well as to the sikas. An impressive animal, Thorold's deer has a white muzzle and chest, and its great antlers are also whitish.

A strange story is that of Père David's deer. In 1865 a French missionary-explorer saw a small herd of these deer, kept in a royal park on the outskirts of Peking; and he was able to obtain a few living specimens for shipment to Europe, where a herd was soon developed. No wild individual was ever discovered, and the Chinese herd was wiped out in the Boxer uprising of 1900. Descendants of the European herd may still be seen in captivity. Père David's deer, which has no close relatives, is the only undomesticated mammal that has never been known in the wild state.

The giant panda, a superficially bearlike animal, is confined to the central and western parts of Szechwan in central China. It lives only in bamboo forests at high elevations. Its teeth are modified for a diet of bamboo, and on its paw is a thumblike structure used in manipulating the bamboo stalks. Obviously the

Azaleas, Weigelas, and Camellias are prime attractions in gardens near Charleston, South Carolina, but were native to the Sino-Japanese Province. (South Carolina State Development Board)

panda's range must be limited by the dietary restriction. The lesser panda, a long-tailed and fox-faced species, is more widespread, from Szechwan to Nepal, and southward in the mountains to northern Burma and northern Laos. In Szechwan, at least, it lives at higher altitudes than its giant relative. Pandas belong to the same family that includes the raccoons and coatimundis of the New World.

The bear family is notably diverse in the Sino-Japanese Province. The sun bear, although essentially tropical, reaches Szechwan. The widespread grizzly and brown bear group ranges southward into the Himalayas and southwestern China. Allies of the American black bear inhabit Tibet and Japan. The moon bear is distributed from Iran eastward to Japan and Formosa. The cat family is even more diverse. Large cats of the Province include the tiger, which reaches a much greater size in Manchuria than in the tropics; the widespread leopard; the snow leopard, a large, shaggy animal ranging from the Tibetan Plateau to Sakhalin Island; and the clouded leopard, most beautiful of cats, distributed from Indonesia northward into Bhutan. Smaller felines include the Asian golden cat, found from the Malay Peninsula northward through the Himalayas; the leopard cat, an arboreal species ranging from the Philippines and Indonesia northward through Manchuria; the fishing cat, a marsh dweller distributed from Ceylon into southwestern China; and the marbled cat, a diminutive replica of the clouded leopard, ranging from Indonesia to the Himalayas.

From the foregoing remarks it will be seen that animal groups enter the Sino-Japanese Province from several directions; and western China, the vicinity of Sikang and Szechwan, lies at the crossroads of many invasion routes. A physiographic map will reveal why this should be the case. In eastern Asia a mountainous uplift, composed of numerous long sierras, extends in an approximately north-south direction from Siberia into the tropics of Indochina and the Malay Peninsula. Another uplift, running more or less at right angles to the preceding, begins with the Pyrenees of Europe and culminates with the Himalayas. Western China is the meeting place of the two uplifts.

From bottom to top a high mountain offers a variety of environments; and a long sierra usually provides a highway for the dispersal of many animal groups, at some elevation or another. Throughout the ages, numerous groups have followed one or another mountain chain as far as western China or beyond. Often an invading group, reaching the complex mountains of western China, has evolved adaptations to the local environment. Most surprising invader of the region may be the monkey genus *Macaca,* which has produced an endemic species in Szechwan. Called the bearlike monkey, it is a huge and shaggy simian that in winter must hunt beneath the snowdrifts for its food.

Ecologists—students of the environment and its impact on animals and plants—have been especially interested in Szechwan, whose high, steep mountains offer a diverse lot of habitats from top to bottom. And in 1945, biogeographic attention was focused on Szechwan with the discovery therein of the "dawn redwood," a surviving remnant of a plant genus that flourished before and during the early Tertiary; it was growing with chestnuts, sweet gum, beech, maples, alders, blue beech, hop hornbeam, and other probable relics of the Arcto-Tertiary forest.

It is with reluctance that we leave Szechwan; but other topics must be considered. One of these is the invasion of the highest altitudes by vertebrates. The griffon vulture, a mountain bird distributed from the Mediterranean to the Himalayas, in the latter region may set an altitude record for vertebrates. An expedition once struggled to the top of Mount Everest, at 29,141 feet the highest peak on earth; and there they saw a griffon vulture soaring overhead. Although a carrion eater, this bird is more nearly related to the eagles than to the New World vultures. The bar-headed goose has also been seen flying over Everest.

The lammergeier, a larger relative of the griffon vulture, is a bird of the high mountains from southern Europe to Africa and the Himalayas. It has been noted at 25,000 feet on Everest. The common name, from the German Lämmergeier, means lamb-vulture, but the bird is mainly a scavenger. The raven —the same species that inhabits North America—follows the Himalayas to heights rivaling those attained by the lammergeier if not by the griffon vulture. The alpine chough, which resembles a small, yellow-billed crow, lives above timberline in Europe and Asia; on Everest it reaches 27,000 feet. The rock pigeon, common throughout temperate Europe and western Asia, also reaches a high altitude in the Himalayas. Domesticated as early as 3000 B.C., it is ancestral to the common street pigeons as well as to many cultivated breeds. Yet another high-altitude bird is the wallcreeper, an inhabitant of rocky cliffs from southern Europe through the Himalayas. Its halting flight and its red and white wings have given it the name of butterfly bird. In winter it descends to low elevations, where in the absence of cliffs it will creep about the walls of buildings. Thus the birds of the Himalayan peaks, like so many of the high-altitude plants, are not confined to the Province, but live also in the mountains of other lands.

We cannot leave the Sino-Japanese Province without remarking on certain of its salamanders. Nearly restricted to the Province is the family Hynobiidae. These are the most primitive of all salamanders, and were perhaps ancestral to several other families. The most primitive species of hynobiid, Keyserling's salamander, is the most northerly of all salamanders. It ranges from Manchuria and Mongolia northward to Verkhoyansk, that cold Siberian locality mentioned previously, and even to Anadyr at the northeastern corner of Siberia. Several other hynobiids live at very high altitudes. Pinchon's salamander, discovered by Père Armand David in cold mountain streams along the Szechwan-Sikang boundary, reaches 13,000 feet. Schmidt's salamander occupies a similar range and altitude. (Near the White Dragon Pool on the top of Mount Omei there is a temple consecrated to the worship of Schmidt's salamander, and the monks say that the animal will mysteriously die, or vanish, if taken to lower altitudes. In truth it dies, but because it cannot stand the warmer temperature of the lowlands. Like many other hynobiids it can live only in a cold environment.) The Yenyuan salamander is known only from an elevation of 14,500 feet in the Yenyuan District of Sikang. No one knows where salamanders arose, or under what environmental conditions; but it may be significant that many of the most primitive ones live at remarkably high altitudes and latitudes, in cold situations.

22 *". . . we observed several large hares, some ducks, and many of the*

cock of the plains . . . also considerable quantities of wild onions . . . while

we stopped, the women were busily employed in collecting the root of a

plant. . . . It is a species of fennel. . . . The valley continues to be a poor,

*stony land, with scarcely any timber, except some pine trees . . . ."**

THE PACIFIC NORTH
AMERICAN PROVINCE

The Boreal Kingdom, which includes the temperate portions of Eurasia and northern Africa, extends also to the temperate lands of North America. Although temperate North America is a very large area, it is divided into two Provinces only. Smaller of the two is the Pacific North American Province. This subdivision embraces extreme southern Alaska, from about Prince William Sound southward to Ketchikan; the southwestern corner of the Yukon and all but the northeastern part of British Columbia; the Pacific coast states of Washington, Oregon, and California; and the upper half of Baja California. It also includes the states of the Great Basin and the Rocky Mountains uplift, from Idaho and western Montana southward through Nevada, Utah, and Arizona. Western Wyoming, western Colorado, and western New Mexico also belong to this subdivision, as does the extreme western portion of Texas. Finally, the Province extends south-

* MERIWETHER LEWIS and WILLIAM CLARK, in the journal of their expedition, 1803–1806.

241

ward beyond the borders of the United States, to include the Mexican highlands. These highlands make up the greater part of Mexico north of the Isthmus of Tehuantepec.

Contrary to what one might expect, the prairies of Canada and the United States are not included in this subdivision. They belong to a more easterly Province.

California has the richest flora of any political subdivision within the Pacific North American Province. The circumstance reflects the long north-south extent of the state, as well as its climatic and topographic diversity.

Very few plant families are endemic to the Province, even by a loose definition of endemism. The Garryaceae, or silk tassels, include but one genus, *Garrya.* There are about 20 species of *Garrya,* concentrated in this Province although some extend also into the tropics. The best-known Garrya is bear brush, which lives on dry slopes of the foothills in California and nearby areas. It is sometimes transferred to the garden, where it is grown for its showy catkins. The Limnanthaceae include but one genus, *Limnanthes,* made up of five species scattered over western North America. Best known of these species is meadow-foam, an attractive little plant restricted to damp spots. The meadow-foam is sometimes taken into the garden, where it is called Floerkia. The Koeberlinaceae include but one genus and species, a spiny little shrub distributed from northern Mexico into Texas.

Although the Province has but few endemic families, it harbors at least three hundred endemic genera. Most of these belong to families represented in both the Old World and the New. Among the endemic genera of the Province are *Eschholtzia,* which includes the California poppy; and *Abronia,* the sand verbenas. These genera are of wide distribution in the subdivision. Almost as widely distributed is *Lewisia,* which includes the bitterroot, state flower of Montana. *Sidalcea,* the checkerbloom and related mallows, is largely restricted to the northern portions of the Province. Of more southerly distribution is *Choisya,* the genus of the Mexican orange. *Darlingtonia* includes some insectivorous plants found in mountain swamps of California and southern Oregon. Darling-tonias are often sold as "cobra orchids," but they belong to the pitcher-plant family, Sarraceniaceae. Among other well-known genera of the Province are *Romneya,* which includes the Matilija poppy of California; *Dendromecon,* the genus of California's tree poppy; and *Umbellularia,* the California laurel.

Sarcodes is an interesting genus. It has but one species, the snow plant, which lives in the montane forests of central and southern California. It grows especially beneath the great Sequoia trees in the Sierra Nevada. Leafless and without chlorophyll, the snow plant is a saprophyte; that is to say, it feeds upon dead and decaying organic matter, in this case the rich humus that collects beneath the forest trees. In early spring the plant sends up a stalk of blood-red flowers, which may be buried by a late snowfall; thus the common name. A related genus, *Pterospora*, is not saprophytic but actually parasitic, usually on the roots of pines along the Pacific coast. Both genera belong to the Monotropaceae, a family represented by saprophytes and parasites in temperate parts of both the Old World and the New.

Most famous endemic genus of the Province is *Sequoia*, with two species. One of these, the redwood, is distributed at low or moderate elevations along the coastal mountains from Santa Cruz, California, northward into southern Oregon. This conifer grows no farther inland than the limit of ocean fog, which provides necessary moisture. Tallest of trees, the redwood often exceeds 300 feet in height. The maximum height seems to be between 360 and 370 feet. The other species of *Sequoia*, called California big tree, is restricted to high altitudes of the Sierra Nevada, where only a few scattered groves exist. Although not as tall as the redwood, the big tree is more impressive; for its trunk is massive, sometimes exceeding 35 feet in diameter. Although the genus Sequoia is now very restricted in its distribution, fossil Sequoias are known from Europe and Asia, as well as from North America as far east as New Jersey.

Bigelovia, the rayless goldenrods, are concentrated in the Province, but two of the species range from Texas to the Atlantic coast. *Yucca*, which includes the Spanish bayonet, Spanish dagger, Adam's needle, and Joshua tree, is concentrated in the southern part of the Province; but there are several species in the eastern United States also. The Mariposa lilies, *Calochortus*, characterize the southern part of the Province, but also range southward into Guatemala. The genus *Zinnia* is most diverse in the southern part of this Province; but its species are also strung out through Middle and South America, all the way to Chile. These genera, and many more, link the Province with other parts of the New World.

Some genera link the Province with portions of the Old World. Several of these genera are especially interesting. The peonies, *Paeonia,* include about two

Three Sisters Mountains and Deschutes River near Redmond, Oregon. (Oregon State Highway Commission)

dozen Eurasian species; but there is one New World species, called the western peony. It lives on wooded hillsides from California northward into Washington and eastward into Utah. *Meconopsis* includes the blue poppy mentioned in the preceding chapter. There are about forty species of *Meconopsis* in the Himalayas or in the nearby mountains of China and Tibet; but an isolated species lives in western Europe, and two more in western North America. The New World species are sometimes placed in a genus of their own, but the procedure does not alter the relationship. *Lithocarpus,* of the beech family, includes trees that are more or less intermediate between the chestnuts and the oaks. Most species of *Lithocarpus* are Asian; but one, called the tanbark oak, grows on hillsides and in ravines of Oregon and California. *Castanopsis,* also of the beech family, has several species in Asia, and one in the Pacific North American Province. The American one, called giant chinquapin, grows on the slopes of the Pacific coastal ranges.

The madroño, a tree of hillsides and canyons from California into Washington, is the only New World member of *Arbutus,* a genus represented in western Europe and the Mediterranean region by the strawberry tree. The beardtongues, *Pentstemon,* include a great number of species in western North America and a dozen or so in eastern North America. One Pentstemon, however, ranges from Alaska across Bering Strait to Siberia and Japan. The genus *Phlox,* like *Pentstemon,* is particularly diverse in western North America, has a dozen or so species in eastern North America, and is represented in the Old World by one species that ranges from Alaska to Siberia. *Castilleja,* the painted cups, are concentrated in the western United States, and only one species of the genus lives in eastern North America. There is also an isolated species in northeastern Asia.

A great many genera, widespread or at least not restricted to the Pacific North American Province, have produced endemic species therein. Such species are particularly numerous in parts of California where the climate is of a Mediterranean type; for special adaptations are often necessary if a plant is to thrive where rain falls only during the coldest part of the year. In western California, from the coast to the foothills of the mountains, nearly half the flora is made up of endemic species. There are few crop plants, but many familiar garden flowers, among the endemic species of the Province. Examples include a tall, evergreen Ceanothus; a Clarkia called Rocky Mountain garland; a large-flowered Collomia, of the Phlox family; and a 75-foot dogwood, the giant of its widespread genus

The Sego lily, state flower of Utah, a species of Calochortus. *(Utah Tourist and Publicity Council)*

(*Cornus*). Others include the flannelbush, a Fremontia; a Gaultheria known as salal; several Godetias, the best known called farewell-to-spring; a small blue-flowered lupine and a much larger yellow-flowered one; and the holly-barberry, a Mahonia. We might also add the muskflower, a Mimulus; a Nemophila and a Phacelia, both of the waterleaf family; a Tolmiea, native to cool forests and called youth-on-age; and three currants: the golden, the giant red, and the California.

Various patterns of distribution are found among the conspicuous endemic or near-endemic species of the Province. The Washingtonia or California fan palm, so abundant around Palm Springs, characterizes the deserts of southern California and northern Baja California; two other species of Washingtonia live farther south, in Mexico. The ocotillo, tipped with scarlet blooms in April and May, is widespread in the deserts of the southwestern United States and northern Mexico; it is the northernmost offshoot of a desert genus (*Fouquieria*) that is more tropical than temperate. Candles-of-the-Lord, a tall Yucca, is nearly confined to southern California; but as noted above, its genus is represented in eastern North America even though concentrated farther west. A common sage-brush, with a silvery-gray stem, ranges from the Pacific coast eastward into Nebraska; its genus (*Artemisia*) is represented over much of Eurasia and the New World.

An extraordinary plant of the Province is the Mormon tea, a shrub with jointed, green stems and tiny, scalelike leaves. It grows along with the grease-wood and prickly pears from Texas into California. The Mormon tea belongs to a genus (*Ephedra*) that is also represented in the Mediterranean region, Asia, and the New World tropics. Botanists, visiting the North American desert for the first time, look forward to seeing the Mormon tea, for it is the only United States member of an ancient family, the Gnetaceae or joint firs. This family is related to none other; in some respects it is intermediate between the conifers and the broad-leaved plants. Besides *Ephedra*, there are only two other extant genera of the Gnetaceae. *Gnetum* includes woody vines or small trees of several tropical lands; *Welwitschia* has but one species, a grotesque plant that sprawls like a flattened octopus on the desert sands of southwestern Africa.

Special attention must be given to the cacti, so conspicuous in the deserts of the Pacific North American Province. Their family, the Cactaceae, includes nearly 2,000 species, all but a few of them confined to the New World. One

On the facing page, a forest of redwoods in Humboldt County, California. (John Lo Buono from Humboldt County Chamber of Commerce) Above left, a Yucca, state flower of New Mexico. (New Mexico Department of Development) Right, Joshua trees in Antelope Valley north of Los Angeles, California. (All-Year Club of Southern California)

genus, *Rhipsalis,* is made up of species that grow on tree branches, and that bear the name of mistletoe cactus. The seed of a mistletoe cactus is sticky, and will adhere to the beak, foot, or plumage of a bird. Mistletoe cacti are widespread in the American tropics, but are also strung out all the way across southern Africa, and thence across the Indian Ocean from Madagascar to Ceylon. It is suspected that birds are responsible for the unusual distribution of *Rhipsalis.* All other cacti are native to the Americas.

The cactus family is concentrated in Mexico and adjacent parts of the southwestern United States; but there is another, albeit lesser, concentration in the drier parts of tropical South America. From these centers some of the genera have variously reached Middle America, temperate South America, or parts of North America far removed from the western deserts. The most widespread genus is *Opuntia,* whose members are often called prickly pear, devil's-tongue, or cholla. The Opuntias are represented from southern Canada to Patagonia. (And man has taken Opuntias to other lands; they are common along the Riviera, and in Australia they have become a serious pest.)

In the deserts of the Pacific North American Province there are cacti of many different shapes, from pebble-like "living stones" through fat, spiny barrel cacti to the saguaro whose fluted columns rise forty feet in the air. Each species, however bizarre, in the appropriate season will bear showy flowers of red, pink, orange, yellow, yellowish-green, purple, or white. The saguaro, state flower of Arizona, is the tallest of the cacti. Except for a few small stands in California and Mexico, the saguaro is confined to rocky slopes and mesas of the Arizona mountains.

Various cacti grow vinelike up the trunks of shrubs and trees. Some large, climbing cacti open their exquisite blossoms only at night—perhaps only during one night of the year—and are known popularly as "night-blooming Cereus." Probably these cacti are all of Neotropical, not Boreal, origin; but several of them have been planted widely on the Mexican highlands, where the blossoming of a very large specimen may herald the start of a fiesta. A gigantic species with a three-angled stem is commonly raised under glass, or in the gardens of warmer lands such as Mexico, southern California, Arizona, and Florida. This species, the one most often dubbed night-blooming Cereus, offers a problem to intrigue biogeographers (and to intrigue those anthropologists who think the prehistoric civilizations of Mexico and Central America once knew some influence from southeastern Asia.) It was first discovered in China, where it had been cultivated for centuries. Chinese stock eventually supplied the gardens of the world, but the plant was never discovered anywhere in a wild state.

The Nopaleas are cacti related to the Opuntias. One Nopalea, widespread in Mexico and introduced into various lands of the New World and the Old, is fed upon by a scale insect whose dried bodies yielded the brilliant red dye called cochineal. Mexican and Central American Indians relied heavily upon cochineal for coloring their textiles. The insects were "planted" upon the cacti, where they multiplied rapidly; and when they were grown, they were brushed off, dried, sacked, and stored. After the Spanish conquest of Mexico, millions of pounds of scale insects were sent annually to Spain, as tribute to the crown; and soon the insect, along with its cactus host, was successfully transferred to parts of the Old World. In the year 1868 the Canary Islands produced more than six million pounds of cochineal. But with the advent of synthetic dyes the cochineal industry declined, and today *Nopalea cochenillifera* grows wild at many localities that once were cochineal plantations. The account of cochineal has a curious postscript. A close relative of the cochineal insect inhabits the Mediterranean Province. The Mediterranean species feeds upon the kermes oak—from whose branches it was gathered by the local residents, for the production of a red dye.

The cactus genus *Lophophora* includes but one species, the peyote. This cactus is protected not by spines but by toxic juices; the radish-shaped body of the plant contains nine narcotic alkaloids, some of them strychnine-like and the others morphine-like in their action. Certain Mexican Indian tribes made a ritual of eating peyote and attached religious significance to the temporary mental derangements that followed. The peyote cult eventually spread northward to many tribes, even reaching Canada; but the plant itself grows from southern New Mexico and western Texas southward to the Mexican state of Puebla.

Cactus distribution could be discussed at length, but other topics remain to be considered. One of these relates to the northward distribution of plants in the Pacific North American Province. The northern extension of the Province is surprisingly warm for its latitude. An oceanic current, originating in tropical waters, passes northeastward off the coast of Japan, where it is called the Kuro Siwo or Black Current. The Kuro Siwo, gradually turning more to the east as it crosses the North Pacific, eventually splits; and a part of it turns northward to

The holly barberry, also called hollygrape, endemic to the Pacific North American Province. (Oregon State Highway Department)

pass off the coasts of British Columbia and southern Alaska. Another part turns southward along the Pacific coast of the United States. The more northerly portion of the Province would become much colder in winter, were it not for the climatic influence of the oceanic currents.

The beneficent effect of the currents is strikingly revealed by climatic data. In Alaska's capital city of Juneau, the average temperature of January—the coldest month—is 27½ degrees above zero. This is higher than the January average for, say, Chicago, Fort Wayne, Detroit, or Buffalo, although Juneau lies more than 1,500 miles farther north than these inland cities of the United States. Thus, plant groups of the far-western United States have been able to invade western British Columbia and southern Alaska in sufficient numbers to warrant the inclusion of these two latter areas in the Pacific North American Province.

A few species from the northern part of the Province have reached gardens. Among them are the Sitka columbine, which ranges from Alaska into California; the Nootka rose, also distributed from Alaska to California, and used as an understock as well as in hybrid combinations; and the Nootka lupine.

Attention must also be given to the Mexican highlands, which occupy a sizable portion of the Province. The greater part of the highland area is very dry. Deserts make up most of the country from Coahuila and eastern Chihuahua southward through eastern Durango and eastern Zacatecas to San Luis Potosí. In places the desert vegetation may be composed largely of cacti and Yuccas. In other places there is a predominance of various coarse shrubs. These latter usually belong to remarkably widespread genera. Examples include *Cordia* (concentrated in the tropics of both the Old World and the New), *Atriplex* (saltbushes, nearly worldwide although spread into some regions by man), and *Buddleia* (butterfly bushes, tropical and temperate parts of America, Asia, and southern Africa). Similarly widespread are *Euphorbia* (spurges, tropical and temperate regions of the world), *Acacia* (Acacias, concentrated in Australia, but also present in Africa, southern Asia, South America, and warmer parts of North America), and *Celtis* (hackberries, many tropical and temperate lands, even Madagascar and New Caledonia).

In still other parts of the Mexican highlands a creosote bush is conspicuous. With it may grow mesquite trees of the genus *Prosopis* (tropical and warm-temperate parts of the New World, Asia, and Africa), and prickly-pear cacti. Often a single species of plant will appear not only in the deserts of northern

Mexico but also in those of southern South America. This is true, for example, of one species of mesquite, and of the aforesaid creosote bush (*Larrea divaricata*).

There are areas of grassland from eastern Sonora and western Chihuahua southward through western Durango and western Zacatecas into Guanajuato. Here the conspicuous grasses (family Gramineae) belong to such genera as *Bouteloua* (grama grasses, widespread in the New World and a few species in the Old), *Aristida* (wire grasses, warmer regions of the world), *Eragrostis* (love grasses, temperate and tropical lands generally), *Andropogon* (broom grasses, temperate and tropical lands), and *Muhlenbergia* (concentrated in Mexico and the southwestern United States). Growing with the grasses may be various cacti, or trees and shrubs such as mesquite, desert ironwood (*Olneya*), palo verde (*Cercidium*), hackberry, ocotillo, Yuccas, and Acacias.

A broad strip of woodland extends along the Sierra Madre Occidental, with a similar but narrower strip along the Sierra Madre Oriental. In the north these two sierras are widely separated by desert and grassland, but toward the south they converge; and the two woodland strips coalesce in the state of Mexico and nearby areas. This woodland is dominated chiefly by pines (*Pinus*) and oaks (*Quercus*), both genera concentrated in but not limited to the north-temperate lands. In places, pines and oaks are intermingled; but near the deserts and grasslands there are large tracts with various scrubby oaks but no pines, while at higher elevations there are extensive pine forests without oaks.

Still higher, on scattered peaks, some of the pines mingle with fir (*Abies*) and alder, or perhaps with aspen and Douglas fir (*Pseudotsuga*). At the highest altitudes grow one or two pines, a juniper, and a heathlike shrub (*Pernettia*), along with coarse grasses belonging to such genera as *Festuca* (fescue grasses, temperate and colder regions), *Agrostis* (bent grasses, temperate and colder regions), and *Muhlenbergia*. The Pernettia, incidentally, is an invader from the south; its genus is concentrated in south-temperate lands, especially southern South America, and is also represented in Australia and New Zealand. A few peaks of the Mexican highlands are so high as to reach timberline, above which may be an alpine meadow.

Some of the more southerly peaks are constantly swept by currents of moisture-laden air, rising from the tropical lowlands. On such peaks are to be found the Mexican remnants of the Arcto-Tertiary forest. Peaks with such remnants are located along the eastern escarpment of the Mexican highlands,

A record-breaking blue spruce in Gunnison National Forest, Colorado; height 126 feet, circumference 15 feet 8 inches. Note man at lower left. (Colorado Department of Public Relations)

from San Luis Potosí to Oaxaca; in scattered parts of the Oaxaca uplands; and in the Sierra Madre del Sur of Guerrero. Arcto-Tertiary relics also turn up on peaks still farther south, outside the Boreal Kingdom.

The Pacific North American Province harbors many animals, distributed in a variety of ways. Some animals, found all across Canada, also range southward thereof in the Rockies, and perhaps in other mountain chains as well. Usually they range much farther south in the Rockies than elsewhere. For example the wolverine, distributed from Alaska eastward to Ellesmere Island and Labrador, also reaches New England and the Great Lakes region, but follows the Rockies into Colorado. The moose, essentially Canadian, has been found as far south as New York in the east, as far south as Utah in the Rockies. The wapiti follows the Appalachians southward into Georgia, and the Rockies southward into New Mexico. (Actually, the respective ranges of the wolverine, moose, and wapiti have been drastically reduced by modern man, but at this point we are concerned with ranges as they were until recently.) The eastern red squirrel, found from Quebec to Alaska, reaches South Carolina in the Appalachians, New Mexico in the Rockies.

Other animals are widespread in the New World tropics, whence they range northward into warm-temperate parts of North America. Several of these follow the Mexican highlands northward as far as the United States border, or even well beyond. Thus the collared peccary, coatimundi, jaguar, and ocelot all range into Texas and Arizona. Two small cats, the margay and the jaguarondi, barely reach Texas. The boa constrictor, an inoffensive snake in spite of popular belief, reaches the Arizona border. Still other animals, found in mountains or deserts of the Province, range also into the prairies that border much of the sub-division on the east. In this category is the pronghorn, a distinctive North American animal with no close relatives. It is distributed from Durango and central Baja California northward to Manitoba and Alberta. The mule deer (including the Columbian blacktail which belongs to the same species) ranges from the Pacific coast eastward to Kansas, and from Cape San Lucas northward into Alaska and the Yukon.

Turning now to some endemic or near-endemic animals of the Province: The mountain goat, really more of an antelope than a goat, is essentially a species of western Canada. It follows the Rockies southward into Idaho and Montana, the Cascades southward into Washington. The bighorn sheep, a true sheep with

Asiatic congeners, is widespread in the mountains of the Province, from northern Baja California and Chihuahua northward into the Canadian Rockies; but it also ranges into southern Baja California and the Dakota Badlands, outside the Province. The mountain beaver is a strange little animal confined to a coastal strip from northern California northward into Washington. Lewis and Clark, in their famous expedition of 1804–1805, discovered this rodent on the Columbia River and were hard put to suggest its relationship. It is now placed in a family of its own, the Aplodontidae.

Another remarkable endemic of the Province is the bell toad, also called tailed frog because the male has a copulatory organ resembling a pointed tail. The only close relatives of this primitive frog live in far-off New Zealand. The range of the bell toad is discontinuous; one population lives in the northern Rockies of Idaho and Montana, while a second occupies virtually the same range as the mountain beaver. The pika is a small animal found from south-central Alaska to New Mexico, especially in high mountains near timberline. It looks somewhat like a guinea pig, but is actually related to the hares and rabbits. The pika is the only New World representative of its family, the Ochotonidae; but it has several congeners in Eurasia.

The largest flying bird of North America, the California condor, is restricted to the Province. Before settlement it was distributed along the Pacific coast from Baja California to the Columbia River. However, in modern times it has died out from much of this range, and today is found only in a few coastal counties of southern California. It is said to have died in large numbers after having fed on poisoned carcasses set out by stockmen in an effort to eradicate coyotes. Yet over the millennia the range of this giant vulture must have fluctuated greatly, for reasons having nothing to do with man. At some time in the Pleistocene, the California condor reached Florida where it left its bones in fossil deposits.

The roadrunner is a characteristic bird of the North American deserts; but its group, the ground cuckoos, is distributed from Arizona and New Mexico southward to Paraguay, and has two more species in the Malay Peninsula and Indonesia.

The desert portion of the Province is an important area of lizard diversification. Especially conspicuous are iguanids, such as the desert iguana, zebra-tailed lizard, collared lizard, leopard lizard, earless lizard, side-blotched lizard, chuckwalla, and various horned "toads." Most of these iguanids represent endemic or

Above, the jaguar, a big cat that ranges into the southwestern United States from Mexico. (Tod Swalm) At top of facing page, a boa constrictor, an inoffensive (but often maligned) snake that ranges from South America to the Arizona border. (Jerry Shaw) Below it, mountain sheep near the Columbia Icefield in Alberta, Canada. (Canadian Government Travel Bureau)

near-endemic genera. The Iguanidae probably were part of an ancient South American fauna, but reached Middle and North America long before the closing of the Panama Portal. (Iguanids live also on Madagascar, Fiji, and Tonga; and there is no fossil evidence to show that the family reached these islands by way of Asia.)

The racerunner lizards, or whiptails, often seen dashing over the desert sand, all belong to one genus, *Cnemidophorus,* of the Teiidae. The family is clearly an ancient South American one, and *Cnemidophorus* is its northernmost genus. There is a species in eastern North America, a dozen or more in the Mexican highlands and the southwestern United States, and a few in the New World tropics where other teiid genera are concentrated.

The lizard family Anniellidae is confined to a strip of country from about San Francisco Bay southward into northern Baja California. It includes but two species, both legless and elongated. The lizard family Helodermatidae is distributed from the Isthmus of Tehuantepec northward into Utah. The family includes but two living species: the Mexican beaded lizard and the Gila monster, the only venomous lizards in the world.

Among snakes, the boa family (Boidae) has one genus that is endemic to the Province, and another that is nearly so. The rosy and the three-lined boas (*Lichanura*) are found from Sonora and southern Baja California northward into southern Arizona and southern California. The rubber boa (*Charina*) is more northerly in distribution, ranging from southern California northward well into British Columbia. Most boas are tropical, and it is surprising to find one that can live as far north as Kamloops and the Kootenay Lakes of Canada. The rattlesnakes, two genera thereof, are confined to the New World. About thirty species are known, most of them found on the Mexican highlands or in the southwestern United States. A few offshoots carry rattlesnake distribution from this homeland northward into southern Canada and southward into Uruguay.

Salamanders are diverse on the Mexican highlands, where there are several endemic genera of the Ambystomatidae and Plethodontidae. The diversity is surprising because the highlands are dry, and salamanders are most characteristic of cool, damp regions. However, life-history specializations have permitted these amphibians to invade Mexico from the north. Salamanders usually have an aquatic larva which eventually transforms into a less aquatic or even terrestrial adult, just as a tadpole transforms into a frog; but in the high-altitude lakes of

Mexico the ambystomatids often fail to transform, remaining somewhat larval and living a fishlike existence. In contrast, the Mexican plethodontids have eliminated the aquatic larva from their life-history. The larval stage is spent in the egg, which is laid in some damp spot, perhaps beneath a log or in an air plant but not in water; and the egg hatches into a miniature of the adult.

The deserts of this Province have not always been as dry as they are today. The uplifting of the Rockies and the coastal ranges brought the first widespread aridity to the region. Many aquatic animals vanished along with the lakes and rivers, but a few persisted in springs and streams that never dried up completely.

A surprisingly large number of fishes live a precarious life about waterholes in the dry grassland or even in the desert. (Precarious especially on account of man's activities: settlement, and the introduction of predaceous bullfrogs and competitive fishes.) Among suckers (the fish family Catostomidae), the genus *Chasmistes* has one species in Utah Lake near Provo, Utah; a second species in Pyramid Lake and the Truckee River near Reno, Nevada; and a third in Upper Klamath Lake near Klamath Falls, Oregon.

Among minnows (Cyprinidae), the genus *Moapa* is represented by just one species, the Moapa dace, confined to Warm Springs, about fifty miles northeast of Las Vegas, Nevada. This leathery-skinned fish does not invade the cooler waters into which the spring run flows. The Mojave chub, a species of *Siphateles*, lives in the Mojave River, a stream of California's Mojave Desert. The spring-fed stream is all that remains of a former lake system. The least chub, only species of *Iotichthys*, is confined to the Bonneville Basin of Utah.

Strangest of desert fishes is the woundfin, only species of the cyprinid genus *Plagopterus*. An elongated, silvery fish without scales, the woundfin lives in a part of the Colorado River system, from the Grand Canyon and the Painted Desert northward into Utah.

Among killifishes (Cyprinodontidae), the genus *Empetrichthys* has but two species, both in the Death Valley region. One, the Ash Meadows springfish, lives only in Ash Meadows of the Amargosa Basin on the California-Nevada border; the other, the Pahrump springfish, is confined to the nearby Pahrump Valley of Nevada. The genus *Crenichthys* includes only the Railroad Valley springfish, which lives in Railroad Valley of southern Nevada, and the White River spring-fish, an inhabitant of Nevada's White River Basin. The killifish genus *Cyprino-don*, of widespread distribution, includes some desert species with highly restricted ranges. The Devil's Hole pupfish lives only in Devil's Hole, a spring of Ash Meadows; the Death Valley pupfish inhabits Salt Creek in Death Valley; and the Owens pupfish is confined to Owens Valley, in California west of Death Valley. Among the live-bearing killifishes (Poeciliidae), the Big Bend topminnow, a species of *Gambusia,* is known only from Graham Ranch Warm Springs, in the Big Bend country of western Texas.

23 *"We never saw it grow in any other place . . . a very singular and unaccountable circumstance; at this place there are two or three acres of ground where it grows plentifully."** **THE**

ATLANTIC NORTH AMERICAN PROVINCE

The account of the Boreal Kingdom concludes with a discussion of the Atlantic North American Province. This subdivision embraces all of eastern and central North America, except the Arctic of Canada and the tropics of southern Florida. In addition, the Province extends westward through Saskatchewan, Alberta, and the Yukon, and then across central Alaska to the shores of the Bering Sea.

The Province is divided into northern and southern sections. The Canadian and Alaskan portion of the Province has comparatively few broad-leaved trees, but has instead an abundance of the needle-leaved conifers such as fir, spruce, larch, and hemlock. Some of these conifers follow the Appalachian uplift southward as far as North Carolina or northern Georgia. Coniferous forests, predominantly of white pine, red pine, and hemlock, are widespread around the Great Lakes and in New England. All of these coniferous forest regions together form the northern section of the Province. The southern section therefore includes chiefly the Great Plains; the Atlantic drainage east of the Appalachians,

* WILLIAM BARTRAM, eighteenth-century botanist-explorer, on his discovery of Franklinia in Georgia.

from New Jersey through central Florida; the lowlands of the Gulf drainage, from southern Texas through central Florida, and including the Mississippi Valley; the Ozark-Ouachita uplift; and various interior plateaus and lowlands. The flora of the northern section has an affinity with that of the Euro-Siberian Province, while the flora of the southern section has an affinity with that of the Sino-Japanese Province.

In the preceding chapter it was noted that very few plant families are wholly confined to western North America. The same is true of eastern North America. In fact, only seven families are endemic to any part of North America, even by a loose definition of endemism; and they are small families, with but a few species and these mostly unfamiliar. In eastern North America the family Leitneriaceae includes only one species, corkwood, a small tree that grows in freshwater or brackish swamps from southern Georgia to Missouri. Its wood, lighter than cork, has been used for floats and stoppers.

Although few families are endemic to the Atlantic North American Province, nearly two hundred genera are so restricted. Many are small genera, with but two or three species and perhaps only one. Among tree genera, *Maclura*, the Osage orange, is distributed from Georgia to Texas and Missouri, but is widely planted elsewhere. It yields a fine bow wood, prized by the modern archer just as it was by the American Indian tribes; and its vernacular name of "bodock" is merely a corruption of the French *bois d'arc*: wood of the bow. *Gymnocladus*, the Kentucky coffee tree, is found mostly in rich woods from Oklahoma and Tennessee northward into Ontario. It is an ornamental whose dried seeds have been used as a substitute for coffee beans. *Pinckneya*, the fever tree, grows from South Carolina to western Florida. Its flowers, surrounded by large, pink sepals, are sometimes conspicuous at the edges of swamps. Pioneers in Florida used a decoction of the bark to treat malaria, and it may be significant that the fever tree belongs in the same family (Rubiaceae) with *Cinchona*, source of quinine. *Planera*, the water elm, ranges from northern Florida and Texas northward to North Carolina and southern Illinois. Usually it grows in places that are subject to flooding.

Among endemic genera of shrubs, *Ceratiola*, the Florida rosemary, is found from Florida to Mississippi and South Carolina. Unrelated to the Old World rosemary, it belongs to the crowberry family, Empetraceae. *Hudsonia*, of the rockrose family, is the genus of the beach heather, a familiar plant on the Atlantic

The Atlantic North American Province extends westward to encompass the short-grass prairie. A cattle drive in western South Dakota. (South Dakota Department of Highways)

coast from North Carolina to New Brunswick, and in the Great Lakes region. Unlike the other endemic genera so far mentioned, *Hudsonia* includes more than the one species. A second species of the genus lives only on Table Rock and nearby peaks in the Blue Ridge of North Carolina, and a third is distributed coastally from North Carolina to Nova Scotia. *Neviusia*, of the rose family, includes only the snow wreath, a shrub limited to shale and limestone cliffs in the general vicinity of Sand Mountain, Alabama, and one locality in Arkansas. *Asimina* includes two species. One is a small tree, the familiar pawpaw, which grows on river banks and in rich woods from Florida and Texas northward into Ontario. Its sweet fruit is often gathered. The other species, called dwarf pawpaw, is a shrub distributed from northern Florida to South Carolina and Mississippi. The genus is the northernmost outlier of an essentially tropical family, the Annonaceae or custard apples.

Rhapidophyllum, the needle palm, ranges from Florida to South Carolina and Mississippi; and *Serenoa*, the saw palmetto, from Florida to South Carolina and Louisiana. They are among the few palm genera that are not confined to the tropics. *Fothergilla*, the witch alders, includes three species. One grows in rich woods of the southern Appalachians, from Alabama to North Carolina. A second inhabits swamps of the lowlands, from Alabama to Virginia; and a third lives in low pinelands from Florida to North Carolina. The genus is lacking from the Piedmont Plateau, the rolling clay-hill country between the mountains and the lowlands. This distribution may reflect the fact that the mountains and the coastal lowlands both are far more humid than the intervening Piedmont. Other groups of plants (and of animals) are distributed much like *Fothergilla*.

Geobalanus, of the plum family, includes two low shrubs known as gopher apples. One species is restricted to Florida, while the other ranges from Florida to Georgia and Mississippi. The fruit, somewhat pearlike in flavor, is fed upon by the gopher tortoise, a big land turtle; and gopher-apple seeds are spread in the droppings of this reptile. *Cliftonia*, the black titi, grows in swamps from northern Florida to Georgia and Louisiana. Its flowers, inconspicuous but fragrant, are an important local source of honey.

Oxydendrum, the sourwood, is usually a shrub but occasionally becomes a sizable tree. It grows from northern Florida and Louisiana northward to Indiana. Sourwood is a member of the heath family, Ericaceae, a group with several endemic genera in eastern North America. Most remarkable of these heath genera is *Ampelothamnus,* whose single species ranges from northern Florida into Georgia and Alabama. *Ampelothamnus* may be a shrub, or a vine that ascends trees by creeping under the bark. Another ericaceous genus, *Leiophyllum,* includes three species called sand myrtles. Two of these live in the southern part of the Appalachian uplift, but the third grows in the low pine barrens from southern New Jersey to Florida. This distribution recalls that of the witch alders.

Among endemic genera of herbaceous plants, *Muricauda* includes only the green dragon, a close relative of the jack-in-the-pulpits. Green dragon grows in moist woods from central Florida and Texas northward into Ontario and Maine. *Orontium,* the golden club, is a beautiful aquatic plant of the Arum family, ranging from Florida and Louisiana northward into Tennessee and Massachusetts. *Cuthbertia,* the roselings, includes three species collectively distributed from Florida to North Carolina; they closely resemble their relatives, the day-flowers, but have pink blossoms instead of blue. *Uvularia,* belonging to the lily-of-the-valley family, includes the plants called bellworts; the genus is distributed from northern Florida to North Dakota and Quebec.

Pogonia, the rose crested-orchid, lives in acid bogs and other damp places from Florida and Texas northward to Newfoundland and Minnesota; and *Cleistes,* the rose orchid, grows in the southern Appalachians as well as in the coastal lowlands from New Jersey to Florida. *Cubelium,* the green violet, lives in rich woods and cool ravines from Georgia and Mississippi northward into Ontario and New York; it is truly a green-flowered member of the violet family. *Bartonia,* of the gentian family, includes four species variously distributed over eastern North America, mostly in low pinelands and swamps, or about lake margins; two of them contain little chlorophyll, and are partly saprophytic. *Monotropsis,* the sweet pinesaps, includes two species of the southern Appalachians and nearby highlands, and a third species isolated in northeastern Florida; members of the Indian-pipe family, they are fully saprophytic.

The Province harbors several remarkable plants that actually catch and feed upon animal life. The most impressive of all carnivorous plants belongs to a genus that is endemic to the southeastern United States. This is *Dionaea,* its only species the Venus flytrap, an inhabitant of bogs and low pinelands of the Carolinas. The plant has paired leaf lobes that close traplike upon a visiting insect. A small, soft-bodied insect is crushed; a larger one is imprisoned by inter-locking spines along the leaf-lobe margins. Glands in the leaf secrete a digestive

Bald cypress in a Louisiana river swamp. Unlike most conifers, the cypresses lose their needles in winter. (Louisiana Tourist Commission)

juice, and the victim is absorbed in a week or two. Then the trap is set again. *Dionaea* has been known to capture a small frog; and, in spite of sensational yarns, no other carnivorous plant can actively ensnare any larger prey.

The Venus flytrap belongs to the sundew family, Droseraceae. Another genus of the family, *Drosera*, the sundews, is made up of species that catch insects, but in a somewhat less active fashion. Fine hairs of the sundew leaf are tipped with sticky drops that lure and entangle small prey. Once an insect is caught, nearby hairs will bend toward it and so begaum it further. Whereas *Dionaea* includes but one species, and it of very restricted distribution, *Drosera* has nearly a hundred species and is represented on all continents as well as numerous islands. However, sundews are lacking from the Mexican highlands and the deserts of the southwestern United States; their genus is more characteristic of the Atlantic North American Province than of the Pacific subdivision.

Sarracenia, of the Sarraceniaceae, is another genus of insectivorous plants, endemic to eastern North America. Its members are called pitcher plants. (The pitcher plants of the Venezuela-Guiana Province, mentioned in a previous chapter, belong to the same family but to a different genus, *Heliamphora*.) *Sarracenia* is an eastern counterpart of California's *Darlingtonia*; but unlike the latter, *Sarracenia* includes a variety of species, some of them widespread. One *Sarracenia*, the purple pitcher plant, ranges from Louisiana and northern Florida northward to Manitoba and Newfoundland. Several other pitcher plants, with flowers of purple, maroon, yellow, or greenish-yellow, are restricted to the lowlands of the southeastern United States. In all these plants the trumpet-shaped leaf, lined with downwardly directed hairs, is a deathtrap for insects that are attracted by a honey-like secretion.

From the above account it will be seen that some plant genera, endemic to eastern North America, are of very limited range therein. One genus of the Province establishes a record for geographic limitation. In autumn of the year 1765 the pioneer botanist John Bartram, with his son William, discovered a grove of small trees near old Fort Barrington on the Altamaha River in southern Georgia. More than a decade later the younger Bartram returned and found several acres of these trees, all in exquisite bloom. He sent seeds and cuttings to his father, who maintained a garden near Philadelphia. The Bartrams realized that the Georgia tree belonged to the Camellia family; and that it was most closely related to *Gordonia*, the loblolly bays, a genus represented in Asia and

Lebanon State Forest in Burlington County, New Jersey. Many Coastal Plain organisms reach the northern limit of their range in this vicinity. (New Jersey Department of Economic Development)

eastern North America. The elder Bartram named the Georgia tree *Franklinia*, after his friend the illustrious Ben. In 1790 the trees were still growing near the old fort; but no one ever saw them thereafter, and no one has ever discovered another locality for the species. Today you can buy a Franklinia for your garden, but it will be a descendant of the specimens that John Bartram carefully nurtured near Philadelphia. Probably the "lost Franklinia" was an Arcto-Tertiary relic, on the verge of extinction when the Bartrams chanced upon the last remaining colony.

Numerous plant genera, not restricted to the Atlantic North American Province, have produced endemic species therein. Many of the species have a close counterpart in the Sino-Japanese or in the Euro-Siberian Province.

Various species of eastern North America have found their way into gardens. Especially popular in Europe are such North American trees as the box-elder maple, the catalpa, the flowering dogwood, the Magnolia, and the black locust. Garden shrubs from eastern North America include the mountain laurel and the staghorn sumac; while among familiar herbaceous plants are the Philadelphia lily, scarlet Lobelia, blazing stars, horsemint, moss Phlox, and spiderwort. Asters, so conspicuous in northeastern North America, were carried to Europe, hybridized and improved there, and sent back to us as "Michaelmas daisies."

Among species native to eastern North America, several are of medicinal value: witch hazel, goldenseal, slippery elm, Indian tobacco, May apple, and Seneca snakeroot. The sugar maple, black walnut, pecan, cranberry, and various blueberries are all of considerable economic importance. Concord and Niagara grapes are cultivated varieties of the plum grape, a native species distributed from Georgia and Mississippi northward to Indiana and Vermont.

The daisy family is generally successful in grasslands, and this is true in the prairies that form the western part of the Atlantic North American Province. Members of this family, native to the prairies but often transplanted to the garden, include a perennial Gaillardia, a Helenium or sneezeweed, and the black-eyed Susan. The Jerusalem artichoke, which is not an artichoke but a sunflower, probably came from the western part of the Province, but has been established at many localities farther east.

It is instructive to review the distribution of certain plant groups that are represented both in eastern North America and in some other part of the world. The broom crowberry, of the Empetraceae, ranges from New Jersey northward through the eastern part of the New England states to coastal Newfoundland. Its genus, *Corema*, includes but one other species, found in Portugal, Spain, and the Azores. *Corema* is said to have an "amphi-Atlantic" distribution; it is present on both sides of the Atlantic but not in Asia. It is by no means certain that the genus crossed the North Atlantic; it might once have been represented all the way across Eurasia and North America (as the black crowberry, an *Empetrum*, is today).

The lily-of-the-valley was once thought of as an amphi-Atlantic species. It grows in the Appalachian forests from South Carolina and Tennessee northward into West Virginia, and in western Europe. However, it has also been found at scattered localities in Asia, and the existing populations are likely to be relics of a formerly widespread range.

Tipularia includes one New World species, the elfin-spur or cranefly orchid, which lives in woods from Florida and Texas northward into Indiana and New York. The other two species of the genus inhabit the Himalayas. This is a distributional pattern comparable to that of the tulip trees, trumpet vines, moonseeds, Croomias, and other Arcto-Tertiary relics discussed in connection with the Sino-Japanese Province. Additional plants of eastern North America having congeners only in eastern Asia are the lizard's-tail (*Saururus*) and the ditch stonecrop (*Penthorum*, placed in a family of its own, Penthoraceae).

Calycanthus, the sweet shrubs, includes four species of eastern North America, collectively ranging from Florida and Louisiana northward into Pennsylvania; and the genus is also represented in southeastern Asia. However, one sweet shrub is endemic to California. A comparable distribution is that of the stinking cedars, *Tumion*; these include one species in northern Florida and nearby parts of Georgia, one in California, and two in eastern Asia.

Sibbaldiopsis, the mountain cinquefoil, is distributed from Manitoba to Newfoundland, and thence northward into the Arctic of Greenland. In addition, it grows on cold, windswept ridges of the Appalachians southward all the way into Georgia. This is not precisely an arctic-alpine distribution, but is comparable in that a species of the high north latitudes also finds a suitable habitat at higher altitudes farther south.

Fagus, the beech trees, includes a few species scattered over the temperate regions of Eurasia and North America. The American beech ranges from Florida and Texas northward into Ontario and Nova Scotia. This same beech turns up again in the Sierra Madre Oriental of Mexico. Beechdrops (*Epifagus*), a plant that grows only as a parasite on the roots of the beech, has accompanied its tree host from eastern North America to the Mexican highlands.

Castanea is represented in North America, Eurasia, and Africa. Its New World species include shrubs called chinquapins, and a fine tree called the chestnut. The latter was a conspicuous tree of the forest, from the uplands of Georgia northward into Maine and Ontario. However, in 1904 a parasitic fungus, the chestnut blight, arrived in North America from the Orient. The Asiatic chestnuts were more or less resistant to this parasite; but not so the American species, which was killed out everywhere within its range. Today some old chestnut stumps may sprout, but the new shoots never attain much size. In just a few years more, the genus *Castanea* will be represented in the New World only by the shrubby chinquapins, which are concentrated in the lowlands of the southeastern United States.

The Atlantic North American Province is very large and diverse. Certain parts of it merit special comment.

The Appalachians are not extremely high mountains, but they do permit the existence of red spruce and balsam fir as far south as North Carolina. The Canadian hemlock ranges even farther southward, into Georgia.

Mount Washington, in northern New Hampshire, is one of the few peaks of eastern North America to rise above timberline. It is 6,290 feet high, a remarkably low peak to have a timberline; but its summit is completely unprotected from icy winds that sweep out of the north, bringing snow in every month of the year. (Yet summer temperatures may reach 70 degrees, as is also the case in parts of the Arctic.)

Mount Washington is an excellent place to study the effect of altitude on vegetation. Speaking very generally, up the flanks of a mountain the temperature falls about three degrees for every thousand feet of elevation; and in eastern North America the temperature at sea level also falls about three degrees for every six hundred miles northward. The latitudinal change of vegetation is paralleled by altitudinal change; and up the slopes of Mount Washington the vegetation passes from a forest into a thicket of stunted trees and low shrubs, and finally into a treeless expanse reminiscent of the Arctic tundra.

In the Appalachians there are higher peaks than Mount Washington, but they are all far to the south, in the Smokies. The high southern peaks do not

Above, black bears at Vermillion Bay, northwestern Ontario, Canada. A species widespread in, but not restricted to, eastern North America. (Canadian Government Travel Bureau) At the right, white-tailed deer in Wisconsin. Another species widespread in eastern North America, but not confined thereto. (Wisconsin Conservation Department)

reach timberline, but on some of them, the so-called "bald" mountains, the coniferous forest stops just short of the top, and the peak is capped only with shrubs of the heath family, especially Rhododendrons and mountain laurel. The environmental factors that produce a heath bald are still a matter for investigation.

East of the Appalachians, from southern New York to eastern Alabama, is the Piedmont, a belt mostly of rolling hills. In many places the Piedmont is badly eroded, especially in Georgia and the Carolinas where a red clay subsoil is now exposed at the surface. Much of the Piedmont is covered with shortleaf and loblolly pines, and a question arises whether these pine forests have developed chiefly as a result of environmental changes induced by man. Early historical accounts suggest that, when the country was first settled, the Piedmont had both pine forests and extensive hardwood stands.

In parts of the Atlantic and Gulf lowlands, the distribution of various plants is controlled by a peculiarity of the deeper soil. Beneath the loose surface soil, at a depth of a yard or so, there is a layer of compacted, claylike material; and this layer, the so-called hardpan, is almost impervious to water. In rainy weather the rainwater does not soak into the ground, while in dry weather the ground water cannot rise to the surface through capillary action. Such areas therefore alternate between very wet and very dry; and only certain kinds of plants can grow under these conditions. Usually the predominant trees are pines, especially slash pine but sometimes other species. Local residents speak of the "pine flats" or the "pine barrens;" but the hardpan country is not barren to the botanist's eye. It harbors specialized trees such as pond cypress, pond pine, sweet bay, red bay, and Ogeechee lime. Characteristic shrubs include wax myrtle, star anise, titi, gallberry, tarflower, wicky, fetterbush, and sandweed. There are vines such as greenbriers and blackberries, and numerous flowering herbs: pitcher plants, zephyr lilies, Irises, fringe orchids, sundew, mallows, meadow beauty, Sabbatias, and milkworts. Among the aquatic plants are arrowheads, water lettuce, mud-Mary, bladderwort, bonnet water lily, and pickerel weed. There is also a variety of grasses, reeds, and sedges.

The Coastal Plain, or Atlantic and Gulf lowlands, forms a distinctive part of the Province, and not solely on account of the species that are limited to the pine flatwoods. Numerous plants are largely confined to the Coastal Plain, ranging from Florida northward through the lowlands of Georgia and the

Carolinas into southeastern Virginia, and westward through the Gulf states into eastern Texas. Many of the same plants also follow the Mississippi Valley northward into southern Illinois or southern Indiana.

In the Province the rainfall diminishes toward the west, and eventually becomes inadequate to support tree growth. Speaking generally, an annual rainfall of at least 20 to 25 inches is needed for tree growth in the northern part of the Province, and 35 to 40 inches in the southern part where evaporation is more rapid. Forest vegetation extends westward roughly as far as eastern Texas, eastern Oklahoma, eastern Kansas, central Iowa, central Minnesota, and eastern Manitoba. Farther west in the Province, the forest is replaced by a grassland, often called the prairie. Within the prairie the grasses become sparser toward the west, and the grassland is terminated on the west by the mountains and deserts of the Pacific North American Province. Prairie grasses mostly belong to wide-spread genera such as *Stipa* (porcupine grasses, temperate lands generally), *Sporobolus* (dropseeds, concentrated in warmer parts of North America), *Koeleria* (June grasses, temperate lands), and *Agropyron* (wheat grasses, temperate lands); as well as *Andropogon* and *Bouteloua*, mentioned in the preceding chapter.

The Ozark-Ouachita uplift is much smaller, lower, and drier than the Appalachians. However, the Ozark flora has Appalachian affinities. In many cases a species of the Appalachians has a close congener in the Ozarks. The Black Hills of South Dakota, and similar uplifts in North Dakota and southern Saskatchewan, are wooded oases arising from grassland. They have received their flora from the coniferous forests of the north, from the deciduous forests of eastern North America, and from the Rockies.

A problem in classification is offered by the Aleutians and the Alaskan Peninsula. Floral elements of the Pacific North American Province straggle out to the eastern side of Kodiak Island and its adjacent mainland, while elements of the Atlantic North American Province grow elsewhere on the base of the Alaskan Peninsula. At higher elevations the vegetation is arctic. Of course an area like the Aleutian Range and Islands, lying near several Provinces, will support a mixed and transitional flora, and will be difficult to allocate precisely.

The fauna of the Atlantic North American Province is rich, and the species are distributed in a variety of ways. Animals that are widely distributed in this Province may also invade parts of the Pacific subdivision, especially the Rockies and the Cascades. The black bear, gray fox, white-tailed deer, ruffed grouse, and red-shouldered hawk, all thought of as characterizing eastern North America, actually reach various western states, mostly in the mountains; and in addition, the fox and the deer range through the Mexican highlands and thence southward into the Middle American tropics.

Although in phytogeography the Mexican highlands are placed in the Pacific North American Province, the fauna of the highlands has strong affinities toward eastern North America. Animal species or genera of eastern North America, represented also in the Mexican highlands, include the common opossum, least shrews (*Cryptotis*), southern flying squirrel, rice rats (*Oryzomys*), eastern cottontail, and armadillo. Among birds are the turkey vulture, anhinga, wood ibis, swallow-tailed kite, purple gallinule, barred owl, and boat-tailed grackle. Reptiles include the mud turtles (*Kinosternon*), glass lizards (*Ophisaurus*) and a variety of snakes: brown snakes (*Storeria*), water snakes (*Natrix*), ribbon snake, yellow-lipped snakes (*Rhadinaea*), western hognose snake, indigo snake, rat snakes (*Elaphe*), coral snakes (*Micrurus*), moccasins (*Agkistrodon*), and pygmy rattlesnakes (*Sistrurus*). Many of these animal species or genera range also into the New World tropics, and some are represented in Eurasia; but all of them are nearly or quite lacking from non-Mexican portions of the Pacific North American Province.

The plains bison is popularly considered a "western" animal, but its range lay chiefly within the more easterly Province: from Alberta eastward to New York, and southward to northeastern Mexico and northern Florida. The wild turkey hails not from Turkey but from America; it is distributed from Maine and South Dakota southward through the Mexican highlands. Only one other living species belongs in the same family as the common turkey; this is the ocellated turkey of British Honduras, Guatemala, and the southern part of the Yucatán Peninsula.

Some animals inhabit only the northern part of the Atlantic subdivision. Thus the woodland caribou is a deer of the northern forests, from Newfoundland to Alberta. The pine marten ranges from Alaska eastward to Labrador, in addition reaching New York and the Great Lakes region. The least weasel is distributed across the Euro-Siberian Province, and also from Alaska eastward to Labrador; it follows the Appalachians southward into North Carolina.

The insectivore genus *Condylura* includes only the star-nosed mole, an extraordinary animal of eastern North America. Although a burrower like other moles, it is restricted to wet ground near lakes or streams, and enters the water readily. It is distributed from Labrador and southern Manitoba to the uplands of northern South Carolina, and it is also found in the swamps of southeastern Georgia. Thus its range is interrupted by the relatively dry Piedmont and upper Coastal Plain of Georgia.

The really characteristic animals of eastern North America are those that live in lakes, ponds, or swamps, and to a lesser extent rivers; for such animals have had the most difficulty in crossing the deserts or dry grasslands that border the Atlantic North American Province on the west and the southwest.

The fish family Amiidae has but one living species, the bowfin. When the dinosaurs were spreading over the land, amiid fishes dominated the seas; but by Tertiary times they were confined to fresh waters of the northern hemisphere. Eventually they died out from Eurasia and western North America, leaving only the bowfin which is distributed from the Great Lakes southward to Texas and Florida. The fish family Aphredoderidae includes only the pirate perch, found from the Great Lakes and New York southward into Texas and Florida. The family was represented in the western United States before the Rocky Mountains were uplifted, but does not appear in the fossil record of the Old World. The important family Centrarchidae, which includes the numerous sunfishes and the freshwater basses, is nearly confined to eastern North America, although there is an isolated genus in California (*Archoplites*, the Sacramento perch).

The darters, beautiful little fishes of the percid genus *Etheostoma*, include about seventy species, all living east of the Rockies. The shiners, members of the cyprinid genus *Notropis*, include nearly one hundred species in eastern North America, a few in northern Mexico, and one in the Pacific drainage of Arizona and New Mexico. The killifishes, the cyprinodontid genus *Fundulus*, include about twenty species confined to eastern North America, one that ranges from the Brazos River to the Colorado, two or three in northern Mexico, and one distributed coastally from southern California into Baja California. Several *Fundulus* are coastal rather than freshwater, and one reaches Bermuda.

The salamander family Amphiumidae includes but one genus, whose members—the so-called "lamp-eels"—are elongate muck-burrowers with tiny limbs. The genus is represented from Virginia southward through Florida, thence westward into eastern Texas, and northward up the Mississippi Valley

Bison in the Nation Bison Range near Moiese, Montana. (Montana Highway Commission)

Pronghorn in Montana. (Montana Highway Commission)

almost to Illinois. The salamander family Necturidae, whose members are called waterdogs, is distributed from southern Canada to northern Florida, and from the Atlantic lowlands westward into the Mississippi drainage. Waterdogs were once thought to belong in the same family with the olm, an aquatic, cave-dwelling salamander of Yugoslavia; but the relationship now seems to be remote, a matter of parallelism. Nowhere in the world are salamander species as numerous as in the southern Appalachians, where the Plethodontidae, especially, have diversified.

Curious animals are the sirens and mud eels, family Sirenidae. They are eel-like in shape, with feathery gills; they have front legs but no trace of hind legs. Sirenids have generally been placed with the salamanders, but there is a growing belief that they were of independent origin. They may be related to an ancient group called the aïstopods, elongate amphibians that lived in Europe and North America at a very remote period. In any event, the collective distribution of the sirenids now forms a familiar pattern: from Florida northward into Virginia, and westward into eastern Texas; also up the Mississippi Valley into Indiana.

24 *"He is not a true man of science who does not bring some*

sympathy to his studies. . . . The fact which interests us most is the life of

*the naturalist. The purest science is still biographical."**

REFINING THE PHYTOGEOGRAPHIC SCHEME

We have now covered the floral Kingdoms and Provinces of the world. So numerous are they that a summary is clearly in order. When summarizing the faunal subdivisions, we found it profitable first to mention some of the people who helped to develop the zoogeographic scheme. In like fashion we should note a few of the people who contributed significantly to phytogeography. Some of them were as amazingly versatile as Sclater, Wallace, Darwin, and Heilprin; and their ideas still command our attention when we proffer a succinct treatment of the phytogeographic scheme.

Phytogeography is generally thought of as beginning with the Swiss botanist Augustin De Candolle and the German nobleman F. H. Alexander Von Humboldt. De Candolle was a lawyer and a physician, but he turned to botany and the academic life. He came to specialize in the native flora and the crop plants of Switzerland and France; and his studies won the acclaim of eminent

* HENRY DAVID THOREAU

French scientists such as Georges Cuvier and Jean Baptiste Lamarck. Beginning in 1824, De Candolle began to publish a series of works on the plants of the world, concerning himself with both taxonomy and distribution. During this period he headed the faculty of natural sciences at the University of Montpelier and directed a botanical garden at Geneva.

As for the Baron Von Humboldt, he, like Wallace and Heilprin, was on the one hand a scholar, on the other hand a rugged adventurer. Educated at several German universities, Von Humboldt wrote his first technical paper on the basaltic rocks of the Rhine. Later, with Aimé Bonpland, he explored the island of Tenerife in the Atlantic; and then the two went to South America where from 1799 to 1804 they rambled over what are now Venezuela, Colombia, Ecuador, and Peru. While in the New World, Von Humboldt also visited Mexico and Cuba; he wrote some political essays about the latter country. Then, returning to Europe, he joined the French chemist Louis Gay-Lussac in a now famous experiment that demonstrated the chemical composition of water.

A favorite at European courts, Von Humboldt often accompanied the King of Prussia, or represented him at gatherings; and in 1829 the Czar of Russia, Nicholas I, sent Von Humboldt to explore remote parts of Asia: the Urals, the Caspian region, the Altai, and the Chinese dependency of Jungaria. Von Humboldt was charged not only with the task of collecting geological and botanical specimens, but also with discovering beds of gold and platinum, finding diamond fields, reckoning his position on the earth by means of astronomy, and making observations on the earth's magnetism!

For a period of fifty years, beginning in 1807, Von Humboldt published various accounts of his travels; and these works presented his phytogeographic observations along with interesting commentaries on animal life, people, and the natural landscape. But Von Humboldt wrote for a more leisurely age than ours; the story of his New World trip ran to thirty volumes. De Candolle and Von Humboldt are important not so much for their phytogeographic conclusions, which were based upon the inadequate knowledge of their day, but rather for the impetus they gave to the study of plant distribution.

Another pioneer in phytogeography was the German botanist J. F. Schouw. In 1823 he attempted to subdivide the whole world on the basis of plant distribution. His effort was ambitious, indeed overambitious for its day; but some of his categories were well founded. Like most early students, both phytogeographers and zoogeographers, Schouw tended to subdivide too finely in the better studied regions of Europe, northern Africa, and the Near East, while subdividing inadequately in other regions.

Schouw partitioned the world into twenty-five plant kingdoms, some of them subdivided into lesser units called provinces. Where possible, he named a kingdom in terms of supposedly characteristic plant genera or families. For example, northern North America was the "Kingdom of *Aster* and *Solidago*." Of course *Aster* is the plant genus to which belong the asters so common in gardens, woods, and fields; *Solidago* is the genus that includes the familiar goldenrods. Both genera are indeed represented by numerous species in northern North America; and both are largely, but not entirely, confined to this country. In Schouw's system, southern North America was the "Kingdom of Magnolias," a name fairly appropriate for the Southeast but not the Southwest. (And Schouw did not know that magnolias are more numerous in eastern Asia than in North America.)

A "Kingdom of Saxifrages and Mosses" included both the Arctic region and the tops of high mountains where arctic conditions prevail. This kingdom was made up of three provinces. The "Province of Sedges" was the treeless Arctic tundra. The "Province of Shrubby Alpine Composites" included the higher peaks of Canada and the United States. (Plants of the daisy family are called composites because the "flower" is really a composite of many small blossoms packed tightly together.) The "Province of Primulaceae" included the higher peaks of Europe and Asia, thought by Schouw to be characterized by members of the primrose family.

Saxifrages, mosses, sedges, shrubby composites, and some primroses are indeed conspicuous at high altitudes and latitudes. It was especially perceptive of Schouw to realize that an essentially Arctic flora may also be found at high elevations outside the Arctic; that the Old World and the New are linked in the far north; and that a plant kingdom may transgress a continental boundary. His method of naming the subdivisions has features that are still appealing.

Phytogeography was given a new direction by the evolutionary concepts of Wallace and Darwin. First of the new phytogeographers was the English surgeon and naturalist, Sir Joseph Dalton Hooker. Between 1853 and 1909 he published studies on the flora of New Zealand, Australia, Tasmania, the Arctic, the British Isles, India, and other regions. In these works Hooker considered plant distribution in the light of evolutionary theory, and showed how useful distributional studies could be in any effort to unravel the mysteries of organic evolution.

Hooker was another of those versatile people who, like Von Humboldt, Wallace, and Heilprin, became fascinated by problems of distribution. Hooker was the son of Sir William Jackson Hooker, who for years directed the Royal Botanical Gardens at Kew, in the London suburbs. While the elder Hooker was building the world's most comprehensive herbarium, young Joseph joined the Antarctic expedition of the *Erebus* and the *Terror*, to spend four years in exploring the far-southern regions of the globe. Joseph Hooker's studies on the flora of New Zealand, Tasmania, Australia, and the Antarctic were based in large part on specimens he had personally collected.

In 1846 the younger Hooker was appointed botanist to the Geological Survey of Great Britain; and soon thereafter he made an expedition into the high Himalayas of Asia, spending three years in this remote and perilous region. An accomplished mountain climber, in 1871 Hooker was the first European to scale the Grand Atlas of Morocco, where rugged peaks rise to a height of 13,000 feet. Needless to say, he brought back a large collection of African plants for study. Hooker eventually succeeded his father as director of the Kew Gardens, and in later life he also turned out several major contributions to plant taxonomy. He likewise did important work on marine biology, discovering that some sea-bottom deposits are formed in considerable part from the bodies of microscopic plants, and that such plants, drifting in the upper waters, are the main producers of food in the sea.

During the latter nineteenth century there developed a German school of phytogeographers; it included such men as O. Drude, E. Goeze, A. H. R. Grisebach, and A. Hansen, each of whom wrote a book about plant distribution on a worldwide scale. But the best known of the German students was Heinrich Gustav Adolf Engler, a professor at Kiel, Breslau, and Berlin, and director of the Botanical Garden at Schöneberg. Between 1872 and 1882 Engler published a series of works dealing with the phytogeography of the world. In these books he made special effort to divide the globe into more or less equivalent units on the basis of the flora. The effort was successful, and later phytogeographic schemes have differed from Engler's more in detail than in principle.

Singly or in coauthorship with others of the German school, Engler continued to produce botanical treatises for decades. He also brought order to a list of the world's plants, which he arranged in a scale from primitive to advanced; and the "Engler sequence" has since been followed in numerous botany texts, with but minor emendations.

In the twentieth century, the progress of botanical collecting has permitted refinement of Engler's phytogeographic units. Phytogeographies of the world have been written by (among others) V. V. Alekhin in Russia, S. F. Blake and A. C. Atwood in the United States, G. S. Boulger and M. E. Hardy in England, Léon Croizat in the Netherlands, H. Gaussen in France, and A. Hayek in Germany. Of course these more recent students have contributed far more to phytogeography than a revision of the floral subdivisions.

A modern phytogeographic scheme, and one that has been widely used, was first presented by the English botanist and professor, Ronald D. Good. His book, *The Geography of the Flowering Plants*, was published in 1947. It is interesting to note that each subsequent edition of this work (1953 and 1964) made a few changes in the scheme, to take into account new information about plant distribution in remoter parts of the world.

In 1957, Pierre Dansereau, at that time Dean of the Faculty of Science at the University of Montreal, produced a book entitled *Biogeography: An Ecological Perspective*. Dealing primarily with plant ecology, this work slightly revised Good's terminology and subdivisions.

Now to our summary.

In Chapter 8 it seemed desirable to reject some longstanding but ambiguous designations for various Realms and Regions, and to name these faunal subdivisions in the simple terms of geography and climate. Fortunately, the floral subdivisions have already been designated in this sensible fashion, and one could scarcely improve upon the names that are used for most Kingdoms and Provinces. In Chapter 10, "Cape Kingdom" was suggested as preferable to "South African Kingdom;" and in Chapter 13, "Region of New Caledonia" and "Region of Micronesia and Melanesia" were rejected in favor of "New Caledonian Province" and "Micronesian-Melanesian Province."

One further suggestion is advanced: It is important to distinguish between flora and vegetation. The former term relates to the species, genera, and families of a given area, the latter to the nature of the plant cover. As we shall see in the next chapter, the world has also been subdivided on the basis of its plant cover.

However, Provinces are recognized on the basis of the flora; and so "rainforest," "steppe," and "desert," terms descriptive of plant cover, might well be deleted from certain Provincial names.

To summarize the floral subdivisions: The Australian Kingdom includes the Australian mainland and the island of Tasmania. The Kingdom has three Provinces. (1) The North and East Australian Province includes the northern and eastern edges of the continent, and Tasmania. (2) The Southwest Australian Province is formed by the southwestern tip of the continent. The remainder of Australia makes up (3) the Central Australian Province.

The Cape Kingdom includes only the extreme southwestern tip of Africa. The Kingdom is all one Province, the Cape Province.

The Antarctic Kingdom, with three Provinces, embraces some widely scattered, far-southern lands. Its (1) New Zealand Province includes the two main islands of New Zealand, plus the Kermadec, Chatham, Auckland, and Campbell islands. (2) The Patagonian Province embraces Tierra del Fuego, southern Argentina, the southern Andes of Chile, and the Falklands. (3) The South Temperate Oceanic Islands Province is made up of small islands, notably South Georgia, Marion, Tristan da Cunha, the Crozets, the Kerguelen Archipelago, Heard, McDonald, St. Paul, Amsterdam, and Macquarie. Antarctica and the South Shetlands may be added to this last Province.

The Palaeotropical Kingdom includes the tropics and near-tropics of Asia and Africa, and the tropical islands of the South Pacific from Sala-y-Gomez westward through Indonesia. The Kingdom is divided into three Subkingdoms. (1) The African Subkingdom includes the islands of the western Indian Ocean; two small islands of the Atlantic, Ascension and St. Helena; the tropics of Africa; and the deserts that border the tropics in both Africa and Asia. (2) The Indo-Malaysian Subkingdom is made up of tropical Asia, plus Pacific islands from Sumatra eastward through New Guinea and northward through the Philippines, Formosa, and the Ryukyus. (3) The Polynesian Subkingdom extends over the small islands of the Pacific, from the Palaus and Marianas eastward through Sala-y-Gomez, and from the Hawaiian group southward through Lord Howe Island.

Each Palaeotropical Subkingdom is further divided. The African Subkingdom has eight Provinces, as follows: (1) The North African–Indian Province, including the Sahara and the deserts that stretch from Saudi Arabia to extreme

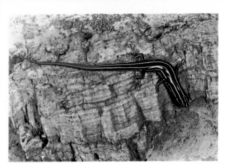

THE MANY PATTERNS OF DISTRIBUTION. *Shown at the upper left is the tuatara, the lone survivor of an ancient group of reptiles, now confined to a few remote islets off New Zealand. At the upper right are spruce grouse in Alaska. Closely associated with the northern coniferous forests, this bird follows the spruce and tamarack about as far into the north as they will grow. The North American otter is pictured at lower left. Semiaquatic, usually in fresh water, otters are widespread in North and South America, Eurasia, and Africa. The blue-tailed skink, lower right, is a lizard of eastern North America. Although its genus, Eumeces, is widespread in North America, eastern and southwestern Asia, and northern Africa, this American species has particularly close relatives in China and Japan. (Upper right, Alaska Division of Tourism; all others, Jim Miller)*

LIFE OF THE NEW WORLD TROPICS. An armored catfish from the Amazon is shown at upper left. It excavates burrows in stream banks, and feeds on algae and other soft substances. Catfishes reach maximum diversity in South America, which has perhaps 1,000 species, most of them freshwater. But some catfishes live in the sea, or in fresh waters of other continents. (Ross Allen's) Below the catfish is the arrow-poison frog belonging to a family, Dendrobatidae, that is confined to New World tropics. A climber like many tropical frogs, it is protected by toxic skin secretions. Bright coloration may warn of inedibility. (Jim Miller) The Caribbean region has a very rich flora. At upper right is the frangipani, Plumeria, growing here in British Honduras. It is a characteristic species of the region, but has been widely transplanted by man to many parts of the New and Old World tropics. (W. T. Neill) Immediately above is a brocket deer from the Amazon forest. Its family, Cervidae, is primarily North American and Eurasian, but moved into South America when the Isthmus of Panama appeared above water. (Ross Allen's)

LIFE OF THE FAR NORTH. *Alaskan plants are surprisingly diverse. During the summer, tiny blossoms may spring up even in stony soil (upper left); grasses sprout and scatter their seeds to the wind (upper right); bog plants appear when low spots begin to thaw (lower left). At lower right are ptarmigan in Alaska. Brownish in summer, this bird whitens in winter to match the snow. Several species of ptarmigan live mostly in far-northern North America and Eurasia, or else farther south at high altitudes. (Alaska Division of Tourism)*

THE MANY FACTORS INFLUENCING DISTRIBUTION. *At upper left is the green turtle, which swims in warm seas but must find sandy beaches on which to nest. (Jim Miller) Arctic waters teem with living things, some of which often help support land life, including man. At upper right, local residents cut nutritious blubber from a small whale on an Alaskan beach. (Alaska Division of Tourism) Plant-eating dinosaurs once lived and nested in what is now the barren Gobi Desert. Shown at lower left is a museum reconstruction based on fossil remains. (Jim Miller) At lower right is a treefrog, Phrynohyas, on Codiaeum leaves in British Honduras. The frog is native, the plant imported from Malaysia. Man often alters distributions, and forms new associations of plants and animals. (W. T. Neill)*

western India; (2) the Sudanese Province, from Senegambia eastward through Mali to the lands of the upper Nile; (3) the Northeast African Province, embracing most of Somalia and Ethiopia, the island of Socotra, and the southern tip of Saudi Arabia; (4) the East African Province, from Kenya and Uganda southward through Mozambique, and westward into Angola and the northern part of Southwest Africa; (5) the West African Province, from the Congo Basin westward to Cameroon, and still farther westward as a narrow coastal strip to Guinea; (6) the South African Province, including Orange Free State and the Transvaal, plus the Kalahari and Karroo deserts; (7) the Madagascan Province, extending from Madagascar and the Comoros eastward through the Mascarenes and Seychelles; and (8) the Ascension and St. Helena Province, including only the two Atlantic islands for which the Province is named.

The Indo-Malaysian Subkingdom has three Provinces. (1) The Indian Province includes most of India, the southern flanks of the Himalayas, the island of Ceylon, and small islands from the Laccadives southward through the Chagos group. (2) The Continental Southeast Asiatic Province embraces the mainland tropics of eastern Asia, from East Pakistan eastward through southern China, and southward through Burma, Thailand, the Indochinese countries, and the upper half of the Malay Peninsula. Among islands appended to this Province are Hainan, Formosa, the Ryukyus, the Nicobars, and the Andamans. (3) The Malaysian Province includes the Pacific islands from the Philippines southward through Timor, and from Sumatra eastward through New Guinea. Also included is the southern portion of the Malay Peninsula. Excluded is a narrow strip of coastal New Guinea, from Hollandia to Madang.

The Polynesian Subkingdom has four Provinces. (1) The Micronesian-Melanesian Province includes the Carolines, Marianas, Bismarcks, Solomons, New Hebrides, and Fiji, as well as tiny islands from the Marshalls to the Ellice group. A coastal strip of New Guinea, from Hollandia to Madang, is also assigned to this Province. (2) The Polynesian Province extends from Samoa eastward to Sala-y-Gomez. (3) The New Caledonian Province embraces New Caledonia, Lord Howe Island, and Norfolk Island. (4) The Hawaiian Province includes the Hawaiian Islands from Hawaii westward through Kure; Johnston Island may also be added.

The Neotropical Kingdom embraces all of South America except the Patagonian Province; all of Central America; the tropical and near-tropical portions of Mexico, including the southern tip of Baja California; the southern tip of Florida; the West Indies; Bermuda; the Galapagos; and the Juan Fernandez group. This Kingdom has seven Provinces. (1) The Amazon Province follows the Amazon Basin from the eastern slopes of the Andes eastward to the Atlantic. (2) The Venezuela-Guiana Province embraces the Venezuelan uplands and the Orinoco Basin, plus the greater part of the Guianas. (3) The South Brazilian Province includes most of eastern and central Brazil, and the Gran Chaco of Paraguay and northern Argentina. (4) The Pampas Province extends from southeastern Brazil and north-central Argentina southward to the northern border of Patagonia. (5) The Andean Province is made up of the slopes and peaks of the Andes, and the land to the west thereof, from southern Chile northward to northern Colombia. The Galapagos Islands also belong to this Province. (6) The Juan Fernandez Province is composed of the Juan Fernandez and Desventuradas archipelagos. (7) The Caribbean Province includes the northern parts of Venezuela and Colombia; all of Central America; all of Mexico except the central highlands and the upper half of Baja California; the West Indies; southern Florida; Bermuda; and small islands off the west coast of Mexico and Central America, from Guadalupe to Malpelo.

The Boreal Kingdom stretches over all the lands north of the Palaeotropical and Neotropical Kingdoms. It thus includes all of Europe; the Mediterranean lands of northern Africa; most of Asia north of its southeastern tropics and southwestern deserts; and all of North America except southern Baja California, southern Florida, and the tropical or near-tropical lands that border the Mexican highlands on the east, south, and west.

The Boreal Kingdom has eight Provinces. (1) The Arctic-Subarctic Province includes all the lands north of tree growth in northern Europe, northern Asia, northern Alaska, and northern Canada; also the islands in or near the Artic Ocean, from Banks and the Sverdrups through Greenland and Iceland, and thence through Spitsbergen, Novaya Zemlya, and the New Siberian archipelago. (2) The Euro-Siberian Province stretches from the Faeroes and the British Isles eastward through most of Europe, and thence still farther eastward through central Asia to the Kamchatkan Peninsula and the Komandorskie Islands. (3) The Mediterranean Province includes southern Portugal, southern Spain, a bit of the French coast around Toulon, central and southern Italy, southern Yugoslavia and southern Bulgaria, all of Greece and Albania, and western and central

Turkey. This Province also encompasses the Asiatic edge of the Mediterranean from Syria into Sinai, and the African edge from northern Egypt westward through Morocco. The islands of the Mediterranean likewise belong to this subdivision. (4) The Macaronesian Province is made up of four archipelagos: the Azores, Madeiras, Canaries, and Cape Verdes. (5) The Western and Central Asiatic Province extends from eastern Ukraine and central Turkey eastward through Mongolia and the Tibetan Plateau. (6) The Sino-Japanese Province includes Korea, most of Manchuria, extreme southeastern Siberia, southern Sakhalin Island, the Japanese archipelago, the Kurils, northern and central China, and most of the Himalayas. (7) The Pacific North American Province embraces southern Alaska, southwestern Yukon, and most of British Columbia; a wide stretch along the Pacific coast of North America from Washington southward through northern Baja California; the Rocky Mountains and the Great Basin; the western portions of Wyoming, Colorado, New Mexico, and Texas; and the Mexican highlands. (8) The Atlantic North American Province includes all of eastern and central North America, except the Arctic of Canada and the near-tropics of southern Florida. This Province also extends westward from Saskatchewan through central Yukon and central Alaska to the shores of the Bering Sea.

25 *"A soil that is adorned by tall and graceful trees is not always a favorable one except of course for those trees. What tree is taller than the fir? Yet what other plant could exist in the same spot?"* *

VEGETATION AROUND THE WORLD

The floral subdivisions having been listed, let us note how the world may be partitioned on the basis of its vegetation or types of plant cover. Roughly speaking, any area is a forest, a grassland, or a desert. Students of vegetation call these three major categories "biochores." A fourth biochore, savanna, is sometimes recognized. Each biochore has subdivisions, called "formation-classes" or "formation-types." The more important formation-types are listed below.

Tropical rainforest occupies lands with constant warmth and consistently high rainfall. Trees are mostly hardwoods. Often they are buttressed, and hung with epiphytes and vines. Tree leaves are evergreen, and there is not much humus on the forest floor. The plants do not exhibit a marked seasonal cycle of growth, reproduction, and rest. The flora is rich in species. This type of vegetation occupies three main areas: the Amazon Basin and eastern Brazil, the western part of equatorial Africa, and a region that includes Sumatra, the southern half of the Malay Peninsula, the Vietnams, Borneo, the Philippines, and New Guinea. There are also smaller areas in southern Mexico, Central America,

---

* PLINY THE ELDER, in the first century.

292

eastern Mozambique, eastern Madagascar, northeastern Australia, the west coast of India, and the southern base of the Himalayas.

Subtropical rainforest is found in a few areas at the edge of the tropics, where summer and winter seasons can be distinguished, and where there may also be a slight drought effect. Trees are mostly hardwoods. They are bedecked with epiphytes, but vines are fewer than in tropical rainforest. Tree leaves are mostly evergreen. A seasonal cycle can be detected. The richness of the flora is reduced by the fall of temperature in winter. This type of forest develops in southern Florida, the Hawaiian Islands, Formosa, certain parts of the Himalayas, extreme southeastern China, eastern Anglo-Egyptian Sudan, and extreme south-eastern Australia; but the largest expanse of subtropical rainforest extends from south-central Brazil and Paraguay southward into northern Argentina and Uruguay. There is also a long stretch of subtropical rainforest along the eastern slopes of the Andes.

Monsoon forest develops where temperatures are tropical or nearly so, although with some cooling in winter. Rainfall is abundant when measured on an annual basis; but most of it arrives during the warmer months, and the cool season is very dry. Trees are mostly deciduous hardwoods; they drop their leaves with the advent of cool, dry weather. Bamboos are often conspicuous, forming thick stands. This type of forest is found mainly in northeastern India and along the southern flanks of the Himalayas, as well as through much of Burma and Thailand; but there are smaller areas of monsoon forest in Indonesia, from eastern Java to the Celebes and Timor, and likewise in western Madagascar.

Temperate rainforest occupies areas of high rainfall just outside the tropics. There is a cool season but no dry one. Trees include hardwoods and conifers, both evergreen. Tree ferns may be conspicuous. Epiphytes are uncommon. The largest expanse of temperate rainforest covers eastern China; there are other areas in southeastern Korea, the western part of Japan's Honshu Island, New Zealand, and southern Chile. In the New World a kind of temperate rainforest is met with on mountains in the tropics, where a high altitude results in some cooling and where humidity is maintained not only by rainfall but also by clouds that drift up against the slopes and peaks. American students usually prefer the name "cloud forest" for this type of plant cover.

Deciduous forest occupies regions with a well-defined winter and with precipitation distributed throughout the year. Usually there is a good bit of snow in winter. Trees are mostly deciduous hardwoods, green only in spring and summer, but there may also be an admixture of evergreen conifers. Epiphytes and large, woody vines are few. In the spring numerous herbaceous plants sprout from the forest floor. A humus layer is thick. There are three main areas of deciduous forest. One extends from eastern Texas and central Mississippi northward about to the Great Lakes and coastal New England. A second occupies northeastern China, much of Manchuria, extreme southeastern Siberia, and the southern tip of Sakhalin Island. The third stretches across Europe, from the Atlantic coast of France northward into southern Norway and southern Sweden, and eastward into western Russia. (Although the fall is proverbially a season when leaves turn red and yellow, most parts of the world have no such autumnal display. Autumn is colorful only in these deciduous forest regions.)

Coniferous forest is characteristic of cold regions with long winters and abundant precipitation. The predominant trees are needle-leaved and cone-bearing; they are usually evergreen. Some of the conifers may reach great size. Bogs are numerous in the forest, and support a variety of sedges, heaths, and mosses. Coniferous forest extends across much of Canada, from Nova Scotia and Ungava Bay southward into the Great Lakes region, thence northwestward into the southern part of the Northwest Territories, and (with interruptions) across Yukon and Alaska. This type of plant cover also extends southward into the Cascades, the Sierra Nevadas, and the humid Pacific coast from southern Alaska into northern California; and it follows the Rockies southward into New Mexico. There are disjunct outliers in the high southern Appalachians. A second great expanse of coniferous forest reaches from the Kamchatkan Peninsula westward across the U.S.S.R. and into Scandinavia. There is a smaller area of coniferous forest in southwestern China, and isolated patches occupy mountain peaks in many lands.

Evergreen hardwood forest characterizes areas with a Mediterranean type of climate, where the summer is warm and dry, and the precipitation is concentrated toward the winter months. Trees are mostly small, their leaves crisp and often spine-edged. There are few epiphytes. Often there are sizable areas of coarse brush or thorny shrubs. The largest expanse of evergreen hardwood forest rings the Mediterranean Sea, with minor interruptions. Other areas are in California and Baja California, central Chile west of the Andes, the Cape of Good Hope, and the southwestern corner of Australia.

Savanna is a puzzling type of vegetation. It is somewhat intermediate between forest and grassland; for the ground is usually covered thickly with tall grasses, but trees are also scattered about. Savanna develops in tropical or near-tropical lands where rainfall on the average is not scanty, but is strongly seasonal. During the long dry season, fires usually sweep over the savanna. There are large expanses of savanna in northern South America, bordering and interrupting the Amazonian rainforest. A lesser area extends from Paraguay into Uruguay. Savanna also covers parts of Central America and southern Mexico, especially on the Pacific slope. In the Old World a belt of savanna borders the Australian central desert on the north, east, and south; and there is savanna in western India also. However, the greatest savanna is one that borders the African rainforest from Senegambia eastward almost to the Red Sea, thence southward into the northern part of South Africa, and then westward into Angola.

Savanna woodland develops in subtropical or warm-temperate regions where the rainfall is seasonal and highly unreliable. The ground is thickly covered with grasses and scattered patches of low scrub; trees are small and mostly scattered. This type of vegetation stretches across northern and parts of eastern Australia. There are also sizable areas of savanna woodland in southern Rhodesia, southern Ethiopia and southern Anglo-Egyptian Sudan, Ceylon, central and southern India, central China west of the temperate rainforest, northern Argentina, and central Brazil.

Thorn forest, also called scrub or scrubland, occupies subtropical or warm-temperate areas where rainfall is both scant and seasonal, with a very long dry season. Trees are small, often thorny; and there may be patches of thorny brush. Numerous annuals spring to life during the short rainy season, flower briefly, and then disappear. Larger herbaceous plants may have leaves or stems that are thickened, fleshy, and resistant to desiccation. A large expanse of thorn forest extends from southwestern Texas southward through the Valley of Mexico; and another occupies a part of northern Argentina, south of the savanna woodland. Thorn forest covers much of southwestern Africa, from southern Angola to the borders of the Cape Kingdom. The southern part of the Red Sea is also rimmed with this type of vegetation. In Australia, zones of thorn forest often lie between the savanna and the desert.

Steppe, often called short-grass prairie in the United States, characterizes temperate areas where the precipitation, rather scanty, is concentrated toward the winter months. Short, tightly bunched grasses are abundant, but do not completely cover the ground. Shrubs are scattered, and trees mostly lacking. This

Temperate rainforest at Lake Alabaster, Otago, New Zealand. (New Zealand Information Service)

Savanna with scattered oaks and pines in British Honduras. (W. T. Neill)

type of vegetation merges on the one hand into prairie, and on the other into semidesert. *Prairie*, or tall-grass prairie as it is called in the United States, develops where the precipitation is more concentrated toward the summer months. It is characterized by tall grasses that die down in winter, and by numerous herbaceous plants that sprout rapidly from bulbs or seeds at the beginning of the rainy season. *Semidesert*, with a dry summer and a severely cold winter, has scattered shrubs, a few trees, and grasses that are rather ephemeral. A vast expanse of steppe and prairie, with some semidesert, lies east of the Rockies in North America, from the southern part of Northwest Territories southward well into Texas, and with an outlier reaching as far east as the western shore of Lake Michigan. An even greater expanse of steppe, prairie, and semidesert reaches from the Black Sea eastward into Manchuria; it is interrupted centrally by desert. Smaller areas occupy the eastern part of southern Argentina and the northern part of South Africa.

In some areas, all at high altitudes or latitudes, there is insufficient warmth to permit the growth of trees; but there is a brief period of thawing that permits the growth of herbs and small shrubs. Grasses and sedges usually predominate, and the vegetation is called *grassy tundra*. The greatest expanse of this vegetation is that of the Arctic, but there are lesser areas in the higher mountains of southern Mexico, western South America, central and southern Europe, and southern Asia.

Here and there in the high north latitudes, there is sufficient warmth to permit the growth of trees, but a coniferous forest does not develop because precipitation is too scanty. Such areas support *cold woodland*. Trees are small and widely spaced, but shrubs may be abundant. Grasses, sedges, and mosses often grow thickly. Muck-filled depressions, called muskeg, become boggy when the ground thaws. Cold woodland is scattered over the northland, but in general is interposed between the grassy tundra and the coniferous forest. *Heath* is nearly confined to Ireland and southern Iceland, where, in spite of the high latitude, winter temperatures are moderated by the Gulf Stream. The shallow, peaty soil is covered with mosses, grasses, sedges, and shrubs belonging to the heath family. Trees are few and scattered.

Deserts are areas with sparse vegetation, but we may distinguish among several kinds of deserts, according to the factors that limit plant growth. *Cold desert* characterizes areas such as Antarctica or most of Greenland, where the

Ice desert high in the Bernese Oberland of Switzerland. Mount Morgenberghorn in foreground, Great Combin and Mont Blanc in background. (Swiss National Tourist Office)

ground is covered with glacial ice or with snowfields that are nearly or quite permanent. Just a few plants can grow around the glacial borders or in the snowdrifts. In other deserts plant growth is limited by insufficient rainfall. In these dry deserts much of the soil is not bound or protected by plant cover; thus it may be eroded by wind or by occasional flooding. In *tropical desert*, especially, the landscape may be one of nearly barren sands; but *subtropical desert* supports shrubs and even small trees, along with many ephemeral herbs. Most plants of the dry deserts show obvious specializations for existence where moisture is in short supply.

Subtropical desert occupies the Great Basin, which lies between the Rockies and the Cascades–Sierra Nevada chain; and it extends thence into northern Mexico and southern California. Subtropical desert also lies east of the Andes in western Argentina. Another expanse of subtropical desert, ringed about with steppe and semidesert, extends from the Caspian Sea eastward through Sinkiang. Much of central Australia is dry desert, varying from subtropical to tropical, and somewhat interrupted by thorn forest. A vast expanse of dry desert, partly subtropical and partly tropical, stretches across northern Africa (the Sahara), and thence eastward through Saudi Arabia into West Pakistan. A smaller dry desert, varying from subtropical to tropical, lies in the southern part of Southwest Africa. There is an exceedingly dry tropical desert west of the Andes in Peru and northern Chile. As noted earlier, a *fell-field* develops at high altitudes or latitudes, where temperatures are arctic and precipitation very scant. A fell-field may be thought of as another kind of desert.

Tropical rainforest, subtropical rainforest, monsoon forest, temperate rainforest, deciduous forest, coniferous forest, and evergreen hardwood forest are all subdivisions of the forest biochore; steppe, prairie, and grassy tundra of the grassland biochore; tropical desert, subtropical desert, cold desert, and fell-field of the desert biochore; savanna woodland, thorn forest, savanna, semidesert, heath, and cold woodland of the savanna biochore. A word of caution: Names relating to vegetation types have been used in a variety of ways, and there is only a rough agreement among authors about what is meant by rainforest, woodland, savanna, steppe, prairie, or desert. Not that the main vegetation types are difficult to recognize; the disagreements are mostly terminological.

Although climate is largely responsible for the nature of the plant cover in a given area, soils and topographic relief are important also. For example, the

Local conditions modify the plant cover. Distant hills are sparsely vegetated, but cottonwoods line the Gunnison River in Colorado. (Colorado Department of Public Relations)

effectiveness of rainfall may be neutralized by a very porous soil, or by marked relief that leads to rapid runoff. Ravines or river bottomlands may offer situations that are cooler and damper than all the surrounding region. Steep and rocky hillsides, recent lava beds, shifting sand dunes, coarse gravels, rocky outcrops, permanently wet depressions, subsoils exposed by erosion—all may inhibit local development of a vegetation type that is widespread in the general vicinity.

Throughout the millennia, temperature and precipitation have varied; vegetation types have shifted about concomitantly; and small enclaves of one type often have persisted in favorable situations within a region occupied mainly by some other type. Here and there in the world, a peculiar combination of environmental factors results in a plant cover that is virtually unique. Such areas are usually small, and of no more than local interest; they need not be listed. However, mention should be made here of the extensive pinelands of eastern North America: the pine barrens and flatwoods that extend from southern New Jersey into Florida and westward into Louisiana; the rolling, pine-clad hills of the Piedmont; the Coastal Plain dunes covered with sand pine, or with a mixture of longleaf pine and turkey oak. Although pines are conifers, surely these forests are not to be included with the coniferous forest of the cold north. In the next chapter we shall have more to say about pinelands.

Zoologists are often much interested in vegetation, for the distribution of an animal species may reflect its restriction to certain types of plant cover. For example, the giant forest hog is confined to tropical rainforest of Africa, the sand-puppy (a burrowing rodent) to tropical desert of Somalia, and the chinchilla to grassy tundra high in the southern Andes. The prairie dog is restricted mostly to steppe, prairie, and semidesert from southern Saskatchewan to northern Chihuahua; the Canadian porcupine mainly but not exclusively to coniferous forest of North America. The kiwi (a flightless bird) is limited to temperate rainforest of New Zealand, the turnstone (as a nesting bird and not a mere visitor) to grassy tundra and fell-field of the Arctic, the wood thrush largely to deciduous forest of eastern North America (although it will also nest in suburban shrubbery). The desert tortoise is restricted to subtropical desert of the southwestern United States and northwestern Mexico, the yellow-lipped snake to pine flatwoods from Florida northward into North Carolina and westward into Louisiana. However, the distribution of a vegetation type is usually much wider than that of any single animal (or plant) that is confined to it.

26 *"I had no idea there was so much going on in Haywood's meadow."** *"... thou canst not stir a flower / Without troubling of a star."*** **OTHER WAYS OF LOOKING AT PLANT DISTRIBUTION**

We would expect considerable agreement between floral and vegetational subdivisions. If an area is distinctive in its climate and soils, it will harbor at least some characteristic floral elements, and it will further be set apart by a vegetation unlike that of surrounding regions. Furthermore, the vegetation of an area in turn modifies some components of the local environment, thus favoring invasion by certain species but discouraging invasion by others.

Consider, for example, some of the effects of a deciduous forest upon its environment. Humus collects beneath the trees; and as it decays it makes a light soil more retentive of water, or a heavy soil more porous. Buried rocks may be converted into fine soil from the chemical action of plant acids and the mechanical action of roots. Above ground, the plant bodies resist currents of air or water, causing them to deposit whatever detritus they may be carrying; while the continuous plant cover prevents the surface soil from washing or blowing away. Dead plants and fallen leaves return to the surface various minerals absorbed by the roots from deeper levels. The denser the forest, the more it shades the

* HENRY DAVID THOREAU, *Walden*.   ** FRANCIS THOMPSON, poet-mystic, 1897.

ground; and so beneath a stand of forest trees the only other plants to grow are those that are adapted for a low intensity of light. (The human eye adjusts to wide variation in light intensity, and the average person is surprised to learn that the light in an open field may be more than five hundred times stronger than that at ground level in a nearby forest.) In some forests the quality of the light may be altered, the tree leaves filtering out certain wave lengths. A forest also decreases wind velocity, moderates the vagaries of temperature, and increases humidity; and there is some reason to believe that atmospheric moisture will condense and fall as rain more readily over forested than over nonforested tracts.

　　Other types of vegetation alter the soil and atmospheric conditions in other ways. Data taken at a weather station may give only a rough idea of the environmental conditions to which the local plants are actually exposed. Ecologists

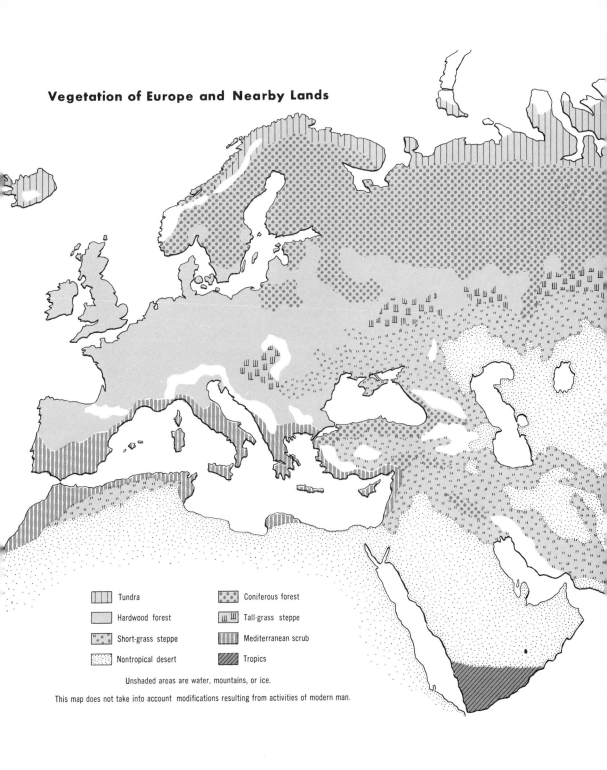

Vegetation of Europe and Nearby Lands

	Tundra		Coniferous forest
	Hardwood forest		Tall-grass steppe
	Short-grass steppe		Mediterranean scrub
	Nontropical desert		Tropics

Unshaded areas are water, mountains, or ice.

This map does not take into account modifications resulting from activities of modern man.

speak of "microclimate," the conditions of temperature, humidity, and wind that actually prevail in a given formation-type. A forest might have a series of microclimates, from the root level in the ground to the tops of the highest trees.

From previous chapters we see that there is indeed some agreement between the floral subdivisions and the geographic extent of certain vegetational types. Thus the Mediterranean Province corresponds to one area of evergreen hardwood forest, the Cape Kingdom to a second area of this formation-type, the Southwest Australian Province to a third. The West African Province corresponds to a large expanse of tropical rainforest, the North African–Indian Province to a vast area of dry desert.

Actually, in all parts of the world there are parallels between floral and vegetational distribution, but we shall pursue this topic further only in the case of North America, north of Mexico. Within this region there are ten areas that are distinctive in their predominating type of vegetation. Unfortunately, such areas are again called "Provinces," and this usage is so well entrenched that a substitute terminology would probably be unwelcome. To avoid confusion we may italicize the vegetational subdivisions.

(1) The *Tundra Province* is circumpolar in the Arctic. ("Tundra" was the name applied by Laplanders to the treeless country of northern Scandinavia. The word passed into Russian and then into English.) It is mostly grassy tundra.

(2) The *Northern Conifer Province* stretches from Newfoundland and Nova Scotia westward into Alaska, and thence across Siberia into northern Europe. It is mainly coniferous forest. In its American portion the most characteristic conifers are white spruce, balsam fir, larch, black spruce, and jack pine.

(3) The *Eastern Deciduous Forest Province* extends from New England and the southern edge of the Great Lakes region southward into central Georgia and eastern Texas. It is mostly deciduous forest, and its scattered areas of conifers are not mapped separately.

(4) The *Coastal Plain Province* includes the lowlands of the southeastern United States, from south-central Florida northward into southern New Jersey and westward into extreme eastern Texas. This subdivision also embraces the lowlands of the Mississippi Valley, as far north as southern Illinois. The *Province* has areas of deciduous forest, but stands of pine usually are more conspicuous. Swamps are numerous.

(5) The *West Indian Province* extends northward from the West Indies to

encompass southern Florida. Its subtropical forest exists mainly as patches, scattered amidst pinelands and swamps.

(6) The *Grassland Province* occupies central North America east of the Rockies, from Alberta and Saskatchewan southward into southern Texas. It is mostly prairie and steppe.

(7) The *Cordilleran Forest Province* is made up chiefly of the Rockies, Cascades, and Sierra Nevadas. It is predominantly coniferous forest, and its characteristic conifers include Douglas fir, Engelmann spruce, western hemlock, western red cedar, ponderosa pine, and lodgepole pine.

(8) The *Great Basin Province* is essentially the Great Basin, which lies between the Rockies and the Cascades–Sierra Nevada chain. It is mostly dry desert.

(9) The *Chaparral Province* extends from southwestern Oregon southward through western California into northern Baja California. It is largely evergreen hardwood forest, although it has some pineland, grassland, and savanna woodland. (In Spain, "chaparro" was a scrubby, evergreen oak, and "chaparral" was an area where such oaks predominated. The Spaniards carried the latter term from their Mediterranean homeland to California, where some of the vegetation has a Mediterranean aspect.)

(10) The *Sonoran Province* extends from southeastern California to western Texas, and thence southward into Mexico. In its United States portion it is mostly subtropical desert, but with thorn forest in western Texas.

It will be seen that this vegetational scheme corresponds well to the floral one advanced in earlier chapters. The *Tundra Province* roughly coincides with the floral Province called Arctic-Subarctic; and the United States portion of the *West Indian Province* corresponds to a similar outlier of the floral Caribbean Province. The *Northern Conifer, Eastern Deciduous Forest, Coastal Plain,* and *Grassland Provinces* are four distinctive areas within the floral Province called Atlantic North American, except that in vegetational studies the northern coniferous forest extends across both the New World and the Old. The *Cordilleran Forest, Great Basin, Chaparral,* and *Sonoran Provinces* are among the distinctive areas recognizable within the floral Pacific North American Province. It is probably safe to say that, within each vegetational *Province,* at least one-half and perhaps three-quarters of the plant species do not range into any other *Province.*

The ten *Provinces* listed above, and about ninety others in the remainder of the world, are more important than might be inferred from the brevity of the foregoing account. Actually they are areas that are distinctive as regards climate, vegetation, and flora. As such, they are interesting to the ecologist, who studies the interrelations of living things with their environment. They are useful to the zoologist, for animals are often restricted not just to one formation-type but to one particular expanse of that formation-type. The *Provinces* are especially important to the phytogeographer whose view is restricted geographically, perhaps to one continent or one country.

Various *Provinces* have been characterized as being "mostly grassy tundra" or "mainly coniferous forest"; but what about the atypical portions of a *Province?* How do they fit into the distributional picture? Usually they fit well, and in a way that is altogether surprising.

Visualize an area with a marked winter and with abundant precipitation at all seasons. The area is covered with deciduous forest. Somewhere in this forest there is a great expanse of outcropping rock, which does not support trees. On the rock face nothing grows but some grayish, crusty patches of lichen. A lichen is a kind of lower plant, or, strictly speaking, two kinds; for under the microscope the lichen body is seen to consist of a green alga whose cells are enmeshed by minute threads of fungus. The fungus mass absorbs water, holds it, shields it from evaporation, and makes it available to the alga; the latter combines the water with atmospheric carbon dioxide to make nourishing carbohydrates, which are shared with the fungus. Dust particles, brought by wind or rain, contain enough nitrogen for the lichen's need. Excreted carbon dioxide, combining with water, forms a weak acid that eats into the rock face, releasing sufficient minerals to satisfy the lichen's requirements. In a few decades or in a few centuries—depending upon the kind of rock and the dampness of the area— the surface of the rock is converted to a fine, dustlike soil.

When this soil has been produced, different kinds of lichens, not crust-forming but more leaflike in shape, can get a foothold. This they do, and soon they shade out the crusts, which die and decay. About the foliose lichens water has a better chance to collect and be absorbed; evaporation is decreased; more wind-borne dust accumulates, along with bits of lichen scrap. Oxidation is reduced, and a "humus" accumulates, especially in minute depressions and crevices of the rock. Acids from living and dead lichens convert more of the rock face into soil.

Soon a different type of lower plant can invade—the mosses. Several kinds, arriving as wind-borne spores, come to overtop the lichens. The little mossy clumps constantly die at the bottom and grow at the top, and an inch of soil may be formed under them. This is sufficient to permit invasion by some ferns and a few higher plants. The first higher plants to arrive usually are herbaceous annuals. They continue the formation of humus and the chemical breakdown of rock into soil; and their probing roots may begin a mechanical breakdown and separation of rock particles. Soon biennials and perennials begin to invade, increasing in numbers as the environment becomes more congenial. If our imaginary outcrop is somewhere in the *Eastern Deciduous Forest Province*, it may now support wire grass, poverty grass, perhaps a blue grass, mullein, alum root, and several ferns, along with drought-enduring mustards, goldenrods, and cinquefoils. The tangled network of rootlets increases, and the thin soil becomes more shaded. Evaporation is further reduced, drought periods are shortened, and temperature extremes are modified. Bacteria, fungi, and microscopic animals come to live in and alter the soil. Eventually the environment has been so modified that it will support woody plants.

The first shrubs to appear may be snowberry, sumac, and ninebark, along with the vine poison ivy. When the shrubby growth becomes dense, the herbs are mostly shaded out. The larger plants, now in possession of the rock, greatly accelerate its conversion into soil. The numerous stems trap falling leaves, abate the wind currents, and accumulate windblown snow. Roots pry deeply into rocky crevices. Humidity is increased above the decaying leaf mold that forms the surface of the soil. The environment has now become suitable for the nurture of tree seedlings.

The first trees to gain a foothold are those requiring comparatively little water; perhaps they will be mostly bur oak and bitternut hickory. Their seeds (acorns and hickory nuts respectively) may well have been brought in by gray squirrels, which recover only a small fraction of the nuts they hide away. When the trees are grown, the rock will be covered at last with deciduous forest. Generations of these oaks and hickories may live and die, eventually altering the soil to a point where it will support hardwoods that need somewhat more moisture—perhaps red oak and shellbark hickory. When these two latter species reach their full growth, they cast a dense shade beneath which the seedlings of the bur oak and bitternut hickory will not thrive. However, various smaller and

Sagebrush vegetation in Rush Valley, Utah; Wasatch Mountains in the distance. Larger, darker plants are conifers: junipers and piñons. (Hal Rumel from Utah Tourist and Publicity Council)

Natural forces, such as erosion, constantly interrupt the course of plant succession. Bryce Canyon National Park, Utah. (Hal Rumel from Utah Tourist and Publicity Council)

more shade-tolerant trees, say hop hornbeam and certain elms, will grow; and these form an understory beneath the canopy of red oak and shellbark hickory. At this point the constant change of vegetation locally comes to an end, but our story does not.

In this same stretch of forest there is a lake. Submerged plants—waterweed, pondweed, water milfoil, and naiad—are rooted in its muddy or sandy bottom, to a depth that depends on the clarity of the water but that rarely exceeds twenty feet. Water buttercup, bladderwort, eelgrasses, and various algae help to fill the lake with vegetation. Streams carry to the lake a burden of eroded material, which is dropped where plants impede the current. As the individual plants die, their remains and those of associated animals all fall to the bottom. Here in the scarcity of oxygen they do not decompose completely but instead form a sort of humus that cements the lake-bottom soil and makes it firmer. In short the sub-merged vegetation renders the lake shallower, and the bottom more suitable for invasion by different plants. When the lake, or a part of it, becomes about six or eight feet deep, it is taken over by floating plants such as various water lilies, lotus, bonnet lily, some pondweeds, and smartweed. These are rooted, but their long stems permit the leaves to unfurl at the surface. As the floating plants increase, their leaves shade the submerged species, which die out (or persist in another and deeper part of the lake).

Currents deposit water-borne material about the numerous stems of the floaters; and, as these plants are often large, their dead bodies rapidly build up the substratum. Soon the area of floating plants, or at least its shallower edge, is built up to a point where swamp plants—as opposed to aquatic plants—can grow. There may develop a reed swamp, perhaps with bulrush, cattail, reed, bur reed, and wild rice. These plants are called emergents for, although they are rooted in saturated soil, their stems emerge from the water and bear the leaves well in the air. The emergents vastly overtop, and overshadow, the floaters; and they advance rapidly for they often sprout not only from seeds but from tuberous rootstocks as well. The floaters disappear, or persist in some deeper part of the lake. The emergents build up the lake shore, which soon becomes suitable for other species, perhaps arrowhead, water plantain, and sweet flag. Lowering of the water level is especially unfavorable to plants growing at the lake edge, and such plants are eventually replaced by sedges, rushes, and spike-rushes, along with marsh marigold, Iris, bedstraw, water hemlock, cottongrass, bellflower, or other herbaceous species. Thus reed swamp has given way to sedge meadow.

All the plants of the sedge meadow bind the soil and accumulate humus; some absorb vast quantities of water from the soil and give it off into the air by a process called transpiration. Eventually a point is reached where the soil of the meadow is saturated only during the spring months. Then shrubs will appear, perhaps small willows, certain dogwoods, or buttonbush. Soon trees appear; they may be larger willows, alders, or cottonwoods. The trees rapidly alter the habitat, building up the soil and drying it through transpiration. The plants of the sedge meadow disappear, to be replaced by more shade-tolerant herbs that can grow beneath the trees. For a time the trees are so scattered that the local plant cover could hardly be termed a forest. However, as the soil is enriched by bacteria and fungi, and as other trees arrive, a deciduous forest develops. At first it may include a mixture of alder, willows, cottonwood, hackberry, elms, ash, oaks, and hickories; but as the forest canopy becomes more dense, trees with very shade-tolerant seedlings will eventually come to predominate. In our hypothetical expanse of forest these are red oak and shellbark hickory. Shade-tolerant shrubs soon form an understory beneath the taller oaks and hickories.

All the stages in a sequence, from submerged vegetation to deciduous forest, may be represented simultaneously as a series of zones within or near the lake. The central or deep-water zone of submerged vegetation is ringed with a zone of floating vegetation, this latter by a zone of emergents, the emergent zone by reed swamp, and so on. In effect the zones are constantly moving in toward the deeper part of the lake, where each in turn vanishes; until at last the erstwhile lake has been converted to deciduous forest.

Now to generalize. At a given locality there will be a predictable sequence of vegetational changes, culminating in a type of vegetation that will persist without further change. This stable type of vegetation is called the climax vegetation of the area, and its nature is determined by the climate. If our lake had been situated farther west, in an area where the rainfall was inadequate to support tree growth, the sedge meadow stage might have passed into prairie and then changed no further; for prairie is climax vegetation under one set of climatic conditions just as deciduous forest is the climax under another set. If our rock outcrop had been in, say, the Cascades, the shrub stage might have been

Desert scrub on hills bordering the Rio Grande in New Mexico. (New Mexico State Tourist Bureau)

Along watercourses in desert areas there may develop plant communities that do not fit into any of the local seres. Cottonwood Creek in Colorado. (Colorado Department of Public Relations)

followed by a tree stage with ponderosa pine, then Douglas fir and larch, and finally white cedar and western hemlock; for in this part of the country there is a coniferous forest climax.

Succession, or plant succession, is the term applied to the passing of vegetation through a number of stages. A particular sequence of stages is called a sere (a word of modern coinage, implying a series). Thus we have described two seres, one beginning on a bare rock and the other in a lake. Several different seres may exist in one area, all leading up to the same climax vegetation. It is curious that each stage alters the environment in a fashion that spells its own doom, but that permits development of the next stage. Only the climax stage is not self-destructive; it functions to perpetuate conditions that are favorable to itself.

It must be emphasized that the development of climax vegetation often is more of a potentiality than a reality. Seres are forever being interrupted, set back, or begun anew. Hurricanes, volcanic eruptions, floods, rock slides, erosion, deposition of soil by wind or water, severe or oft-repeated fires, the collapse of sinkholes, glaciation, the meandering of rivers, elevation or depression of the land by geologic processes, changes of sea level—these and many other natural occurrences may alter the local vegetation, or expose new land on which a sere may be initiated. And of course, all over the world man has been profoundly modifying or removing the vegetation, especially through timbering, burning, farming, and the pasturing of domestic herbivores.

For all practical purposes, some areas of the world will never develop the climax vegetation that is called for by the climate. Repeated disturbance may bring succession virtually to a halt at some particular seral stage. An exceptionally vulnerable stage is that just preceding the climax, and fire is most often the factor halting succession at this stage. In the southeastern United States some stands of pine might pass into hardwood forest, were not the humus and the hardwood seedlings destroyed almost every year by fire. In parts of New England, potentially wooded areas have been burned over repeatedly to hold them in a shrub stage with valuable blueberries. The term "subclimax" is often applied to a stage that precedes the climax, and that is relatively permanent in consequence of disturbance.

Within any *Province* the local vegetation is likely to be diverse because several seral stages are represented. The vegetation is thus more unified than might at first appear, since many of the stages clearly lead toward the same climax. The *Provinces* listed above have been delimited chiefly on the basis of their climax vegetation, and of seral stages that are particularly widespread or long lasting.

There are seven main kinds of climax vegetation in North America, north of Mexico. Each kind can be subdivided, but here we shall not do so in any formal way. (1) Tundra Climax characterizes the lands north of tree growth, and certain montane areas above timberline. (2) Forest Climax extends over much of Canada, eastern United States, and mountainous portions of western United States. This climax includes both deciduous and coniferous forest, each varying from region to region. It also includes subtropical forest. (3) Grassland Climax extends over most of central North America, east of the Rockies; and there are scattered areas of this Climax, either prairie or steppe, in the northern part of the Great Basin and in central California. (4) Woodland Climax, with juniper, pines, and oaks, occupies scattered areas of the southwestern United States, from Texas to California. The largest expanses are on the Colorado Plateau and in the higher ranges of southern Arizona and northern Mexico. (5) Chaparral Climax characterizes the foothills of the Cascades, the Sierra Nevadas, and the southern Rockies. (6) Sagebrush Climax typifies the central Great Basin. (7) Desert Scrub Climax extends from southern Nevada southward into southeastern California, southwestern Arizona, and northern Mexico. This list is simplified to the highest degree.

A region may have patches of vegetation that do not fit into any local sere. The most common example is a strip of forest along a river bottomland in an area of Grassland Climax; here forest vegetation can thrive because it need not rely on local rainfall for water. Within the Forest Climax country of eastern United States there are isolated areas of prairie, where porous soils negate the effectiveness of precipitation. During past millennia there have been marked shifts of climate, at times permitting forest to extend westward into what is now grassland, at other times permitting grassland to strike far eastward into what is now forested country. Vegetation developed by a certain climatic regimen has often persisted as relict patches in areas of suitable environment, long after that regimen has given way to another.

The concept of plant succession is important in the study of animal distribution. An animal species may be confined to climax vegetation, or to some one seral stage. For example, the scrub lizard of Florida (*Sceloporus woodi*, an iguanid) is restricted to a type of vegetation known locally as sand pine scrub. This scrub is part of a sere that begins on dunes of shifting sand. Remains of the lizard, and of other scrub-restricted animals, have been found in a fossil deposit a good fifty miles from any scrub, in an area that now supports hardwood forest. As a result of plant succession, the scrub and its animals are today less widespread than they once were. As another example the meadowlark, a familiar bird, is distributed from New Brunswick southward through Florida, and westward into Minnesota, Nebraska, and Texas. Throughout most of its range the climax vegetation is forest; yet the bird feeds and nests only in open, grassy places, and will never be seen where trees grow so thickly as to form a forest. The abundance of this bird, and quite possibly its range, has been increased by man, who is responsible for most large, grassy tracts in areas of forest climax.

So far, we have been concerned with the distribution of flora, and of vegetation. Fresh conclusions emerge from an investigation into the distribution of the so-called "life-forms" of plants. Christen Raunkiaer, a Danish phytogeographer, first advocated the life-form approach in 1903. He recognized and named five general types of plants:

(1) Phanerophytes (the name means "exposed plants") are perennials, living for several or many years. Usually they are over a foot tall, and may be very tall. Their buds, or points of growth, are continually exposed to the weather. Several kinds of phanerophytes are recognizable: stem succulents, such as cacti and many spurges; epiphytes, such as most bromeliads and many orchids; evergreen plants, some trees and others shrubs; and deciduous plants, again either trees or shrubs.

(2) Chamaephytes (meaning ground plants) are small shrubs, usually under four feet in height; their buds are located so near the ground as to be protected by fallen leaves or snow. The common periwinkle is a familiar chamaephyte.

(3) Hemicryptophytes (half-hidden plants) are perennial herbs. Their buds for next season's growth are located at the surface of the ground, and their aerial parts die down to the ground at the end of the growing season. An example is the dandelion.

(4) Cryptophytes (hidden plants) are perennial herbs with buds well below the surface. Some cryptophytes grow from bulbs, tubers, or rhizones that are

beneath dirt; lilies and cannas are well-known examples. Other cryptophytes, all swamp plants such as the cattail, have their buds beneath dirt and shallow water. Still others, the aquatics such as water lily and eelgrass, have the entire plant body in the water or at the surface.

(5) Therophytes (summer plants) are annuals, passing the adverse season only as seeds. Examples are many garden flowers such as Zinnia, marigold, and Petunia.

Raunkiaer estimated that swamp and aquatic plants together make up only 1 per cent of the world's flora. Stem succulents account for another 2 per cent. Three per cent of the species are epiphytes, and another 3 are the soil-rooted cryptophytes. Trees make up 6 per cent, surprisingly outnumbered by chamaephytes with 9 per cent. Therophytes or annuals contribute 13 per cent to the total. Tall shrubs make up 17 per cent, small shrubs 20. Most abundantly represented are the hemicryptophytes, with 27 per cent. These percentages constitute the "normal spectrum" of plant life-forms. When we investigate the spectrum that obtains in a limited portion of the world, we may find it to depart widely from the normal spectrum. For example, all the phanerophytes together make up 47 per cent of the world's flora, but they are below this figure everywhere except in tropical lands with moderate to heavy rainfall. In parts of the Arctic, phanerophytes are lacking, while in the wet tropics of Queensland they make up 96 per cent of the flora. Chamaephytes, forming 9 per cent of the normal spectrum, far exceed this figure in the Arctic, where in some places they make up two-thirds of the flora. Annuals, forming 13 per cent of the normal spectrum, in hot deserts make up 40 to 50 per cent.

It will be seen that Raunkiaer was classifying plants chiefly according to the position of their regenerating parts, and that the local predominance of some particular life-form often reflects the local climate. Thus phanerophytes, with fully exposed buds, predominate in the better-watered tropics, where there is no danger of drought or freezing. However, stem succulent phanerophytes are drought-resistant, and so they are numerous in warm, dry lands. Epiphytic phanerophytes are most common in tropical forests, where the plants compete fiercely for light.

Chamaephytes, mostly small and with protected buds, are particularly well adapted for areas where the growing season is short, and where the ground is snow-covered in winter.

Aspen in Colorado. A species of poplar, the aspen is quick to invade disturbed areas, such as burned-over tracts, even where the forest is primarily coniferous. (Colorado Department of Public Relations)

Hemicryptophytes thrive in temperate lands where the plant's leafy parts are not maintained over the winter, and where the bud, just at the ground surface, is in a position to sprout rapidly with returning warmth and rain. Hemicryptophytes also thrive under arctic conditions, the buds sprouting when the ground thaws. Yet the abundance of hemicryptophytes in the Arctic may also reflect the absence of trees there; for most hemicryptophytes do not compete well with the much larger trees.

Cryptophytes, if rooting only in soil, are well adapted for life where the rainy season is brief, and the rest of the year dry. Aquatic and swamp-dwelling cryptophytes are common where rainfall, topography, and soil permit the formation of numerous ponds, bogs, and lakes.

Therophytes or annuals, mostly small and quick growing, are especially characteristic of deserts, where they can flower rapidly during the brief and unpredictable rainy season, and spend the unfavorable part of the year as a drought-resistant seed.

There is some correlation between spectrum and vegetation type, since each is determined largely by climate. However, the correlation is not perfect, for one highly successful species might usurp most of the situations suitable for its particular life-form. Also, in any area of vegetation certain plant species, called dominants, actually control the environment and so limit invasion by other species. Thus in any mature forest of North America, a few tree species are dominant. According to locality these may be certain oaks and hickories, or beech and maple, or spruce and larch, or white cedar and hemlock; but in any event their presence will exclude that of most other trees. Many hemicryptophytes are grasses, whose roots form a thick sod and whose stems and leaves form a dense ground cover; and where grasses are thickly clustered, other plants cannot readily seed in.

27 *"It is interesting to contemplate a tangled bank, clothed with plants of many kinds, with birds singing on the bushes, with various insects flitting about, and with worms crawling through the damp earth, and to reflect that these elaborately constructed forms, so different from each other, and dependent upon each other in so complex a manner, have all been produced by laws acting around us."** **PLANTS AND ANIMALS TOGETHER**

Obviously there is some correlation between the distribution of plants and that of animals. The Coastal Plain lowlands are the home not only of the bald cypress and the cabbage palm but also of the alligator and cottonmouth moccasin. A mention of the Great Plains will call to mind not only waving grasses but also prairie dogs and jackrabbits. The deserts and near-deserts of our Southwest are famous as much for the Gila monster, chuckwalla, roadrunner, and sidewinder rattlesnake, as for prickly pear, cholla, and Joshua tree.

The first students to investigate the distribution of biota (flora plus fauna) were primarily concerned with the fauna. They felt that animals were better biogeographic material than plants. And it is true that two areas, separated by some kind of a barrier, may well be more different faunally than florally, because spermatophytes surmount barriers more readily than do vertebrates. But not all

* CHARLES DARWIN, *The Origin of Species.*

higher plants disperse with such ease, and the early workers assumed—and correctly—that an area with a distinctive fauna would harbor at least some characteristic floral elements as well.

Efforts had already been made to subdivide the faunal Regions and Subregions into lesser units, which were called (not surprisingly) provinces. Eventually these subdivisions became elevated to the status of "Biotic Provinces," but not without considerable change in the related concepts.

The early workers were taxonomically oriented; to them, an area was distinct if it harbored characteristic species, genera, and families. In theory a Biotic Province is still to be defined on the basis of its flora and fauna, but now it is said to be characterized also by peculiarities of formation-type, climax vegetation, physiography, and soils.

As now visualized, the Biotic Provinces of North America, north of Mexico, number two dozen, as follows:

(1) The Eskimoan Biotic Province is essentially the Arctic. It is named for the Eskimo, the primitive people who were scattered from northeastern Siberia eastward through northern Alaska and northern Canada to Greenland. The name is misleading; man's distribution cannot be used to help define any biogeographic subdivision. This and several other provincial names were simply intended to honor various aboriginal peoples. South of the Eskimoan is (2) the Hudsonian Biotic Province, which corresponds roughly to the extent of the northern coniferous forest in the New World. The name is not highly appropriate, for the greater part of Hudson Bay lies farther north.

(3) The Canadian Biotic Province involves southern Quebec, southern Ontario, the New England states, most of the Great Lakes region, most of New York, and the central part of Pennsylvania. (4) The Carolinian Biotic Province extends from Lake Erie and Southern New Jersey southward to central Georgia and extreme northeastern Oklahoma.

(5) The Austroriparian Biotic Province includes the Coastal Plain, from extreme southern Virginia southward through all of Florida, thence westward into eastern Texas and southeastern Oklahoma, and up the Mississippi Valley into extreme southern Illinois. "Austroriparian" means "of southern riverbanks;" the name directs attention to the numerous streams and swamps of the southeastern United States. In the Biotic Province scheme, southern Florida is not set apart, for it has only a few faunal elements that are not typically Austro-

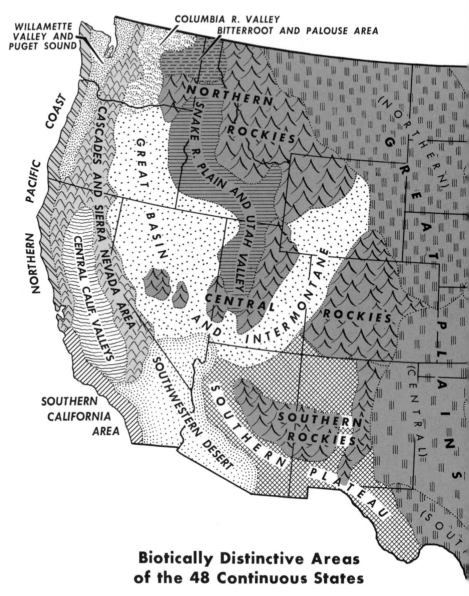

Biotically Distinctive Areas
of the 48 Continuous States

The map shows areas that are distinctive in flora, vegetation, fauna, climate, soils, and topographic relief. Most biogeographic schemes combine these areas in varying fashions, depending upon the schemer's outlook. The areas are named informally here.

GREAT LAKES AND ST. LAWRENCE AREA

(NORTHERN)

(WESTERN)

(CENTRAL)

NORTHERN

PRAIRIES

(CENTRAL)

OHIO AND TENNESSEE R. VALLEYS

APPALACHIAN AREA

THE PIEDMONT

OZARK AREA

UPPER COASTAL PLAIN

COASTAL PLAIN

CENTRAL FLORIDA HIGHLANDS

LOWER COASTAL PLAIN

SOUTH FLORIDA

Scale of Miles

0   200   400

riparian. (6) The Illinoian Biotic Province occupies central North America, from southern Manitoba southward through western Kansas, and eastward to include northern Illinois.

(7) The Texan Biotic Province is a strip running through east-central Oklahoma and east-central Texas. (It corresponds well to what Texans call the "farm belt.") The southern tip of Texas falls in (8) the Tamaulipan Biotic Province, which is largely Mexican. Central Texas and southwestern Oklahoma make up (9) the Comanchean Biotic Province.

(10) The Kansan Biotic Province extends from northern Texas northward into western Nebraska. (11) The Saskatchewan Biotic Province reaches from southeastern Alberta and southern Saskatchewan to extreme northern Nebraska. (12) The Montanian Biotic Province follows the Rockies from southeastern British Columbia southward into northwestern Wyoming. (13) The Coloradan Biotic Province lies farther to the south in the Rockies, from southern Wyoming southward in a strip through west-central Colorado and into northern New Mexico. (14) The Navahonian Biotic Province includes southeastern Utah, central and northeastern Arizona, and western and central New Mexico, with minor extensions into southern and western Colorado and western Texas.

(15) The Chihuahuan Biotic Province extends from Mexico into western Texas and southern New Mexico, (16) the Sonoran Biotic Province from Mexico northward into southwestern Arizona and southeastern California, and (17) the Apachian Biotic Province from Mexico northward into extreme southwestern New Mexico and southeastern Arizona. (18) The Mohavian Biotic Province embraces the Mojave Desert of southern California, as well as southern Nevada, the extreme southwestern tip of Utah, and a part of western Arizona. (19) The Californian Biotic Province includes most of California, as well as a small portion of southern Oregon. The subdivision also extends southward into Baja California.

East of the Californian is (20) the Artemisian Biotic Province, involving most of Nevada, the northeastern tip of California, southeastern Oregon, western Utah, and southwestern Idaho. *Artemisia* is the genus that includes the sagebrush, a bitterly aromatic plant of the daisy family. Just north of the Artemisian is (21) the Palusian Biotic Province, including northeastern Oregon and southeastern Washington, plus a bit of western Idaho. The name is derived from the Palouse River, which arises in western Idaho and flows into the Snake River in southeastern Washington. (22) The Oregonian Biotic Province involves part of western British Columbia, western Oregon, western Washington, and north-western California. (23) The Sitkan Biotic Province extends from northwestern British Columbia northward into southern Alaska. Finally, (24) the Aleutian Biotic Province includes the Aleutian Range and Islands.

It is unfortunate that the Biotic Province concept took the course that it did. Biogeography would have profited if two lines of investigation had developed, one subdividing the land into areas distinctive as regards families, genera, and species of plants and animals, the other concerning itself with formation-types, plant succession, environmental factors, and animal distribution in relation to formation-types. In the first case the point of view would be purely taxonomic, in the second ecological. The first approach would discover areas of endemism, the second areas of distinctive environment. Both the taxonomically oriented and the ecologically oriented investigator would subdivide parts of the world in about the same way, for an area that is distinctive in environment is also likely to harbor a characteristic assemblage of plant and animal species, genera, and perhaps families. However, agreement would not be perfect, for the ecological approach restricts its view to the modern environment, while areas of endemism may reflect the different environment of some past age.

For example, during a very warm, dry period between glaciations, when glacial ice melted away almost completely, sea level rose and reduced Florida to a series of small islands. On these islands some animals evolved differences from their respective mainland stocks, and others persisted as relics without living kin. When glaciers reformed and sea level fell, the islands came to form a central part of the Florida peninsula. Today central Florida harbors an interesting as-semblage of endemic animals, but they all live in vegetation types that extend outside the area of endemism. Thus the short-tailed snake, the only known species of a genus (*Stilosoma*) endemic to central Florida, lives almost always on sandhills covered with turkey oak and longleaf pine; the vegetation type extends northward well into central Georgia but the snake only reaches north-central Florida.

Australia, so long and so distantly isolated, affords the best evidence that an ecological approach will not reveal all that we need to know about distribution. The Australian fauna is especially important in being an ancient one preserved in isolation, and this topic can be discussed without reference to formation-types or the climatic conditions that they reflect. True, there are animal species confined to desert, others to grassland, and still others to forest; but the relict fauna characterizes all environments. Even many Australian plants are biogeographically significant in ways that do not have to do with modern ecology. Thus the tree genus *Eucalyptus* has species in the deserts, in the grasslands, and in the forests of Australia; but more striking is the fact that nearly all the species of this large, diverse genus are confined to Australia.

While faunal Districts were being elevated to Biotic Provinces, another approach to the study of distribution was being developed by the American naturalist C. Hart Merriam, who became chief of the United States Biological Survey (now the Fish and Wildlife Service). Merriam felt that the faunal similarities or differences between two areas were best expressed in terms of the number of animal genera they held in common. He found that there was a comparatively rapid change of genera from north to south in North America, but considerably less change from east to west. He felt that temperature, which also changes from north to south, was the most important factor controlling generic distribution, and that faunal subdivisions should stretch across the country just as do temperature zones.

In 1889 Merriam made an expedition into the San Francisco Mountains of Arizona. There he noted altitudinal zonation of plant and animal communities: as he went higher and higher into the mountains, desert gave way to a zone with piñon pine, this to a zone with yellow pine, then balsam fir and above it spruce. Next came timberline, and finally an alpine zone. Merriam saw that zonation exists because temperature decreases with altitude, and he was impressed with the fact that comparable zones existed in flatter country but were there arranged latitudinally. As a result of his discovery, Merriam extended his ideas to encompass both plants and animals, and first published his work in 1892. His later study, "Life Zones and Crop Zones in the United States," is better known; it appeared in 1898.

Merriam recognized (1) an Arctic Life Zone, with a southern boundary drawn quite far north. This Life Zone extended from the Atlantic to the Pacific, but was broken by Hudson Bay and again by Ungava Bay in northern Quebec. (2) The Hudsonian Life Zone stretched from Labrador westward through the James Bay region, and thence northwestward through the Great Bear Lake country. From here it passed westward (interrupted by a tongue of Arctic Life

Zone in the Mackenzie Mountains) to the shores of the Bering Sea. Merriam's (3) Canadian Life Zone reached from New Brunswick westward through the Lake Huron and Lake Superior region to the Pacific coast of British Columbia; it was interrupted by a tongue of Hudsonian Life Zone in the northern Rockies. (4) The Transition Life Zone extended from the New England states westward through Lake Ontario, Lake Erie, and Lake Michigan; and thence northwestward into southern Saskatchewan. Farther west it was nearly broken by tongues and outliers of Canadian Life Zone in the Rockies, but continued westward to extreme southern British Columbia. The Transition Life Zone also sent tongues southward, through the Cascades and Sierra Nevadas in the west and through the Appalachians in the east.

Merriam's (5) Upper Austral Life Zone extended from southern New Jersey southward in the Piedmont to central Georgia. Swinging around the Appalachian tongue of Transition Life Zone, it reached from western Pennsylvania and northern Alabama westward to the Rockies, where it was much broken by isolated areas of other Life Zones at varying elevations. West of the Rockies, the Upper Austral occupied the Great Basin and extended well down into the Mexican highlands. Farther west it was much interrupted, but reached the Pacific coast in parts of California and northern Baja California. (6) The Lower Austral Life Zone extended from southeastern Virginia southward through central Florida, thence westward through the lower Mississippi Valley into eastern Texas, and southward into Mexico. Farther west, considerably interrupted, it reached central California, central Baja California, and the deserts that extend from Nevada into Sonora. (7) The Tropical Life Zone included the tropical lowlands that border the Mexican highlands, along with southern Baja California and southern Florida.

Merriam's scheme was immediately attacked on many different counts, and on several of these it was indeed vulnerable. European workers rejected it, but in the United States it became more popular than any other biogeographic system, for in most areas it worked. Today we can see that it worked for reasons Merriam might not have suspected. In his time, as today, few people dealt at first hand with the biogeography of a very large area; it was contribution enough to work out patterns of distribution within a single state. Let us apply the scheme to one state, and see how it fits the facts of distribution therein. According to Merriam, the state of Georgia is parceled out over three Life Zones: the Transition, Upper

Austral, and Lower Austral. Within the state, each Life Zone indeed has many characteristic faunal and floral elements, and the local worker could scarcely improve upon Merriam's mapping. His boundary lines were said to have been determined from temperature data assembled by weather stations; but the Georgia expanse of the Transition Zone also happens to coincide with the extent of the Appalachians in that state. The Upper Austral corresponds in Georgia to the Piedmont Plateau, and the Lower Austral to the Coastal Plain. In short, when the view is restricted to this one state, the three Life Zones are really as many physiographic divisions, each with its peculiarities of temperature, rainfall, soils, topography, and vegetation. No wonder each has a characteristic flora and fauna.

But the Life Zone concept was not advanced primarily as a tool for describing distribution within a state or other limited region. Merriam held each Zone to be characterized by genera that followed the subdivision all the way across North America. Certainly in the Arctic most genera, if not most species, are distributed in this way. The Hudsonian and Canadian Life Zones are also moderately homogeneous from eastern end to western. And as we have already seen, at least a few organisms that range widely across northern North America also follow the Appalachians and the Cascades–Sierra Nevada chain; their distribution helps demonstrate the reality of the Transition Zone. But the farther south in the United States, the less useful becomes the zonal concept. For example, the easternmost portion of the Lower Austral is the Dismal Swamp region of southeastern Virginia and adjoining North Carolina, while the westernmost is the Vizcaino Desert of Baja California; the opposite ends of this Life Zone are not markedly alike in the plant or animal genera that they harbor.

As time went by, the original Life Zones were considerably revised by many people, including Merriam himself. The most obviously needed emendation was a north-south line separating the United States into damper eastern and drier western portions. The line was drawn from southern Manitoba southward through the eastern part of the Dakotas, central Nebraska, west-central Kansas, western Oklahoma, and east-central Texas. Of course the line was an admission that rainfall, too, is important in distributional studies.

The Transition, Upper Austral, and Lower Austral Zones were each split into eastern and western portions by the aforesaid line. The new units thus formed were called Divisions. The western portion of the Transition Life Zone was called the Transition Division, the eastern portion the Alleghanian Division.

The Upper Austral Life Zone was divided into the Upper Sonoran Division in the west, the Carolinian Division in the east. The Lower Austral Life Zone was split into the westerly Lower Sonoran Division and the easterly Austroriparian Division. Canadian, Hudsonian, and Arctic Life Zones were not split. Additional modification of Merriam's scheme involved a more detailed mapping of boundaries, especially in topographically diverse areas where the Life Zones are arranged altitudinally. It also proved desirable to extend the Tropical Life Zone of eastern Mexico northward to include the extreme southern tip of Texas.

In its modern form the Life Zone map of North America, north of Mexico, is a useful presentation. However, the Life Zone concept is not readily extended to South America, where rainfall differences are more important than temperature differences in explaining the existence of biotically distinctive areas; nor to Eurasia, where several great mountainous uplifts, trending east and west, have never been connected among themselves.

The Life Zone concept is ecological in that it is based chiefly on an environmental factor, temperature; but it is also partly taxonomic, in that it was intended to analyze distribution at one taxonomic level, the genus. Now we come to another biogeographic approach, one that is strongly and frankly oriented toward ecology.

"Ecology"—the term is derived from Greek root words meaning "a study of the home;" in other words a study of the environmental factors that exist wherever an animal or a plant happens to live. But the Greeks did not have a word for it; Ernst Haeckl, a brilliant and widely traveled zoologist and philosopher, coined the term in 1869. Today ecology is defined as the science that deals with interrelations between living organisms and their environment, including both the physical and the biotic environment, and emphasizing not only the relations between different species but also the relations between individuals of the same species.

Almost anything ecological might at times be grist for the biogeographer's mill, but certain aspects of ecology are particularly important when the world is subdivided into biogeographically significant units. In an ecological approach to distribution, the world is partitioned into Biomes. A Biome is an area with its own characteristic maximum temperatures, minimum temperatures, rainfall (and in places snowfall), seasons, and changes of day length. The interaction of these climatic factors gives the Biome a characteristic vegetation, and this in turn

shelters a distinctive assemblage of animals. Each Biome not only has a characteristic flora and fauna, but each also passes through a series of developmental
stages and reaches a characteristic equilibrium with its environment.

From this definition one can see that the Biome approach brings together
many topics that have previously been parceled out over several other approaches.
From earlier chapters, and especially those relating to vegetation and plant succession, one might guess the location of many Biomes. Thus the major Biomes of
North America, north of Mexico, are (in something less than formal terminology): (1) the Arctic tundra; (2) the northern coniferous forest; (3) the eastern
deciduous forest; (4) the central grasslands; (5) the chaparral or evergreen hardwood forest of California; and (6) the southwestern deserts. Each of these areas
has already received attention; and to avoid repetition we shall not say much
more about these Biomes. However, Biomes are grouped into Biome-types, and
we shall briefly consider these latter and larger categories; for a review of the
Biome-types not only brings out some fresh facts but also shows the breadth and
usefulness of the ecological approach to biogeography.

(1) The Tundra Biome-type is circumpolar in the far north. Most of the
area is a gently rolling plain, with depressions occupied by lakes, ponds, or bogs.
The characteristic vegetation includes Sphagnum moss and various lichens, especially the lichen called "reindeer moss;" also grasses, sedges, and other herbaceous
higher plants, the last of which grow mostly in sheltered places on drier hillsides.
Characteristic animals of the Tundra Biome-type include the musk ox, the reindeer and caribou group, the Arctic wolf, the Arctic hare, and several lemmings.
Birds are abundant in the brief summer, especially waterfowl which nest on the
tundra. However, in winter most of the birds and some of the mammals move
southward, out of the Tundra Biome-type. Some of the animals, such as the
polar bear and snowy owl, are white at all seasons; while others, such as the
Arctic hare, Arctic fox, and ptarmigan, turn white for the winter. The Diptera,
an insect group including flies, midges, and mosquitos, are surprisingly abundant.
(The Biome approach, unlike previous ones, takes into account the lower animals
and lower plants, where these are sufficiently well known.) There are also large
bumblebees whose hairy coat retains heat generated by vibrating wings.

The floating polar ice, the northernmost Arctic islands, and the Greenland
ice cap may be included in the Tundra Biome-type; they are cold deserts, but not
wholly lacking in animals. Some of the animals, such as the nomadic Arctic fox,

are partly dependent on the tundra for subsistence; others, especially the polar bear, take their food largely from the sea. In the Tundra Biome-type the topsoil may thaw in summer, but the subsoil is permanently frozen, often to a depth of many yards. (At Point Barrow, the northern tip of Alaska, the Navy drilled wells through 1,000 feet of frozen subsoil; and in Greenland, scientists penetrated 2,000 feet without reaching the bottom of the frozen zone.) The southern extent of the "permafrost," the permanently frozen subsoil, marks the southern limit of the Tundra Biome-type. In general, permafrost exists where the mean annual temperature does not exceed 23 degrees Fahrenheit. Thus the Tundra Biome-type is sharply defined. Nevertheless, the Sphagnum bog component of the tundra extends southward into other Biome-types, in bog-captured lakes; and conversely, there are scattered enclaves of low birch forest well out in the tundra.

The Tundra Biome-type is markedly homogeneous throughout. There is no need to divide it into Biomes. It is all one Biome, the Tundra Biome. There are areas of alpine tundra on mountain peaks outside the Arctic; but they do not have the fauna of the Tundra Biome-type, and no more than a small portion of the flora. The alpine environment offers no nearby sea rich with food, and no "midnight sun." Areas of alpine tundra are not to be included in the Tundra Biome-type. On the edges of the Antarctic continent, and on some of the Antarctic islands, there is a counterpart of tundra. However, these areas are not included in the Tundra Biome-type either, for they differ from the Arctic tundra in both fauna and flora. Also, Antarctic vertebrates are nourished almost entirely from the sea, and should be considered in the light of marine ecology.

(2) The Taiga Biome-type is the coniferous forest belt that stretches across North America and Eurasia, just south of the tundra. ("Taiga" is the Siberian name for this forest.) It is delimited by the extent of the forest. In North America the most characteristic tree is the white spruce, and east of the Rockies the balsam fir is common. Damper sites may support larch and black spruce. Stands of jack pine may develop on drier sites, especially after fires. Jack pine is usually replaced by spruce and fir, except on thin, poor soils where it may persist almost indefinitely. Moist, disturbed sites may support three broad-leaved trees: paper birch, quaking aspen, and balsam poplar; but in the course of succession these are replaced by spruce and fir. In the Eurasian portion of this Biome-type the characteristic trees belong to the same genera as do the American ones, but may be different species thereof. The most characteristic animals of the taiga are

the moose and its Old World counterpart the true elk. Also common are certain rodents, rabbits, and members of the weasel family.

The lakes, ponds, and bogs of the taiga are not strongly differentiated from those of the tundra. In places there is a transition from taiga to tundra, the trees becoming progressively fewer, smaller, and more widely scattered until at last only shrubby vegetation remains. Just south of the taiga, isolated areas of this Biome-type may exist either in grassland or in deciduous forest; these taiga enclaves are associated with relatively minor differences in physiography or soil. In northern parts of the Great Plains, taiga may occupy cold, damp ravines, although the country round about supports grassland. More so than the tundra, the taiga is differentiated into New World and Old World portions, but the differentiation is insufficient to warrant recognition of two taiga Biomes. The Taiga Biome-type is also the Taiga Biome.

The coniferous forest of the Cascades and the Sierra Nevadas is made up in part of conifers not represented in the main east-west expanse of taiga; and some extraordinary animals live in these mountains also. The biotic peculiarities of the mountains are largely a reflection of the strong oceanic current that warms the region and leaves it dripping with fog between the frequent showers; the most characteristic elements of the biota require consistently high humidity, or else situations that are permanently wet, such as bogs, mountain rills, and cascading streams. However, the western mountains do have enough in common with the east-west conifer belt to warrant inclusion of both in the same Biome. The higher Appalachian peaks, although crowned with forests of spruce and balsam fir, are excluded from the Taiga Biome-type, for they harbor only a small fraction of the taiga biota.

(3) The Deciduous Forest Biome-type is more heterogeneous than taiga or tundra. As noted in a previous chapter, there are three main areas of deciduous forest, respectively in Europe, eastern Asia, and eastern North America. All these once were parts of a single Biome, but this was fragmented by climatic changes of the Tertiary as described in Chapter 21. The breaking up of the old, continuous Biome took place long ago, and the three main remnants have differentiated biotically as the result of extinctions, invasion, and evolution. The three now form as many different Biomes.

In eastern North America the Deciduous Forest Biome-type is extended to include the pinelands of the Coastal Plain. In the latter region there are many sites with deciduous forest, and the pinelands, although more conspicuous, are often subclimax. The fauna of the pinelands is not significantly different from that of the eastern deciduous forest. Puzzling is the forest of the Great Lakes region; in places it is deciduous, but in many areas there is a predominance of conifers, such as white pine and hemlock. The Great Lakes region is really intermediate between taiga and deciduous forest, harboring floral and faunal elements from both Biome-types.

(4) The Grasslands Biome-type is represented in all continents. Its local manifestation may variously be called steppe, prairie, plain, campo, llano, pampa, or some other term. Savanna is also included in this Biome-type. The world's major expanses of grassland are very different biotically. In other words they constitute a series of very distinct Biomes. As usual the differences are especially marked in the fauna; contrast the game herds and predators of the African grasslands with those of the North American prairies and steppes. Nevertheless there has been faunal interchange especially among the South African, East African, Sudanese, Central Asia, and Mongolian Grassland Biomes. The interchanges reflect both climatic shifts and the ability of many grassland animals to cross desert barriers. The North American grassland, called the Great Plains Biome, has interchanged animals with both Asia and South America. Of course the most isolated and distinctive grassland is that of Australia.

The animals of one grassland Biome may resemble those of another grassland Biome in habits and superficial appearance, without being actually of close relationship. Thus the grasslands of southern South America are inhabited by the mara, a nervous, long-legged, diurnal, vegetarian rodent that has a white rump patch for signaling to others of its kind; the tuco tuco, a rodent that burrows molelike in sandy places; and the vizcacha, a burrowing rodent that lives in large colonies, and whose holes are often taken over by burrowing owls and snakes. Does not this assemblage call to mind the North American grasslands, where the nervous, long-legged, diurnal, vegetarian pronghorn flashes its white rump patch; where the pocket gopher burrows in sandy soil; and where the prairie dog lives colonially and digs holes that may be taken over by burrowing owls and snakes?

A grassland usually is sharply differentiated from any adjoining coniferous or deciduous forest. In North America small patches of deciduous forest form enclaves within the Great Plains Biome, and "peninsulas" of forest may extend into grassland along river valleys; while relict areas of the Great Plains Biome exist well within what is otherwise deciduous forest country.

(5) The Desert Biome-type has representative Biomes in every continent. A chain of deserts, stretching from northern Africa eastward to West Pakistan, is fairly homogeneous in biota; but otherwise, each major desert area of the world is a very distinct Biome. The Desert Biome-type is noted for the adaptations of its biota to a scarcity of water. These adaptations may be physiological and behavioral as well as morphological. Desert plants include thorny bushes, perennial succulents such as cacti, sparse and tightly bunched grasses, and fast-growing annuals. Desert animals are often fleet, keen of senses, and with a coloration that matches the desert sand or rock. Eyes, ears, and nostrils of desert animals may have special structures to keep out sand or dust; and locomotor organs are often modified to facilitate movement across shifting sand.

Desert grades into grassland, or into thorn forest, in very complex fashion. In the area of transition there may be an intermingling of communities from each Biome-type, their local distribution governed by minor differences in physiography or soils; or there may be an actual intermingling of species from both Biome-types.

Difficult to classify are the world's two high plateaus: the Tibetan and the Bolivian. Each is somewhat intermediate between grassland and desert, and each harbors distinctive herbivores dependent upon grassland produced by melting snows. (Tibet has the yak, various wild sheep, and the goat antelopes; Bolivia has the llama, alpaca, vicuña, and guanaco.) Animals of the high plateaus are adapted to withstand oxygen deficiency, extreme cold, and biting winds. It is probably well to recognize (6) a High Plateau Biome-type, with separate Tibetan and Bolivian Biomes.

(7) The Tropical Forest Biome-type has three major components. One of these occupies parts of central and western Africa, with smaller and closely similar outliers elsewhere on the continent. A second covers the Amazonian, Orinocan, and Guianan regions, east of the Andes in South America, with an extension into Middle America. A third extends from southern India through the East Indies and New Guinea into Queensland. This last or Australasian expanse may be split into two Biomes, chiefly on the basis of marked faunal differences.

In the Tropical Forest Biome-type some trees are of great height. There is a complex stratification of the vegetation: a ground layer of herbs, overtopped by shrubs; a layer of sparse, narrow-crowned young trees, up to about 60 feet in height; a layer of taller and sturdier trees, from 60 to 120 feet high, whose rounded crowns compete with each other for light, and interweave to form a closed canopy that shades the levels below; and a discontinuous layer of forest giants, towering from 120 to more than 200 feet in the air, and with crowns fully exposed to the sun. The distinctiveness of the layers is blurred by young, growing trees, as well as by epiphytes and vines. Arboreal animals are abundant, and may be confined to some one particular layer. Many animals show modifications for arboreal life: perhaps extreme agility, a prehensile tail, hands and feet that grip tightly, expanding membrances that permit gliding, or a greenish coloration that renders the animal well-nigh invisible amidst the leaves. The Tropical Forest Biome-type cannot be extended into southern Florida. As we have already seen, some tropical floral elements are present in that area along with temperate ones; but the fauna is largely of temperate affinity.

Here and there in the world are small areas that must stand alone in a scheme of Biomes and Biome-types. For example, temperate South America has a flora related to that of New Zealand, but the fauna is mostly of American origin. Long chains of small islands, as from the Palaus to Sala-y-Gomez, are hard to classify because the biota is derived from larger land masses but is often nourished from the sea. Here we shall omit discussion of minor areas.

From the foregoing account of the major Biome-types, it will be seen that the Biome concept seeks to integrate a vast amount of data. Because the Biome approach is so broad, by a little more stretching it can be applied to distribution in the sea; and so much cannot be said of any other approach considered in previous chapters.

28 *"Sometimes I rambled to the pine groves . . . or to the cedar wood beyond Flint's pond . . . or to swamps where the Usnea lichen hangs in festoons from the white-spruce trees . . . ."** **FURTHER BYWAYS OF BIOGEOGRAPHY**

Sir Joseph Dalton Hooker once remarked that of all branches of botany, plant distribution required the greatest amount of study. By now the reader will surely agree with him. We have already described several ways of looking at plant distribution, and additional ways exist. Nor have we exhausted the possible ways of analyzing the distribution of animals, or of plants and animals together.

A useful biogeographic approach is the Biotic Community concept. This approach defines "habitat" as the environment normally occupied by an organism within its geographic range. The habitat is most conveniently described in terms of vegetation types, which result from an interaction of many environmental factors. Each area of distinctive vegetation harbors a characteristic assemblage of animals and plants, among which numerous interactions take place. Each area of distinctive vegetation is therefore a community of organisms.

The Community is not completely closed from other Communities, yet to some degree is independent of them. A Community may be climax or seral. It is important to consider all Communities separately, be they climax or seral. Be-

* HENRY DAVID THOREAU, *Walden.*

338

tween two Communities there may be a zone of transition or intermediacy, called an ecotone. An ecotone may be so large and distinctive as to harbor organisms that do not live in either of the bordering Communities; therefore an ecotone may also constitute a Community. Two Communities may be represented as scattered enclaves within the ecotone that separates them. Some Communities correspond closely to Biomes, but others do not. In the Community system, there are nine Major Biotic Communities in temperate North America, north of Mexico. These are respectively the Tundra, Coniferous Forest, Deciduous Forest, Grassland, Southwestern Oak Woodland, Piñon-Juniper Woodland, Chaparral, Sagebrush, and Scrub Desert. Most of the subdivisions are already familiar to us from other approaches.

The system recognizes both Arctic and alpine tundra. The southern limit of the Arctic Tundra is drawn very far north; it corresponds in area to the Arctic Life Zone of Merriam. The dominant plants—the ones that control the habitat and so are highly important in the Community's ecology—are lichens, grasses, sedges, and occasionally dwarf willows and other small shrubs. The tundra as a distinctive land is already familiar to us, but here is a new idea: south of the Arctic Tundra is a broad ecotone, or transitional area, between tundra and coniferous forest. This corresponds closely to the Hudsonian Life Zone of Merriam. Recognition of a hyphenated floral Province, the Arctic-Subarctic, and the explorers' insistence upon a "high" Arctic and a "low," hint at the existence of this subdivision. The ecotone is added to the Major Biotic Community that includes the tundra proper.

Still farther south is the Moist Coniferous Forest, a Biotic Community equaling in extent the Canadian Life Zone of Merriam. Its dominant plants are coniferous, evergreen trees: pines, spruces, hemlocks, firs, and cedars; also Douglas fir in the Rockies, Cascades, and Sierra Nevadas. The puzzling forest that stretches from the Great Lakes region eastward to Nova Scotia is considered to be an ecotone, a transition between moist coniferous forest and deciduous forest. This ecotone is added to the Major Biotic Community that includes the Moist Coniferous Forest.

The Deciduous Forest extends from Massachusetts and the southern part of the Great Lakes region southward into Louisiana. Dominant plants are broad-leaved, deciduous trees; among them are beech, maple, basswood, oaks, hickories, walnut, and (until recent extinction) chestnut. The Piedmont Plateau, and the clay-hill lands of the upper Coastal Plain in Alabama and Mississippi, are considered ecotonal between the deciduous forest and the subclimax pinelands of the lower Coastal Plain.

The Southeastern Pine Woodland Subclimax is a Biotic Community extending from Florida northward along the lowlands to northeastern North Carolina, central Georgia, north-central Alabama, and southeastern Mississippi. Dominant trees are pines and sometimes oaks, as well as grasses and shrubs where trees are widely spaced. Extreme southern Florida might be thought of as ecotonal between the Southeastern Pine Woodland Subclimax and a tropical Community that does not reach the United States.

Central North America is occupied by a Grassland Biotic Community, as we might have guessed. In the eastern part of this subdivision the dominant plants are tall grasses such as bluestem, Indian grass, and switch grass; in the western part, short grasses such as buffalo grass and grama grass. The present scheme borders the grassland with ecotones of which we have previously said but little. There is an ecotone between grassland and deciduous forest, narrowly bordering the latter on its western side, and widening markedly in eastern Texas. A second ecotone, also between grassland and coniferous forest, forms a moderately wide belt just south of the moist coniferous forest from Alberta to northwestern Minnesota. A third ecotone, between grassland and sagebrush, forms a wide but interrupted band from central Wyoming westward to eastern Oregon.

A Mesquite Subclimax is confined in the United States to western Texas and southern New Mexico, but extends far down into Mexico. (It was mentioned in Chapter 25 simply as an expanse of thorn forest.)

The Southwestern Oak Woodland develops on hills and mountain slopes in parts of Utah, Nevada, California, New Mexico, Arizona, Colorado, and Oregon. The dominant trees are nearly all oaks. Usually they are spaced close enough for their branches to overlap, but sometimes they are widely spaced and then there are intervening patches of grasses and shrubs.

The Piñon-Juniper Woodland occupies parts of the Great Basin and the Colorado River region, as well as the eastern side of the Cascades–Sierra Nevada chain in California. As the name suggests, the dominant plants are the piñon pine and several species of junipers. Yuccas are often conspicuous.

The Chaparral is the evergreen hardwood forest that develops in California under the impact of a Mediterranean type of climate. Dominant plants, in one situation or another, include snowbush, chamise, scrub oak, mountain mahogany, coffee berry, manzanita, poison oak, and a cottonseed bush (*Baccharis*).

Sagebrush—the Biotic Community, not just the sagebrush plant itself—occupies higher parts of the Great Basin, above the level of the valley floors. Dominant plants are sagebrush, shadscale, rabbit brush, greasewood, winter fat, and sometimes grasses.

Scrub Desert includes the lowlands and the valley floors from western Texas westward to southwestern California. The dominant plant is creosote bush. Also conspicuous are mesquite, palo verde, catclaw Acacia, western ironwood, ocotillo, Agaves, cacti, and Yuccas; and occasionally grasses.

Finally, the Biotic Community approach recognizes such seral Community types as shores, which are poorly vegetated areas bordering large bodies of water; freshwater marshes and salt marshes, usually inundated and with grasses, sedges, reeds, and other herbaceous plants; and swamps, which are inundated areas with trees and shrubs. Also recognized are savannas, rather level grassland with widely spaced trees and shrubs; shrublands, with a predominance of low, woody plants; and riparian woodlands, mostly of deciduous trees, near streams or in bottomlands where ground water is readily available to plants.

There have been other approaches to the study of distribution, not differing markedly from the Biome and the Biotic Community schemes. Each biogeographic scheme represents a somewhat different approach to the analysis of distribution. If two schemes partition the land into similar areas, so much the better; for the distinctiveness of the area is all the more firmly proven. For example, you will arrive at the approximate outlines of the vegetation *Provinces* whether you subdivide the land on the basis of climatic, floral, or vegetational differences; and so the individuality of a *Province* is triply assured. Furthermore, the real goal of any biogeographic approach is not a map but an understanding of the forces that have brought about the various patterns of animal and plant distribution. Each approach contributes something to this understanding.

The various biogeographic schemes have all partitioned the world into rather large areas; but what about a finer subdivision? Any of the major approaches, if carried far enough, will produce the identical arrangement of small divisions. If the world were subdivided to the utmost on the basis of its vegetation, what would the final units be? They would be your woodlot, Farmer Brown's pasture, the old millpond, the creek swamp, Neighbor Jones's fields, the patch of woods on the edge of town, the Boy Scout lake, the pinelands in the State Park. But if the world were subdivided on the basis of its flora, the final units would be the same; for there are trees and shrubs and herbs in your woodlot that you cannot find in Farmer Brown's pasture or Neighbor Jones's fields; while there are certain grasses in the pasture, different ones in the fields, and perhaps none of them in your woodlot or in the woods just outside of town. Each of the small areas is at least slightly distinct as regards fauna, also. Furthermore, within the woodlot there is a stirring of life that has little to do with surrounding areas. Leaves fall and make humus in which insects burrow; shrews eat the insects, and in turn fall prey to some owl or snake; caterpillars with strict dietary requirements eat certain plant leaves, transform into butterflies, and are ensnared by spiders who spin their webs from twig to twig; berries, caterpillars, and spiders lure some birds but not others. In short the woodlot is a small-scale community, to some degree independent of the many others that exist nearby.

And so it is time in our narrative to leave the large areas, the Biotic Provinces, Biomes, and the like; and to turn our view toward distribution within some small area, say a county. Not that a county is a unit of biogeographic significance; it is simply one of convenient size for study and discussion. Marion County, in north-central Florida, is a good area to choose, for distribution there, while exceptionally complex, has also been intensively studied. In this county the small areas of differing vegetation form a bewildering mosaic. In twenty miles of driving you can see more than twenty different types of vegetation, some of them so distinctive as to have received vernacular names. (Each type may be called an "association," a term implying only that a number of plant species habitually appear with each other. Often, but not invariably, an association is named from two of its most conspicuous plants.)

Many sand dunes in the county are covered with little sand pines, Florida rosemary, and dwarfed oaks; local people call this type of vegetation the "scrub." On other dunes, scattered longleaf pines tower above small turkey oaks. Here and there are pure stands of enormous live oaks, whose branches are decked with Spanish moss and whose overlapping crowns shade the ground below. Local residents speak of "live-oak hammocks." ("Hammock" is a word adopted from the language of the Timucua Indians; it means an island-like stand of hardwoods surrounded or bordered by some other type of vegetation.) There are also stands of mixed hardwoods, some deciduous and others evergreen; among them are

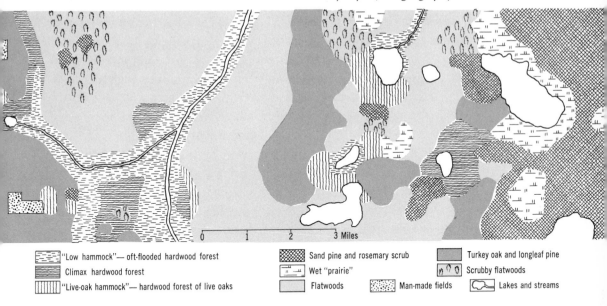

≡ "Low hammock"— oft-flooded hardwood forest	░ Sand pine and rosemary scrub	▓ Turkey oak and longleaf pine
☰ Climax hardwood forest	Wet "prairie"	Scrubby flatwoods
⫿⫿⫿ "Live-oak hammock"— hardwood forest of live oaks	Flatwoods	Man-made fields	Lakes and streams

Vegetation in Detail. A portion of Marion County, Florida, showing mosaic of vegetation.

Magnolia, blue beech, hop hornbeam, American holly, laurel oak, basswood, various hickories, Florida sugar maple, winged elm, and red mulberry. This mixed hardwood forest is the climax vegetation of the county.

There are flatwoods with wire grass and widely spaced slash pines; or with fetterbush and black pines; or with thickets of palmetto, gallberry, and tarflower, and with dense stands of pond pines in numerous depressions. In places, dune sand has blown over the flatwoods, producing a scrubby flatwoods; some of its plants are invaders from the sand pine scrub.

In river bottomlands there is a forest with cabbage palms, red bays, sweet gums, hackberries, and red maples. At the edge of a river, or even extending out into its shallows, is a forest mostly of gigantic bald cypress trees. The river bottom itself supports a distinctive community of aquatic plants such as eelgrass, a naiad, a Sagittaria, pondweeds, a coontail, and many algae.

At the heads of small streams, especially in flatwoods, there are tangled thickets of sweet bay, red bay, wax myrtle, black gum, red maple, dahoon holly, and loblolly bay, all bound together with greenbriers and other vines. In local parlance these thickets are "bayheads." Here and there are freshwater marshes that present a solid expanse of sawgrass, actually a coarse, gigantic sedge whose leaf blades will cut flesh and clothing alike. Other marshes support little or no sawgrass, but a mixture of pickerel weeds, yellow bonnet lilies, white water lilies, arrowhead, cattail, and other herbaceous species.

Many shallow basins are covered with tall, waving grasses that overtop and obscure the other herbaceous plants. Local residents call this association "prairie," and from a distance the aspect is indeed prairie-like; but walk out into a Florida "prairie" and you will soon be knee deep in mud and water.

Some basins, mostly in flatwoods and only intermittently filled with water, are muck-bottomed and are covered thickly with pond cypress, a dwarfed relative or variant of the giant bald cypress. Other basins, often in the same flatwoods, are sand-bottomed; they are grown over with a head-high shrub called sandweed, a member of the same genus that includes the little Saint-John's-wort.

Almost anywhere in the county may be expanses of nearly barren sand, on which just a few pioneering herbs are beginning to appear. These expanses were recently made by man, who removed the original vegetation in order to haul off the sand for fill. Also widely scattered are fields covered with broomsedge and dog fennel; the angular outlines of the broomsedge fields reveal that they are tracts cleared some years ago by man.

There are open woodlands with scattered redhaws that sprout from thick broomsedge; and there are a few hills, all with a reddish and somewhat claylike soil, that support a puzzling mixture of live oaks, laurel oaks, bluejack oaks, longleaf pines, grape vines, blackberry tangles, and sassafras bushes.

Along the main watercourses are great mounds of freshwater clam and snail shells, the accumulation of prehistoric Indian villages. Several acres may be piled with shell to a depth of twenty feet or more. Some of the higher and drier middens are covered with giant cacti and twisted junipers. Larger middens in the river swamp support a climax forest, but the smaller ones may bear only a dense stand of cabbage palms.

Suburban areas offer what might be called a "horticultural forest." It includes native plants that man thought to be ornamental, but also a great variety of introduced trees, shrubs, and herbs. Plantings are spaced in such a way as to leave many open spots, where introduced grasses are encouraged to grow.

Vast expanses of rolling sandhills, which might otherwise be covered with turkey oak and longleaf pine, have been converted into pastureland by the complete removal of trees and the sowing of non-native grasses. Herds of race-horses, of Black Angus cattle, and of fat, red Santa Gertrudis cattle, help to keep the tracts in grassland, for they eat or trample any tree seedling that happens to spring up. Many dunes, which might otherwise be covered with sand pine and rosemary scrub, have been converted into orange groves; the dark green trees march in neat ranks over the low hills of whitish sand.

This complexity of vegetation types can be resolved into seres. As an illustration, when Florida was nearly inundated by a rising sea level between glaciations, dunes formed along what was then the coastline, and were left there when the sea retreated. We may suppose that these dunes, like the ones just back of the present coast, were eventually captured and stabilized by certain vines, grasses, and other herbs adapted for life on shifting sands. The sand pine scrub is the first timber stage to develop after dunes are stabilized. In the course of succession this scrub changes into live-oak hammock, and the latter into the climax forest of the area.

Within historic times some streams have vanished, and their beds, which no doubt supported aquatic plants, are now overgrown with bald cypress. The cypress or river-swamp forest in time changes to bottomland forest, and this latter to climax vegetation. Other seres can be detected but need not be described.

The local student of distribution may be interested not so much in succession but in the fact that each type of vegetation has characteristic animals and plants. In the county you could discover the short-tailed snake only in the turkey oak and longleaf pine stands, and here too is about the only habitat in which to find the interesting little plant called green-eyes. The turkey-oak tree itself rarely appears in any other association, but the longleaf pine may turn up in a variety of situations. For the yellow-flowered and the orange-flowered milk-wort you would have to visit damp, sunny spots in the flatwoods; while shadier parts of the flatwoods, with an abundance of fallen pines, would be about the only place to find the yellow-lipped snake. The glossy watersnake lives only in flatwoods ponds; it finds the southern limit of its range in this county, for it did not inhabit the islands that preceded the Florida peninsula, and has not pushed very far southward in the state.

To discover the few local populations of the burrowing owl you would seek out fields of very low grass, especially the pasturelands where horses and cattle graze. Not every pastureland will harbor the owl, however; for it lives in holes dug by the gopher tortoise, a reptile restricted to dry, sandy places. The indigo snake trails its eight-foot length beneath the live oaks, but also moves into the turkey oak and longleaf pine association, and just rarely into other situations. The armadillo, introduced into Florida and now very common there, is most characteristic of the drier situations, but will venture into many of the wetter ones to dig for grubs and earthworms. The big red-headed skink, here at the approximate southern limit of its range, is virtually confined to bottomland forest and nearby climax forest. Very few reptiles live in the climax forest, which seems to be taken over by mammals and birds. But among amphibians the slimy salamander, which hides beneath rotting hardwood logs, is common in this association.

The trees of the climax forest are mostly lime-loving species, and their presence usually indicates available lime in the soil. A species of juniper, or red cedar, often grows in the climax forest but obviously is not dependent on the forest itself. It will grow wherever there is lime, even on the dry, sunbaked boulder fields around limestone quarries.

The scrub jay is limited to the sand pine scrub, and one patch of scrubby flatwoods that is sufficiently scrublike for the bird's needs. Occasionally, flocks of scrub jays fly through other types of vegetation, but they are simply moving from one scrub patch to another. The whiptail scorpion or vinegaroon, which (like the burrowing owl and the scrub jay) lives also in the arid southwestern United States, is nearly limited to the driest plant association, the turkey oak and longleaf pine.

The corn snake is far more abundant in the horticultural forest than in any natural situation. It feeds upon the introduced rats and mice, which are generally associated with man. Venus's looking-glass and several other plants are only to be found in lawns, where their seeds were probably introduced by accident along with the seeds of imported grasses.

The narrowmouth toad lives under rocks and logs, eats ants, and can breed in very transient rainwater pools. It may be found in any of the vegetation types except the fully aquatic ones and the driest parts of the sand pine scrub. The saw palmetto, a small palm, is just about as adaptable as the narrowmouth toad. It will not grow in places that are permanently or frequently flooded, nor is it one of the first plants to invade a devegetated tract; but otherwise it may appear just about anywhere in the county.

To summarize, the local investigator finds some animals and plants to be restricted, at least in his area, to one or a few vegetation types. Other species inhabit many types. Individual animals, belonging to exceptionally mobile species, may pass through several associations in a day. An organism that is confined to a certain type of vegetation may or may not be present in most individual patches of this vegetation. A species may occupy several vegetation types but be outstandingly successful in only one or a few. The local distribution of a species is occasionally limited by some historical factor; that is to say, by a geologic or climatic episode of past ages.

An organism may occupy only a limited situation within the vegetation type to which it is restricted. Thus in the climax forest of Marion County the slimy salamander never ventures farther above ground than the top of some rotting log; while in the same forest the flying squirrel hardly ever sets foot on the ground. The towhee, a small bird, scratches henlike in dead leaves, and feeds also in low brush; but almost never does it fly higher than ten feet off the ground. Serpent fern rarely grows anywhere but in the tops of the cabbage palms; Spanish moss decorates trees to the very top, but seldom dangles so low as to drag the ground.

The ultimate biogeographic subdivision—the so-called "biotope"—is the place or the situation in which you will find any particular kind of plant or animal.

29 *"There is nothing constant in the universe / All ebb and flow, and*

every shape that's born / Bears in its womb the seeds of change."*

THE CHANGING
ENVIRONMENT

For all its importance in biogeography, the modern environment does not always explain modern distribution. The three great faunal Realms exist for reasons that have little to do with the environment of today: it has generally been easy for animals to pass between Eurasia and North America or between Eurasia and Africa; harder for them to pass between North America and South America; and hardest for them to pass between Australia and Eurasia. And at a much lower level of distributional study, ecological data will not explain why the redheaded skink and the glossy watersnake should find the approximate southern limits of their respective ranges in Marion County, Florida, when the plant associations they frequent are not so restricted.

Many plants have transgressed barriers to reach areas of suitable environment, and so the floral and vegetational subdivisions correspond to areas that are environmentally distinct; but animals are slower to spread, and their distribution may still reflect some barriers that have long since vanished. Also, animal distribution may imply that some present barrier has not always existed; thus the wood frogs have extended their range from Siberia to Canada, although today there is no possibility of their crossing the cold and salty Bering Sea. In short, the

* OVID, *Metamorphoses,* in the first century.

348

faunal subdivisions tell more about the geologic and climatic history of the earth than do the floral ones.

This situation concerned the Russian phytogeographer E. V. Wulff, who felt that geologic and climatic changes must have affected plant distribution, and in ways that could not be deduced from the outlines of the floral and vegetational subdivisions. Wulff reasoned that, among plants, at least some cases of widely discontinuous distribution have resulted not from the "leaping" of barriers but rather from an environmental change that wiped out a particular group from a large part of its range. If numerous plant groups exhibit roughly the same pattern of discontinuous distribution, some major environmental change may have been responsible for the discontinuity. Thirteen significant, oft-repeated patterns of disjunct distribution were found by Wulff to exist among the higher plants. Of these we shall list only the five that involve North America.

One is the "Arctic-alpine" pattern. We have already noted its existence in Chapter 17. Plant seeds might be carried by winds or birds from the Arctic to distant high mountain peaks; but glacial periods reduced the gap between areas of arctic climate. Actually, it would be well to recognize within a single pattern not only the occurrence of tundra plants on distant mountains above timberline, but also the occurrence of taiga plants in isolated communities on distant mountains at climatically suitable elevations above the deciduous forests.

Studies made in southeastern Pennsylvania show how climatic change is involved with the occurrence of high-latitude plants at high elevations elsewhere. The analysis of fossil pollen from a bog near Valley Forge revealed that during a peak of glaciation the locality was on the border between tundra and taiga. At such a time it must have been relatively easy for taiga plants (and animals) to move into the southern Appalachians, where some of them still persist on scattered peaks.

The "North Atlantic" pattern of distribution we have elsewhere called amphi-Atlantic. The amphi-Atlantic plants inhabit eastern North America (east of the central plains) and western Europe (west of the Eurasian central plains). Some students have postulated former land connections across the North Atlantic, a topic to be considered later.

The "North Pacific" pattern involves restriction to eastern North America and eastern Asia, with occasional representation also in Pacific North America. The pattern characterizes the remnants of the Arcto-Tertiary forest.

The "North and South America" pattern involves restriction to deserts or semideserts of both these continents, and absence from Middle America. Creosote bush is the best known plant to be distributed in this way.

There are very dry areas along the Pacific slope of Middle America, which receives less rain than does the Caribbean slope; and elsewhere in Middle America there are some remarkably dry valleys in the rain shadow of sierras. Past periods of reduced rainfall further dried all these areas, and permitted many of them to coalesce. Dry corridors were formed through Middle America, facilitating the north-south distribution of desert and semidesert plants.

The "Tropical" pattern has several subpatterns, only one of which concerns us. This one involves restriction essentially to tropical Africa and tropical America. (Some tropical American groups send offshoots well into North America, and so command our attention here.) The genus *Annona*, the custard apples, is the best known plant group to be distributed in this way. Once again, trans-Atlantic land connections have been postulated; how and when they might have existed is a topic for subsequent discussion.

Fortunately, a number of sciences deal directly with the geologic and climatic events of past ages, and the dating thereof. The findings of these sciences may be scanned for information of possible use in biogeography.

In the seventeenth century, when man took an egocentric view of the universe, Lightfoot and Ussher held the world to be scarcely 6,000 years old. Today we believe its age to be at least a million times greater than this. The first half of earth's history is scarcely of biogeographic concern, for the planet was still lifeless.

We do not need to say much about the remote time when life was first dawning in the sea. It is perhaps worth noting that many present-day investigators find no place to draw a firm line between living and nonliving. They feel that, given the environmental conditions of the ancient sea and the ancient atmosphere, carbon molecules would have combined and recombined in complex fashion, and on many occasions would have produced giant molecules with the power of duplicating themselves. Carbon-chain molecules sufficiently complex to be called "organic" may have evolved many times, and the living things of today might not all be of common descent. The so-called Monerans—the bacteria and the blue-green "algae"—may well have developed independently of other organisms. The Protista—the true algae, slime molds, fungi, and protozoa—may

Cliffs at Beachy Head, Sussex, England, are formed largely from shells of tiny marine organisms. Such deposits reveal that land and sea were not always disposed just as at present. (British Travel Association)

ERA	MILLIONS OF YEARS AGO		PERIOD
			Recent
	1		Pleistocene
CENOZOIC	13		Pliocene
	25	*Tertiary*	Miocene
	36		Oligocene
	58		Eocene
Paleocene: General erosion	63		
			Cretaceous
	135		
MESOZOIC			Jurassic
	181		
			Triassic
	230		
General erosion			
	280		Permian
	345 *Carboniferous*	{	Pennsylvanian Mississippian
PALEOZOIC	405		Devonian
	425		Silurian
	500		Ordovician
	600?		Cambrian
Uplift and general erosion			
PROTEROZOIC			
Uplift and general erosion			
ARCHEOZOIC			
Uplift and general erosion			

The Changing Environment. The major environmental changes of the past. Dates are approximate. Earlier topographic developments are omitted, since their effects on

TOPOGRAPHIC DEVELOPMENTS IN NORTH AMERICA	BIOLOGIC DEVELOPMENTS IN NORTH AMERICA
	Dominance of man
Elevation of California coastal mountains; glaciation	Extinction of many large mammals at end of period
Elevation of Rocky Mountains, Sierra Nevada, and Cascade Range	Development of mammals
Beginning of Rocky Mountains Submergence, Gulf of Mexico to Alaska	Rise of mammals and birds Extinction of dinosaurs Development of flowering plants and deciduous trees
Appalachian Mountains base-leveled; submergence, Colorado to Alaska. Drifting continents?	Dinosaurs diverse; first appearance of birds and mammals Many insects of modern types
Large land area; widespread aridity	Reptiles diverse: walking, flying and swimming types Forests mostly coniferous
	Decline of tree ferns, rise of conifers
	Rise of reptiles and insects Amphibians diverse
	Fishes diverse; rise of amphibians
	Rise of fishes; first land plants
	Early fishes; no land life
	Marine invertebrates
	Primitive marine life; fossils scant
	Dawn of life
	No living organisms

distribution are no longer recognizable. Biologic developments reveal frequent extinction, accompanied by the rise of new groups to positions of dominance.

represent a second line of development, and it has even been suggested that the Protista themselves are of multiple origin. The remaining organisms—the true plants and true animals—evidently evolved from Protistan stock.

Earth's history begins with an Azoic Era; the phrase signifies a vastly long age of lifelessness. The Azoic was succeeded by the Archeozoic Era, a great span of time during which life was first developing in the sea. We know little about the Archeozoic. It may have begun about four and a half billion years ago. The very earliest organisms, presumably one-celled and delicate, have left no discernible traces in the Archeozoic rocks. Alga-like fossils, over three billion years old, have been found in South Africa; apparent algae and fungi, around two billion years old, in Ontario. About a billion years ago, certain algae and bacteria were forming protective shells about themselves, out of limestone extracted from the water; and their remains supply us with additional fossils of great age. (Our "billion" is the American one, a thousand millions; not the English one which is a million millions.)

There is some indication that the Archeozoic was terminated by a period of marked geologic and climatic changes. It was succeeded by the Proterozoic Era. Proterozoic fossils are scarce, but one deposit in Australia has produced them in fair abundance. This fossil bed yielded remains of jellyfishes, jointed worms, and soft corals, as well as two curious invertebrates akin to nothing that lives today. This diversity of marine life implies a lengthy evolution of which only traces have so far been found. A few other Proterozoic fossils have been discovered in South Africa and England.

The Proterozoic, like the preceding Archeozoic, was terminated by profound geologic and climatic changes. There came a period of widespread glaciation— apparently not the first the world had known, and surely not to be the last; and when this early glacial age was over, new forms of life soon became abundant in the sea. A new Era, the Paleozoic, had opened.

The fossil record of the Paleozoic is good, revealing constant change in living things. It is therefore convenient to divide this Era into lesser units, called Periods. The first Period of the Paleozoic is the Cambrian, which began 600 million years ago.

The Cambrian sea teemed with protozoa, sponges, mollusks, brachiopods or lamp-shell "mollusks," sea cucumbers, starfishes, and sea urchins. Graptolites, resembling triangular bits of latticework, fastened themselves to the sea bottom.

Dominating the Cambrian sea were the trilobites: hard-shelled invertebrates with jointed legs, looking somewhat like the horseshoe crab of today. Trilobites were the most advanced organisms of their time, and more than a thousand fossil species of them are known. A few swam free in the water but most were sea-bottom scavengers, and none exceeded two feet in length. Before the Cambrian had ended, all the main invertebrate or lower animal types had come into existence; but vertebrates had not yet appeared.

The Cambrian ended 500 million years ago, and was succeeded by the Ordovician Period. The latter saw the appearance of the first fishlike animals—primitive, jawless types somewhat like modern lampreys and hagfishes but provided with a bony armor. Yet in the Ordovician the fishes played a minor role. Some deposits of this Period represent sediments washed down by rivers, and contain the fragmentary remains of armored fishes; so it may be that these, the earliest known vertebrates, evolved in fresh water while the sea was still ruled by invertebrates. The sea was full of snails, squidlike animals, corals, sponges, and the bryozoa or moss animals. The most highly evolved plants were seaweeds (larger algae), and no organism had yet invaded the land.

The Ordovician ended 425 million years ago, passing into the Silurian Period. In the latter, trilobites still existed but were no longer dominant; nautiloids were now in the ascendancy. These were carnivorous, many-armed mollusks, resembling a squid in a cone-shaped shell; they were related to the paper Nautilus and pearly Nautilus that live today. The largest of the Silurian nautiloids reached a shell length of 16 feet. Snails, bivalve mollusks, brachiopods, graptolites, jellyfishes, and corals still thrived. Sea lilies became abundant; they looked like flowers growing from jointed stalks, but actually were carnivorous animals related to starfishes.

Heavily armored sea scorpions, equipped with claws, also became abundant in the Silurian. Some of these invertebrates reached a length of nine feet. In spite of their name, the sea scorpions may have been more characteristic of estuaries and the lower reaches of streams than of the ocean proper. The early fishlike vertebrates may have developed armor as protection against the great, clawlike pincers of the sea scorpions.

The primitive, fishlike vertebrates persisted into Silurian times, but were still unimportant. The fossil record of this Period also includes an animal related to the modern lancelet. Lancelets are not included among the vertebrates, yet are related thereto; their body is stiffened by a rodlike structure that runs from head to tail, but the structure is not divided into joints or vertebrae. Such primitive animals, which are somewhat on the borderline between invertebrate and vertebrate, probably arose before the fishes; but being small and soft-bodied (and perhaps restricted to a habitat that was not conducive to fossilization), they left few traces in the Paleozoic rocks.

The latter Silurian was a time of mountain-making. Great ranges arose in Europe. (All that remains of them today are the Scottish highlands and the Scandinavian ranges.) In the streams of these uplands, several types of armored, fishlike vertebrates diversified; but by the time their bodies had been washed downstream and deposited in delta sediments, only broken bits of armor plate remained.

Sea plants had been evolving, and some had developed an internal system of tubes through which nutrients and waste products could flow. They had further diversified their structure into a root that anchored, and a stem that held the rest of the plant to the light. In the Silurian, certain of the plants invaded the shore and finally the land itself. This was a step of great consequence, for plants are producers while animals are only consumers; and animal life could not leave the water permanently until plants had done so. Before the Silurian was out, at least one invertebrate, a scorpion-like animal, was living on land.

The Silurian ended 405 million years ago, passing into the Devonian Period. In the latter, the most significant evolutionary advance took place among the primitive fishes. In earlier types the mouth was but a weak, jawless opening; but certain Devonian types modified a pair of bony gill supports into jaws. The modification permitted fishes to invade the sea, and to compete successfully with the invertebrates that teemed there. During the Devonian the fishes diversified in, and established their dominance over, the sea as well as the fresh water.

Six main lines of fish evolution were represented in this Period. One, the armored, jawless fishes, persisted from the Silurian. A second line, armored but provided with jaws, was not destined to survive much past the Devonian. A third, with a skeleton of cartilage rather than of bone, was a forerunner of the sharks and rays. A fourth line persists today as the lungfishes. A fifth, with a bony skeleton and with fleshy, lobed fins, was ancestral to *Latimeria*, the "living fossil" mentioned in Chapter 2. The sixth line of Devonian fishes was ancestral to the bony fishes that predominate in all waters today.

Devonian deposits are often of reddish sandstone, believed by many students to have been deposited under arid conditions. A famous Devonian locality, at Scaumenac Bay in Quebec, has yielded thousands of fish specimens, about 95 per cent of them being either lungfishes, lobe-fins, or jawed and amored fishes. All three of these types were supplied with lungs, and presumably could, like the modern lungfishes, survive long droughts by burrowing in the mud. The Devonian environment may have put a premium on the ability to exist away from water. The Devonian also saw a proliferation among land plants. There were kinds resembling the modern Equisetums or horsetails; others that were fern-like, and that reproduced by wind-drifting spores; still others with a kind of seed.

Before the Period was over, several animal groups had taken to the land. There were mites, millipedes, and insects that ate the plants. Vertebrates were represented on land by amphibians that looked like oversized salamanders, and that had developed from lobe-finned fishes. Little modification was necessary to evolve stumpy legs from lobed fins, and the earliest fishlike amphibians may have used their appendages chiefly to leave drying mudholes in search of ponds that still held water.

The Devonian ended 345 million years ago, passing into the Carboniferous Period. The latter name means "coal-bearing"; many coal deposits were formed from the swamps that existed during the latter part of this Period. The Carboniferous has two subdivisions. The earlier of these is called the Mississippian; it is characterized by limestone deposited in shallow seas. The limestone was produced especially from the minute shells of Foraminifera, a protozoan group. Brachiopods, sea lilies, and mollusks were still abundant in the Mississippian, but trilobites had nearly vanished, along with both the jawed and the jawless armored fishes. Sharks flourished.

The latter part of the Carboniferous, called the Pennsylvanian, saw a general shifting of the land; beds of shallower seas were uplifted, while many land areas sank about to sea level. Vast swamps were formed. In them grew scale trees, some of them one hundred feet tall. Horsetails reached thirty feet. Seed ferns diversified. True ferns and conifers appeared, but played only a minor role. Nowhere on earth was there yet a flower blossom.

Carnivorous insects appeared, preying on their herbivorous kin; among the carnivores was a dragonfly with a wingspan of twenty-nine inches. Spiders arose. In the swamps the amphibians diversified, and during the Pennsylvanian they

Rivers may cut into and expose ancient strata. Notable in this regard is the Colorado River, here photographed near Dead Horse Point in Utah. (Utah Tourist and Publicity Council)

The Scotch highlands are remnants of the world's oldest mountains. Glencoe, Argyllshire, Scotland. (British Travel Association)

reached their peak. One amphibian line, becoming fully emancipated from the water, evolved into reptiles. A coal mine at Linton, Ohio, yielded several types of early reptiles, as well as amphibian-reptile intermediates. The deposit dated from the middle of the Pennsylvanian, and we may suppose that the first reptiles were of early Pennsylvanian age.

About 280 million years ago the Carboniferous passed into the Permian Period. The latter was a time of violent changes. Mountain chains arose, in about the same places where the Alps, the Rockies, and the Appalachians were later to rise. Deserts formed; glaciers waxed and waned. As the millennia passed, the mountains eroded down, and their material was deposited far and wide. Many organisms vanished. Tree ferns declined. Conifers thrived in the northern hemisphere; a fernlike plant called Glossopteris predominated in the far-southern lands. Amphibians persisted, but in reduced numbers and variety. Reptiles diversified, and some of them began to develop mammal-like characteristics. A small, heavily armored reptile from the Permian of South Africa may be ancestral to the turtles. Toward the end of the Permian there was a long episode of geologic and climatic disturbance. When this came to an end, the flora and fauna of the land had been radically altered, the old marine fauna had been decimated, and a whole new Era was ushered in. The Permian was the last Period of the Paleozoic Era.

About 230 million years ago the Mesozoic Era began. Its first Period was the Triassic, a time of widespread aridity. Forests were mostly coniferous. The arthropods—the invertebrates with jointed legs, such as spiders, insects, and the like—continued to diversify. Amphibians persisted, and even established one new and successful line, the frogs. Reptiles achieved a position of dominance in the Triassic; there were walking, running, and swimming types. The first beak-headed reptiles appeared in this Period; their lone survivor is the tuatara of New Zealand. Another reptilian group, the ichthyosaurs, took to the sea. Reptiles probably took also to the air in the Triassic, for highly evolved flying types are abundant in the succeeding Period. In the Triassic two reptilian groups were of minor importance yet held great promise. One of these was the dinosaurs, the other the mammal-like reptiles.

Some 181 million years ago the Triassic passed into the Jurassic Period. Climate was mild, rainfall generally abundant. Conifers, ferns, and cycads spread widely, along with maidenhair trees. (The Asiatic ginkgo tree, often planted as an ornamental, is a lone survivor of this last group.) Some relatives of the cycads evolved flower-like structures. As plants diversified, so did the insects.

Dinosaurs stalked the Jurassic land in a variety of shapes. Especially numerous were long-necked, ponderous dinosaurs that lived a semiaquatic existence. Other dinosaurs, like the herbivorous *Stegosaurus*, were protected all along the back by bony plates. Protection was needed, for some of the Jurassic dinosaurs were carnivorous types that strode about on their hind legs. In the sea, the ichthyosaurs had become highly adapted for marine life; their streamlined form suggests a porpoise although there is no relationship. An occasional ichthyosaur fossil is that of a female with fully developed young inside her. Evidently these marine reptiles had evolved the live-bearing condition. And in the Jurassic another group of reptiles took to the sea; these were the plesiosaurs. They were stout-bodied, paddle-limbed, and long-necked.

The first lizards appeared in the Jurassic, and the first true crocodilians. One group of the latter joined the ichthyosaurs and plesiosaurs in the sea. Flying reptiles, the pterosaurs, were abundant. They were not on the line of bird ancestry; birds developed in the Jurassic from quite a different reptilian line.

Perhaps the most interesting fossil of the Period is *Archaeopteryx*, the reptile-bird intermediate. It looks much like one of the small, swift-running, bipedal dinosaurs, complete to long tail and sharp teeth; but the body is densely feathered, the tail supports a series of vanelike feathers, and the dinosaurian forelimb (with three long-clawed digits) bears wing feathers. We are fortunate that *Archaeopteryx* was fossilized in sediments that preserved a detailed imprint of the feathers; the bones alone suggest little more than a specialized type of small dinosaur. (It is strange to think that the dainty mockingbird of the garden is a cousin of the dinosaurs; only its scaly legs and its slender, long-clawed toes are suggestive of its reptilian ancestry.)

Mammal-like reptiles were still unimportant in the Jurassic; but they were continuing to evolve, and some types of that Period were already more mammalian than reptilian.

The Jurassic is a convenient Period in which to interrupt the chronological account, in order to make a point that is of considerable importance in biogeography. Time and again, some particular group will appear, flourish and diversify, then decline and perhaps die away completely. Extinctions are especially numerous during times of marked geologic and climatic changes. But conditions unfavorable to one group may permit another group to rise and spread. Decline or total extinction is but the dull side of a coin whose bright and shiny face is the rise of some better-adapted group to a position of dominance.

30 *"Till the slow sea rise and the sheer cliff crumble / Till terrace and meadow the deep gulfs drink."** **THE LAST 181,000,000 YEARS**

The Jurassic is also a convenient time at which to pause and ask a question that is of biogeographic interest. We have mentioned the existence of land organisms as early as the Silurian; but just where might the land have been located?

It has recently been learned that the continents are underlain by granite, the ocean beds by a different rock called basalt. The basalt dips beneath the granite of the continents, which latter might be thought of as granite chunks floating atop the heavier basalt. The arrangement suggests that continental material was segregated from other material at a very remote time, presumably when the earth itself was in a formative stage. But we must not assume that continental material was always divided into the same continental masses that exist today. Much attention has been given to the possibility that there once existed a single continental mass, a super-continent dubbed Pangaea; that this mass broke into pieces which drifted apart and to their present locations.

The notion of drifting continents is not new. It was first proposed in 1857 by Richard Owen, then a professor at the University of Nashville (not Sir Richard Owen, of British fame in paleontology). In 1915 the German meteorologist Alfred Wegener devoted much attention to the subject, and his name is commonly associated with the theory of continental drift; but more recent proponents of the theory have put forth much evidence that was not available to

*The Theory of Continental Drift. Some geologists believe all conti-
nents orginally formed one land mass. This map shows postulated
former relationships of major land masses in an early stage of the
drift. Note the former clustering of land masses in the southern
hemisphere, the separation of India from Asia, and the existence of
a seaway between Eurasia and Africa.*

Wegener. Especially welcome was the demonstration that convection currents,
deep in the earth's molten interior, might have the force actually to rip continents
and push them about.

Take a globe of the earth, or even a map, and imagine that the continents
are movable over the face thereof. If Africa, the Saudi Arabian peninsula, and
Eurasia were shoved together, they would fit rather well. If the Americas were
pushed eastward across the Atlantic, the eastern coast of South America would
fit nicely into the western coast of Africa. Europe, Greenland, and North America could also be made to coincide like so many pieces of a jigsaw puzzle.
Madagascar could be pushed westward against the Mozambique coast. New
Zealand would fit against the east coast of Australia, Tasmania against the southern coast, and so on. Most of the land masses of the world, including Antarctica,
could all be fitted together, and in such a way that many mountainous uplifts

and other geologic formations, now widely separated, would align themselves neatly. For example, areas along the coast of western Europe and the British Isles, where mountain ranges once existed, closely match the coastal ends of ancient mountain chains in northeastern North America; the Paleozoic and early Mesozoic rock sequences of southern South America are strikingly like those of southern Africa in composition, character, and grain size. Proponents of the drift theory see no reason why coastlines and geologic formations should correspond across wide seas, unless the land masses had once been united.

Interestingly, the rocks and formations of India ally it not to the remainder of Asia but to the southern continents, and India is held to have drifted against Asia from a location much farther south. Proponents of continental drift also point to the Great Rift Valley which scars Africa from north to south. Seemingly, eastern Africa came close to being split off from the remainder of that continent. In Permian times, glacial ice spread from centers in the southern continents and India, a curious circumstance indeed unless these areas were at that time much nearer the pole. And apparently the same lands were again glaciated in the Devonian.

Certain magnetic minerals, present in many kinds of rocks, are aligned in a fashion that reflects the orientation of the earth's magnetic field at the time the rocks were formed. The magnetic poles of the earth have not always held the same position, but some of their wanderings have been charted. Yet the rocks of Europe and contemporaneous rocks of North America are not quite properly aligned; the discrepancy could be accounted for if North America had been gradually drifting to the west, away from Europe. Alignment in India could be accounted for only if that region once lay thousands of miles farther south—a position that would explain its glaciations.

The continental drift theory should stand or fall on geological and geophysical evidence, but the biogeographer can at least comment.

Acceptance of the theory would clarify many problems of biogeography, such as the existence of an amphi-Atlantic distribution not only among plants but also among some animals; the presence of chelid turtles in South America and Australia; the presence of leiopelmid frogs and a beak-head reptile on far-isolated New Zealand; and the former existence of horned turtles in several lands of the southern hemisphere but not of the northern. Especially would the distribution of many long-extinct groups be more intelligible if continents had once been closer to each other. Various dinosaur types attained a surprisingly wide range in the Jurassic. Thus *Brachiosaurus*, greatest of all the dinosaurs, is known from Africa and North America; its smaller relative, *Bothriospondylus*, from Europe and Madagascar. Much earlier, in the Permian, the fernlike *Glossopteris* and several associated plants lived in South America, Africa, India, Australia, and Antarctica, but have left no trace of their presence in more northern lands. The distribution of the "*Glossopteris* flora" bolsters the notion that India and the southern continents had all been connected. (It is now known that *Glossopteris* did not reproduce by wind-blown spores but by a sizable, seedlike structure.) In the latter Pennsylvanian or early Permian, small reptiles called mesosaurs inhabited South America and southern Africa, but left no sign of existence in any northern continent.

In 1937, Alexander L. Du Toit suggested that there had been two primordial continents, separated by a seaway. One of these continents lay in the southern hemisphere and was eventually split into Australia, New Zealand, South America, Antarctica, and Africa, plus India which drifted northward. The other, in the northern hemisphere, split into North America proper, Greenland, and Eurasia (minus India). Time will soon tell whether Du Toit's view is the correct one, or Wegener's, or neither; for the possibility of continental drift is currently being investigated through a variety of scientific techniques.

Our chronological account was interrupted in the Jurassic by a discussion of continental drift, because the Jurassic is the Period in which the drift is thought by many students to have taken place. It may be significant that living organisms, distributed in such a way as to suggest continental drift, usually belong to groups that are probably or certainly of Jurassic age if not older. For example, the living lungfishes, thoroughly confined to fresh water, inhabit South America, Africa, and Australia. Their group dates back to the Devonian, and the Australian species belongs to a family that is known from the Triassic. The tuatara, New Zealand's beak-head, represents a family that had arisen at least by early Triassic times, and types very similar to the modern survivor were in existence by the Jurassic. The New Zealand frogs, and their one relative in western North America, represent a distinctive group within the jumping amphibians, a group whose other known members are of Jurassic age.

Most vertebrates that live today belong to groups that arose after the Jurassic, and zoogeographers have therefore concerned themselves but little with problems of continental drift. But the higher plants average much older than the

Mouth-plates of an ancient ray in a block of limestone from Marion County, Florida. Both fish and rock attest to the former presence of the sea in that area. (W. T. Neill)

higher animals, and phytogeographers have often accepted the drift theory with gratitude. Not only would drift explain some minor puzzles of plant geography (for example, the amphi-Atlantic and some "Tropical" patterns of discontinuous distribution); it would also throw considerable light on two major problems. The respective floras of Australia and the Cape Kingdom are far more alike than would be expected from the great distance that separates the two regions; the respective floras of Australia and New Guinea are far more different than would be anticipated from the proximity of the two, even granting the climatic differences. Both these problems would be solved if Australia today were closer to New Guinea and farther from Africa than it had been in the past. And of course the existence of an Antarctic floral Kingdom, embracing New Zealand, southern South America, and various scattered islands, is less surprising if its geographic components had once been closer to each other.

Now back to the narrative. The Jurassic ended 135 million years ago, when the Cretaceous Period began. Cretaceous climate was mild on the whole. During at least a part of the Period, Bering Bridge stood above water, linking Asia and North America; and the climate did not restrict the use of the bridge only to cold-hardy organisms. But in this Period, shallow seas began to encroach upon the land. One such split North America in two; it extended from the Gulf of Mexico northward through the present Great Plains, and on through western Canada to the Arctic Ocean. Another arm of this sea isolated western North America from the lands that lay farther south.

The Cretaceous was notable for the diversification and spread of the flowering plants. By the middle of the Cretaceous there were trees, shrubs, and herbs very similar to ones that live today. Some of the trees and shrubs were deciduous. Dinosaurs continued to dominate the land, but the great, long-necked, semi-aquatic ones were declining in North America. The stegosaurs of the Jurassic had vanished, to be replaced by flat-headed, spine-armored types. There was a diversification of the dinosaurs that walked on their hind legs; some were slender runners, rather ostrich-like in build, while others were heavily muscled predators. One of the latter, *Tyrannosaurus,* was the most fearsome carnivore that ever walked the earth; it was 47 feet in length, and its massive jaws were armed with serrated teeth, each six inches long. It lived in the western United States and Canada.

In the Cretaceous sea, the plesiosaurs still flourished, but the ichthyosaurs were declining and the marine crocodiles had vanished. Several lines of sea turtles developed, and one group of lizards also took to the salt water. These latter, called mosasaurs, evolved into elongate, fierce-looking predators.

Beak-heads had vanished from most parts of the world, and so had many of the flying reptiles.

Cretaceous mammals were few, primitive, and insignificant. (The Powder River fossil beds of Wyoming, discussed in Chapter 9, are of late Cretaceous age.) Of the mammals, insectivores held greatest promise; they were the first placentals. Birds had lost such reptilian characters as teeth and a long, bony tail; some of the Cretaceous birds would attract little attention in a modern aviary. Salamanders first appear in this Period. The bony fishes were diversifying in both fresh water and salt, and had already come to resemble many modern types.

The Cretaceous was the last Period of the Mesozoic Era. Like preceding Eras, the Mesozoic terminated with geologic and climatic changes that brought about the extinction of many groups, both plant and animal. The dinosaurs were among the groups that did not survive past the Cretaceous. Their extinction has captured the popular fancy, but was not especially remarkable. Some of them (not all by any means) reached a great size, and were variously possessed of long teeth, powerful claws, sharp horns, and spiny ornamentation. Man is impressed by size, teeth, claws, horns, and spines; and so he places undue emphasis on the disappearance of the dinosaurs. At the same time that these reptiles vanished from the land, ammonites vanished from the sea; yet little do we hear of ammonites, widespread but unspectacular mollusks. The disappearance of the dinosaurs is neither more nor less significant than the disappearance, at one time or another, of trilobites, early armored fishes, sea scorpions, scale trees, ichthyosaurs, plesiosaurs, sea crocodiles, flying reptiles, mosasaurs, or hundreds of other animal and plant groups that we have not mentioned.

The Cretaceous Period, and so the Mesozoic Era, ended 63 million years ago, to be succeeded by the Cenozoic Era. The isolation of Australia from the rest of the world probably came about in the late Mesozoic; the isolation of South America from North America took place in fairly early Cenozoic times.

Between the Mesozoic and the Cenozoic there was a long Period of geologic and climatic disturbance. Called the Paleocene, and of five million years duration, this interval is commonly included with the Cenozoic. It was marked by widespread erosion; and in consequence, Paleocene fossils are comparatively few, and are often badly worn or broken. The world's best hunting ground for Paleocene fossils is a broad stretch of western North America, from New Mexico northward through Colorado, Utah, Wyoming, Montana, and Alberta; for in this region the valleys and plains were covered deeply by material washed down from the Rockies.

The most significant biological development of the Paleocene was the diversification of the placental mammals, from a Cretaceous insectivore stock. In the Paleocene were established trends that would eventually lead to the carnivores, and the hoofed animals, and the rabbits. There were even Paleocene primates, standing midway between insectivore and lemur. Several of these earliest primates had chisel-like front teeth, very suggestive of rodents.

About 58 million years ago the earth quieted down again, and the Cenozoic proper got under way. The first four Periods of the Cenozoic (excluding the Paleocene) are collectively called the Tertiary. Previous chapters have shown the Tertiary to have been, in overall view, a time of gradually deteriorating climate; but the first Period of the Tertiary, the Eocene, was mild, warmer in fact than the preceding Paleocene. The Cretaceous sea had fallen back from North America, but many areas now land were still under water. Among such areas were the present Coastal Plain, the lowlands of the Mississippi Valley, and much of California. Land was continuous from western United States southward through what are now the Mexican highlands.

In North America the shores of the sea were bordered with tropical mangrove swamp, and tropical forest extended northward to what is now southern

Virginia, southern Illinois, southern Kansas, and southern Nevada. Subtropical forest, somewhat like that now found in parts of the Florida peninsula, covered much of the United States, and extended northward along the Pacific coast to southern Alaska. Figs, laurels, avocados, cinnamons, and Magnolias grew in Oregon and Washington where the Cascades would later rise, and in Nevada where the Great Basin would later develop. Marine animals of tropical affinity left their remains in coastal sediments of Washington.

Breadfruits, laurels, figs, and Magnolias grew in Wyoming, mixed with less tropical redwoods, hickories, maples, and oaks; and the lakes of Wyoming harbored alligators. Evidently Wyoming lay near the boundary between a tropical and a warm-temperate environment. In the east, Vermont occupied a similar position; and tropical marine organisms swam along the New Jersey coast. Palms and cycads reached British Columbia and the islands of southeastern Alaska. A temperate forest—the Arcto-Tertiary forest of which we have spoken—blanketed most of Canada, Alaska, and the southern half of Greenland, as well as northern Eurasia; it included dawn redwoods, maples, beeches, sycamores, ashes, alders, and basswoods. (The fossil beds of Disko Island, mentioned in Chapter 17, are of Eocene age.) Spruces, pines, willows, hazels, and birches grew still farther north, as far as land extended. There was no polar ice cap of any magnitude, and no Greenland glacier.

In the Eocene there appeared mammals that were ancestral to the even-toed herbivores such as deer and cattle; others that foreran the odd-toed herbivores such as horses, rhinos, and hippos. One mammalian group took to the sea, where they eventually became whales and porpoises; another sea-faring group became the manatees. Mammals also took to the air in the Eocene, forerunning the bats.

It must not be inferred that the entire Eocene assemblage existed on any single continent. The Eocene ancestors of the elephants were African; of the tapirs, Asian; of the horses, North American; of the armadillos, South American. Many groups were not destined to pass between continents until later Periods (if at all). But evidently there was a great deal of movement across Bering Bridge, for Eurasia and North America had many faunal elements in common during the Eocene.

Even back in Cretaceous times the Rockies were beginning slowly to rise; but by the end of the Eocene, 36 million years ago, they still had attained no great height, and cast no significant rain shadow. The Great Plains region, and other North American areas exposed by the retreat of the Cretaceous sea, developed extensive swamps and numerous stream drainages, in which lived turtles, alligators, and a variety of fishes.

In the next Period, the Oligocene, the Rockies continued to rise. Climate became cooler. Tropical forest fell back; in North America it reached northward only to central Mexico. Subtropical forest reached the southern half of the United States (except the Coastal Plain, much of which was still under water). Nevada and Oregon no longer supported figs, cinnamons, avocados, and Magnolias; instead there were dawn redwoods, alders, maples, pepperwoods, dogwoods, tan oaks, hazels, and sycamores. The Rockies were casting a rain shadow, and east of them the first grasslands and open woodlands were beginning to develop. In the sea the reef-building corals, characteristic of warm water, reached the coast of Labrador.

During this period, the Tethys Sea began to vanish. The Alps and the Himalayas arose. Seas finally disappeared from the Ural region, and Eurasia attained roughly its modern outline. The Bering Bridge probably was submerged; for the Oligocene saw comparatively little interchange of animals between Eurasia and North America. South American mammals, both marsupial and placental, were evolving in isolation. The characteristic South American rodent stocks—the ones culminating in porcupines, agoutis, pacas, and capybaras—appear in the Oligocene.

This period, ending 25 million years ago, passed into the Miocene Period. Western and central North America were becoming drier, and extensive grasslands were forming. The Miocene saw a temporary reversal of the trend toward a cooler climate. Palmettos, avocados, mahogany, and lancewood pushed northward to Oregon and Washington. Drought-resistant shrubs entered southern California and Nevada from northern Mexico. Tropical fishes swam off the California coast. But toward the end of the Miocene the cooling trend was resumed. In the Great Plains, grassland was spreading at the expense of forest; lakes, ponds, bogs, and swamps were becoming progressively fewer; aquatic and swamp-dwelling animals were vanishing from the region. In contrast, eastern North America was well watered, and covered with subtropical forest as far north as Virginia.

In the Miocene, numerous animals entered North America from Asia, across Bering Bridge which again stood above water. The proboscideans or elephant-like mammals, appearing first in the Eocene of Africa, and reaching Eurasia in the early Miocene, invaded North America across the Bridge in the late Miocene. Toward the end of this Period, or perhaps early in the next, the Isthmus of Panama appeared above water, permitting faunal interchange between North and South America. The retreating sea still had not uncovered the lower Coastal Plain of the southeastern United States; but a sizable island appeared not far offshore, and soon this became connected with the mainland to form a peninsula, the forerunner of Florida.

The Miocene ended 13 million years ago, to be followed by the Pliocene Period. During the Pliocene a great many aquatic and swamp-dwelling animals disappeared from the central and western United States. Grassland was widely established over the Great Plains; and among herbivorous mammals some large, swift-running grazers had evolved from smaller browsers of the early Tertiary. The Cascades–Sierra Nevada chain was rising west of the Rockies, leaving a dry basin between the two great uplifts. Eastern North America was still forested, and also continued to provide abundant habitats for aquatic and semiaquatic organisms. Most of the Coastal Plain had been uncovered by the sea, and the Florida peninsula had enlarged although not to its present size.

Each Period of the Cenozoic had seen the disappearance of species, genera, even whole families of animals. By the Pliocene the fauna of North America and Eurasia had attained an essentially modern aspect. There would be more extinctions at the end of the Pliocene and in even later times; but most of the Pliocene animals belonged at least to families that survive today. (This is not to say that the animal groups were distributed in the Pliocene just as they are today. In that Period, for example, North America had elephants and rhinos but no bison or wapiti—just the reverse of the present situation.)

There was some movement over Bering Bridge in the Pliocene, but during most of that Period the bridge was submerged; and at best it offered a cold-temperate environment, which prohibited its use by animals intolerant of cold climate. For instance, the giraffes and gazelles spread from Asia into Africa but did not reach the New World. South American animals were moving northward; by mid-Pliocene a ground sloth had reached North America, and by late Pliocene a tortoise armadillo. Parts of the old South American mammal fauna were vanishing, but others were thriving. A South American marsupial stock evolved into a great, saber-toothed carnivore, superficially resembling the saber-toothed cats that had been developing in northern continents. But as noted in Chapter 5, much of the old South American fauna was slated for extinction as North American invaders became established. By late Pliocene times, the cats, saber-tooths, bears, raccoons, horses, peccaries, and deer coming from North America had pushed southward to Patagonia.

The Pliocene was the last Period of the Tertiary but not of the Cenozoic Era. The Pliocene ended about a million years ago, when vast glaciers formed in the high latitudes of both the northern and the southern hemispheres, and lesser glaciers on many high mountain peaks. The new Period, ushered in by glaciation, is called the Pleistocene. This is the glacial age whose biogeographic effects were mentioned in earlier chapters.

In the northern hemisphere the ice sheet reached far down into Europe and the northern United States. In the latter country, the southern edge of the glacier followed roughly a line from New York City westward through Pennsylvania to the Ohio River, then down this river, and thence westward in the Missouri River basin to west-central Montana. From this last region the line ran westward, paralleling the Canadian border, to the Pacific, with small lobes of ice extending into the northern Rockies and the Cascades. Farther south in the mountains of western North America, separate glaciers formed on peaks as far south as Arizona and south-central California. Ice caps also developed on some higher peaks of Mexico and Central America.

In Europe the ice came down roughly to Stalingrad, Kharkov, Dresden, Frankfurt, Bonn, Brussels, and London. A large, separate glacier covered the Alps, and another the Pyrenees. Lesser glaciers formed on dozens of peaks from Portugal eastward through Corsica and Italy to Greece and Romania. Glaciation extended even to a few peaks of the Atlas uplift in northern Africa. The Grand Atlas of Morocco, especially, developed a sizable ice cap.

The ice sheet that covered much of Europe also reached into northwestern Asia, as far as the tip of the Taimyr Peninsula. For reasons to be given later, in northeastern Asia the glacial ice was confined chiefly to the mountains. In southern Asia, long and narrow glaciers capped the various mountain ranges as far south as the Hindu Kush and the Himalayas, with lesser glaciers on isolated peaks as far east as Korea and Japan. There was glaciation in southeastern Australia, Tasmania, and of course New Zealand, which still has ice-cap remnants. The higher peaks of New Guinea developed glaciers, as did Mauna Kea in

Hawaii. It is not known how much of Antarctica was ice-bound before the Pleistocene. At present the Antarctic ice covers 4,400,000 square miles, and is well over a mile in average thickness. In South America, ice covered Tierra del Fuego, southern Chile, and southwestern Argentina. Farther north, more or less interrupted glacial chains followed the Andes. There were isolated ice caps on higher peaks even in the equatorial latitudes. In Africa south of the Sahara, the only glaciers were those formed on the isolated mountains that lie in the eastern part of the country near the equator.

By their very weight the main ice sheet of the northern hemisphere and its Antarctic counterpart actually dented the earth's crust beneath them; yet the glaciation lowered the sea level throughout the world. Normally, moisture evaporates from the ocean into the air, is wafted over the land, condenses and falls as rain or snow, and eventually reaches rivers which carry it back to the sea. But in a glacial age a vast and increasing amount of moisture is being frozen into

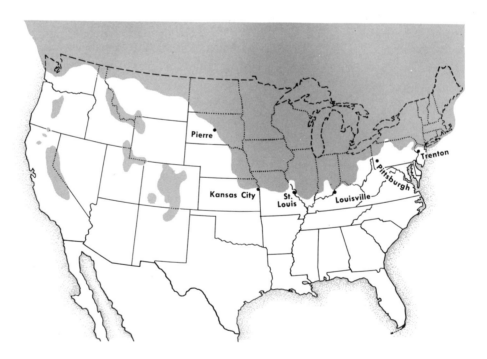

Glaciation in the United States. The maximum extent of glacial ice in the United States during the Pleistocene.

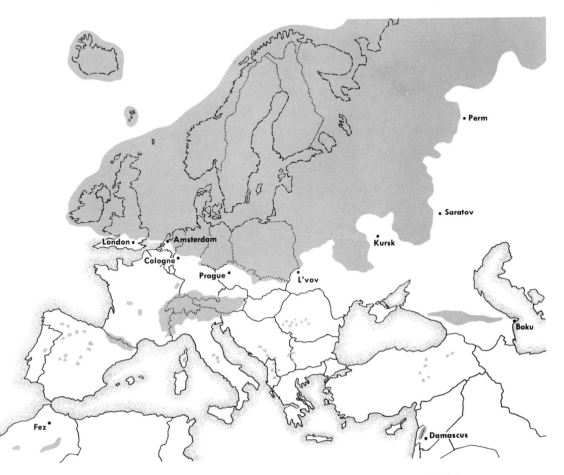

Glaciation in Europe and Northern Africa. The maximum extent of Pleistocene glaciation in Europe and northern Africa. Note the major ice sheet covering northern Europe; lesser sheets in the Alps, Pyrenees, and Caucasus; small, mountaintop glaciers in lands bordering the Mediterranean.

the ice sheet. Thus as the first Pleistocene glaciers grew and spread, sea level dropped by more than three hundred feet.

At a peak of glaciation the Life Zones of North America were telescoped to the south. A narrow belt of tundra bordered the ice, and south of the tundra lay a wide belt of coniferous forest. Remains of spruce, fir, larch, arborvitae, and northern pines have been found in fossil beds as far south as the Gulf States and northern Mexico. And in western Europe, other trees of the boreal forest have been found fossil along the shores of the Mediterranean.

Although the effects of glacial chilling were not strongly marked in the low latitudes of the world, the effects of a falling sea level were very noticeable. Various islands of the West Indies became interconnected, also becoming more alike faunally. The same thing happened in the East Indies, the Philippines, and other archipelagos. Some offshore islands became connected with a nearby mainland.

At last the climate became warmer. The ice sheet halted its spread and began to melt. It melted away almost completely, leaving far less ice in the world than there is today. Sea level rose a good three hundred feet above the present stand, inundating many islands and lowlands. Some animals and plants pushed northward again. Precipitation decreased; western organisms pushed eastward along newly formed tongues of grassland.

Then climatic trends reversed again. The high latitudes grew colder, rainier; for a second time great ice sheets formed. Sea level fell, but not as far as before. Then came another reversal; the ice melted, but not as thoroughly as before. Then came a third glaciation, followed by a third warming period; and finally a fourth glaciation. This last encroachment of ice, which brought glaciers southward as far as Wisconsin, came to an end only some eleven or twelve thousand years ago. Since then the overall trend (with some brief reversals) has been toward warming, a melting of the ice, and a rising of the sea level.

During the Pleistocene a few animals reached North America from Asia. Most important of these were the moose, wapiti, caribou, white-tailed deer, bison, mountain sheep, and mountain goat. Movement across Bering Bridge was facilitated by the absence of glacial ice from much of northwestern Alaska and northeastern Siberia, even during peaks of glaciation. Ice sheets did not cover this region because glaciers, being but compacted snow, cannot form where precipitation is scanty. Cold-hardy animals were able to inhabit northeastern Siberia during at least a part of the Pleistocene, and so were in a favorable position to reach North America whenever Bering Bridge was above water and climatically tolerable.

Through a curious chain of circumstances, we know something of the environmental conditions that once prevailed near the Asiatic terminus of Bering Bridge. In the year 1900 a Lamut tribesman wandered into Sredne Kolymsk, in northeastern Siberia, with an ivory tusk he had chopped from a "great, hairy devil" that lay dead in a crevasse near the Beresovka River. Two scientists, O. F. Herz and E. V. Pfizenmeyer, visited the locality the next year, and found the frozen remains of a mammoth elephant. Obviously the animal had fallen into the crevasse; its pelvis and one leg bone were broken, and its mouth was full of unchewed food. Its struggles had brought an avalanche of snow down on it, and its position showed that it had died in a piteous effort to free itself. Preserved for millennia in nature's deep freeze, the Pleistocene beast was in good condition; its mouth and stomach contents could readily be identified. The mammoth had fed upon the cones and branches of larch and pine; upon sedges, mosses, wild thyme, Alpine poppy, buttercups, and a variety of grasses. The plants suggest that the shaggy elephant was living near the taiga-tundra border, under climatic conditions duplicated in northeastern Siberia today. The animal had eaten many plants that were in fruit, and so we may be sure that it died in the autumn—a season when the crevasses are often masked by new snow.

In the Pleistocene many mammal stocks came to reach an enormous size. Deer, bison, felines, sabertooths, bears, elephants, rhinos, camels, ground sloths, tortoise armadillos—all produced Pleistocene giants. There was a giant beaver in Eurasia and North America. During this Period the Australian marsupials produced an enormous kangaroo, a ponderous wombat, and a lion-sized carnivore. Even the duck-billed platypus developed an oversized offshoot in the Pleistocene. On Madagascar a lemur grew as large as some of the great apes.

Most of the present-day mammals were living in the Pleistocene along with the giants. Some paleontologists hold the Pleistocene Period to have ended, and a Recent Period to have begun, when the giant mammals became extinct. In general, they survived the first three glaciations but vanished during or soon after the fourth—at the same time that man, then a primitive hunter, began to spread over the world. This does not necessarily mean that Pleistocene animals were exterminated by man, any more than dinosaurs were exterminated by their successors the marsupials and insectivores. But it is clear that early man, in both the New World and the Old, preyed upon the giant beasts. Even mammoths and mastodons were felled by spears and stones.

It is also clear that a major fraction of the earth's biota is destined for rapid extinction as man multiplies and converts all natural resources to his own needs. In broad view the spread of man, beginning about the time the last ice sheet began to retreat, heralded not only the end of the Pleistocene Period but also of the Cenozoic Era. We are well started on a new Era, the Psychozoic, when the survival of any animal or plant will depend chiefly upon the whims of man.

In the meanwhile, climate is still warming. In the United States the mean average temperature has risen three and a half degrees Fahrenheit in 40 years, the rise being due especially to the amelioration of winter temperatures. In Spitsbergen the mean winter temperature has risen 15 degrees in the last half-century; the seaports, once ice free for only three months, are now open for seven. In Alaska the melting back of the glaciers has been evident for 200 years; one glacier near Juneau retreated two miles in ten years. Trees are advancing into the tundra in Alaska, Quebec, Labrador, and Siberia. In the Canadian prairie the agricultural crop line has shifted as much as 100 miles northward, and the growing season has been lengthened by ten days. In northern New England and eastern Canada the cold-adapted birch trees are beginning to die off; the spruces and firs are beginning to suffer from the higher summer temperatures. In Sweden the timberline has moved 60 feet up the mountains since 1930.

The cardinal, the turkey vulture, the tufted titmouse, the blue-winged warbler, the opossum, the armadillo—all have recently spread much farther north in the United States. In the last 40 years or so, 25 species of birds have pushed into Greenland from the south. Codfish moved into Greenland waters as the seals moved out; and Greenland Eskimos, once seal hunters, have become cod fishermen. Tuna have pushed north into New England waters, and flying fishes may be seen off New Jersey. A few areas of the world have not warmed, perhaps because of shifts in the ocean currents; but the general picture is one of climatic amelioration and the poleward spread of many organisms.

31 *"A worm is as good a traveler as a grasshopper or a cricket, and a much wiser settler."** **FACTORS INFLU-ENCING DISTRIBUTION**

Biogeography seeks to explain why any organism, or any taxonomic group, is distributed in the way we find it to be. Previous sections of the book reveal distribution to have been affected by a complex interaction of factors, and it is now time to examine some of these factors separately.

Obviously the distribution of a species or group is influenced by its time and place of origin. Thus the placental mammals evolved in the northern continents at a time after Australia had been isolated. Therefore the only placental mammals to reach Australia were the ones that were adapted for swimming, flying, or drifting on rafts of vegetation. As another illustration, the rattlesnakes evolved on and dispersed from the Mexican highlands, in comparatively late times; within their whole span of existence there may never have been a period when Bering Bridge stood above water and also offered an environment suitable for rattlers. Thus these snakes are lacking from Asia. Distribution is affected by what may be called evolutionary factors.

Organisms react to the environment in different ways, and a barrier to the spread of one species may be a broad highway for the dispersal of another. As an obvious illustration, the Isthmus of Panama permits movement of tropical land animals between North and South America, but prohibits movement of tropical marine animals between the Caribbean and the Pacific. Or one species may have

* HENRY DAVID THOREAU

its spread blocked by a mountain range, while another species, finding a suitable habitat at some particular elevation, may follow the range for a thousand miles. In short, distribution is affected by what may be called innate factors—the anatomical, physiological, and behavioral specializations of the organism.

Innate factors merit further discussion. The mammals and birds differ from the reptiles and amphibians in having an internal mechanism to maintain the body at a given temperature. Thus mammals and birds can, on the average, withstand cold better than reptiles and amphibians. This does not mean that all mammals and birds are cold tolerant, for some cannot abide the slightest chilling; but it does mean that a good many mammals and birds can live in the high Arctic, while no reptile or amphibian can do so. In fact only a small percentage of reptiles and amphibians can range into taiga.

Among mammals, the placental method of reproduction usually conveys an advantage over the marsupial. Whenever placental mammals have been introduced into Australia, the marsupial fauna has suffered. (Yet an occasional marsupial can withstand placental competition; the best example is the opossum, a common animal in North America. It has thrived probably because it is unspecialized; it can live in high country or low, forage on the ground or in trees, and eat almost anything from sweet corn to a dead mule. But the significant point is the very limited number of marsupial species that can survive where placentals are diverse.)

The hoofed, herbivorous mammals fall into two main types as regards distribution. They may be grazers, feeding mainly upon grass; or they may be browsers, eating other herbaceous vegetation and the leaves of shrubs and trees. The distinction is more important than might be guessed. Most grasses contain some silica, a very hard substance. (Quartz is a form of silica.) Exclusive subsistence upon grasses always requires great modification of tooth structure and growth. The sharply cusped teeth of a deer typify browsing dentition; they are very unlike the ridged, flat-crowned teeth of the grazing horse. A browsing animal, forced to subsist upon grasses, would soon wear its teeth down to the gums. Grazers and browsers can live together in some areas, where grassland is dotted with trees and shrubs; but in general, forests are inhabited by browsers and grasslands by grazers.

Birds, being fliers in most cases, have been able to reach remote islands. But land birds are mostly rather specialized in their requirement for life, and only a small minority of the species have attained very wide ranges. Birds are generally quite limited in diet and foraging grounds, as well as to nesting site and nesting material. They do not disperse across barriers as well as one might think. On the main land masses, birds are distributed in a way that reflects the components of the modern environment and of the different environments of the past.

Many birds of temperate and arctic lands migrate at the onset of winter, covering a vast area in their migratory flights. However, biogeography is mainly concerned with the breeding range of a migratory bird. One might think the migratory habit to favor extension of a bird's breeding range; but it must be remembered that the migratory bird has an impulse to return to its starting point, not to take up residence in some land that it flies through.

Reptiles have no internal mechanism for maintaining a constant body temperature; but in spite of popular belief they are not "cold-blooded," not usually cold to human touch. Reptiles must take on the temperature of their environment; but snakes and lizards (which make up 95 per cent of living reptiles) maintain a surprisingly high and a remarkably constant body temperature by restricting activity to places where, and times when, a certain temperature prevails. Most snakes and lizards of the United States are active at temperatures between 65 and 85 degrees Fahrenheit; hardly any will venture abroad if the body temperature falls below 60 degrees, or rises above 108. A temperature above 97 degrees is fatal to many reptiles, especially snakes, which on the average are less heat tolerant than lizards. Both snakes and lizards take their temperature mostly from the substratum on which they rest, not the circumambient air; a lizard basking on a rock may raise its body temperature to 95 although the air may be no more than 50. Nevertheless, temperature is a major factor limiting the distribution of reptiles.

Most reptiles deposit eggs, which are covered with a hard or leathery shell. Reptile eggs are somewhat resistant to desiccation; at least they do not have to be laid in water, but instead are deposited in the ground, beneath rocks or other ground debris, in humus, or in the pulp of rotting logs. A minority of reptile species have become live-bearers, and these are the ones that establish records for distribution into high altitudes and latitudes, where ground temperatures would endanger an egg clutch. In North America, for example, the harmless garter snakes and the venomous rattlers, both groups live-bearing, range farther north into Canada and higher into the Rockies than any other snakes. Some groups

Ability to fly does not necessarily result in a wide distribution. The bird family of ibises and spoonbills includes 20 genera, only one represented in both the Old World and the New. Shown is the Hagedash ibis, confined to Africa. (South African Tourist Corporation)

of snakes are represented in Asia by both egg-laying and live-bearing species, but in the New World only by the latter; these are groups that passed from Asia into North America at a time when Bering Bridge had grown fairly cold, cold enough to filter out most of the egg-layers.

The amphibians—the caecilians, frogs, salamanders, and sirenids—take their body temperature from their surroundings; and in addition, most of them rapidly lose water through the skin if exposed to a dry environment. Amphibian eggs desiccate easily; in most cases they are laid in the water, and hatch into a larva that for a time lives an aquatic existence before transforming to an adult. Amphibians are therefore most abundant in well-watered lands, and only a few specialized groups have been able to invade the very dry regions or habitats.

Freshwater fishes engaged the attention of early zoogeographers. If a freshwater fish is dropped into salt water, it usually shrivels and dies; if a saltwater fish is dropped into fresh water, it usually bloats and dies. Oceans were therefore presumed to be barriers to the dispersal of freshwater fishes, and the presence of these latter on an island was taken to be evidence of a former mainland connection. But in 1938 George S. Myers, an ichthyologist at Stanford University, divided freshwater fishes into several groups, according to their tolerance for salt.

Some fish families are made up of species that are wholly intolerant of saline waters. Such fishes indeed do not disperse through the sea. They are said to make up the "primary" division of freshwater fishes.

Other families may be composed largely of freshwater species, but also include species that live occasionally or frequently in brackish water. A member of such a family may be confined to an island where it lives only in fresh water; but we may suspect that it, or its ancestor, was able to pass through at least a narrow stretch of sea. These families, known or suspected to have salt tolerance, form the "secondary" division of freshwater fishes. The secondary families rarely cross a wide sea barrier.

Islands that are well separated from any mainland may receive no members of any primary or secondary group. On such islands, aquatic habitats are open for the taking, and they may come to be occupied by some freshwater offshoot of an otherwise marine stock. Such offshoots are said to be "vicarious." A few marine stocks have vicarious representatives that successfully compete with primary fishes in continental streams.

Certain fishes move periodically between the fresh water and the salt, usually

(Facing page) Snakes that lay eggs are restricted to regions where low temperatures will not endanger the clutch. Shown above at right is a hognose snake from Marion County, Florida, and her recently deposited eggs. (Ross Allen's) The coyote, at lower right, has one litter of pups a year; its prey, various rats and rabbits, usually has several litters annually. Fecundity has some bearing on dispersal. (Canadian Government Travel Bureau)

Mammals and birds withstand cold better than do other vertebrates. A baby roe deer, below, photographed in the Grisons of Switzerland, tolerates cold that would kill a reptile or amphibian.

in connection with spawning activity. They may live most of the time in fresh water but go to the sea to spawn; or may live in the sea but push upstream to lay their eggs. Fishes that move into or out of the ocean in order to spawn are said to be "diadromous." This term is derived from Greek rootwords that mean "running through."

Finally, there are "sporadic" fishes that live mainly in the sea, and just occasionally make their way upriver into fresh water.

The vicarious, diadromous, and sporadic fishes can spread across wide seas, and their distribution is of little significance in the biogeography of the fresh waters.

A local fish fauna may include a mixture from several divisions. A stream system in eastern North America might simultaneously harbor primary types, such as basses, sunfishes, darters, minnows, suckers, pikes, and ameiurid catfishes; secondary kinds such as gars and killifishes; the glass minnow, a vicarious offshoot of a marine family; the common eel, a diadromous fish that goes out into the Atlantic to spawn; and a sporadic mullet.

Not all freshwater fishes can be categorized beyond any possibility of argument; but the foregoing scheme merits consideration when the biogeographer puzzles over the existence of an isolated sucker in China, of killifishes in the West Indies, or of characins in the tropics of Africa and the New World.

Turning now to some innate factors affecting the distribution of plants, a distinction should be noted between conifers and broad-leaved trees. Many huge conifers grow only where the climate is cool and moist. But certain other conifers seem to take over where some environmental factor is unusually rigorous, leaving the more favorable situations to the broad-leaved trees. Consider, for example, the situation in eastern North America. Coniferous forests range northward beyond the limits of broadleaf forests; and on cold mountain peaks, conifers grow above the last broad-leaf stands. Pines cover the clay hills of the Piedmont, where the topsoil has eroded away; they cover the flatwoods, which alternate between very wet and very dry. Pines also predominate on sandhills where a porous soil and quick runoff reduce the effectiveness of precipitation; and on tracts that burn over frequently. Pond cypress and pond pine grow in shallow, oft-flooded basins; bald cypress in the river shallows and swamps; junipers on windswept rocky ledges or about limestone outcrops.

As noted in Chapter 26, plants are divided into life-forms on the basis of innate characteristics, all of which have some effect on distribution. The heat requirements of trees have been described in Chapter 18, the water requirements in Chapter 23. Previously discussed topics will not be reviewed; let us turn instead to some plant adaptations that have not as yet been considered.

Many plants have special adaptations for dispersal. In tumbleweeds the whole plant body breaks off near the ground, and, driven by the wind, rolls for long distances and drops seeds along the way. (The name "tumbleweed" is popularly applied to several unrelated plants; common in the central and western United States is the Russian thistle, an imported species.) In the morning campion, blue squill, and many poppies the seeds are shaken from their dry pods by the wind. Many orchids have fine, dustlike seeds that drift in the air. Witch hazel, touch-me-not, Oxalis, and several others have a seed capsule that ruptures abruptly, spraying the seeds. In some plants the fruit is provided with a wing, as in the ashes, or two wings as in the maples, or four wings as in the Terminalias; up to a dozen wings in some other groups. In the Catalpas and the yams the fruit is not winged but the individual seeds are. In Anemones, mountain Avens, and Clematis the fruit is plumed in such a way that it will drift in the air; in milkweeds the fruit is not plumed but the seeds are. Coconut and Nipa palm have fruits that can float for a long while in sea water. Cockleburs and beggar-lice have fruits that will cling in the hairs of a mammal; mistletoe has sticky seeds that adhere to the beak and plumage of a bird. Persimmons, blackberries, and many others have a fruit that is eaten by animals which later evacuate the hard and still viable seeds. Acorns and hickory nuts are stored by birds and rodents; often the stored nuts are not recovered, and sprout where they were hidden. The big, round fruit of the Osage orange frequently rolls to some point far away from the parent plant. Willows, wandering Jew, and some Philodendrons can grow from a sprig torn off by wind or water; life plant can sprout from a mere leaf.

These and other mechanisms for plant dispersal have not evolved primarily to increase the range of the species. Seeds are dispersed chiefly to prevent the seedling from having to grow in competition with, and perhaps in the shade of, its parent. A great majority of seeds or other viable plant parts are not carried very far from the parent plant, and species with the most elaborate dispersal mechanisms are not necessarily the ones that have attained the widest ranges. The chief phytogeographic effect of this slow dispersal, generation by generation, may be to permit invasion of territory opened up by a gradually changing climate.

Some marine fishes have a larval stage that drifts with the current, but otherwise the vertebrates do not have a life history stage that is adapted for passive dispersal by wind, water, gravity, or other animals. However, many vertebrates have an impulse when young to leave the locality where they were hatched or born. Thus, when young spadefoot toads transform from the aquatic tadpole stage, they leave the pond where the larval life was spent; they move overland so far and in such droves that guillible people insist upon "a rain of froglets." When spotted salamanders transform, they are under a compulsion to leave their pond, and in their wanderings they turn up in situations where the settled adults will never be found. Newborn watersnakes are found rambling through dry woods where the adults are never seen. In several familiar garden birds the adults drive off the well-grown young if the latter do not leave voluntarily.

A young animal in its emigration may find no spot where it can settle and thrive, but at least it has a good chance of doing so; whereas if it did not emigrate, it would surely have to compete with adults of its own kind, and with its litter mates. The situation is analogous to that in plants, whereby numerous seeds fall upon unsuitable ground, but there is a chance that some will reach a congenial spot at a fair distance from the parent plant. And in animals, just as in plants, this slow dispersal, generation by generation, helps a species to spread into territory opened to it by environmental changes.

The mortality rate is usually very high among seedling plants and young animals; but once the critical juvenile stage is passed, the organism has a good chance of reaching old age. For example, in an oak forest, thousands of acorns and oak seedlings must succumb each year, but only rarely is a sizable oak tree destroyed by anything but the decay that accompanies old age. Or to take an example from the animal world, in Florida some clutches of alligator eggs die while still in the nest; the little 'gators that do appear, hatching in the late summer or fall, are decimated by the following spring; but the 10 per cent that do survive the first winter are not often imperiled by anything but man. In short the survival of a species in any area may be determined more by the requirements of the young stage than of the adult.

The circumstance is of obvious importance to biogeography: if in any given species the requirements of the young stage are more exacting than those of the adult, then the range of the species will be limited to areas where the needs of the young are met. This conclusion comes as no surprise to the gardener, who knows that he can transplant to his garden many species that he could never raise there from seeds; but biologists in general have devoted insufficient attention to the requirements of young organisms as contrasted with those of the adults.

In addition to evolutionary and innate factors affecting distribution, there are also environmental factors. Configuration of land and sea is important in distribution, as shown in the discussion of the faunal Realms and of continental versus oceanic islands. Also important are the climatic values: temperature, precipitation, wind, and light. We need not review what has already been said about temperature and precipitation.

Wind is important chiefly in modifying temperature, humidity, and the distribution of rainfall. As a force in itself, wind may occasionally limit the distribution of certain plant life-forms; for example, tree growth in the Falkland Islands is inhibited by the frequency of strong winds. Here and there in the world, some plant species are very tender and susceptible to wind damage; they grow only in regions where sheltered situations are available. And as noted in connection with innate factors, some plant fruits or seeds are provided with devices that facilitate dispersal by wind.

Of light as a factor involved with distribution, we have noted only that the reduced illumination in a forest helps to determine what plants can grow there. But light also affects distribution in more subtle ways. What stimulates so many plants to bloom only at a certain time of year? In some cases it might be a warming of the ground, or the arrival of a rainy season; but often it is day length. Numerous plants, especially of the warm regions, will bloom only when the days shorten. (A Poinsettia, growing under a bright street light, may never flower; while in a greenhouse it may be brought into bloom at a desired time by controlling the length of daily illumination.) Such plants could reproduce themselves only where weather conditions are favorable during the short days when the blossoms are forming and setting seed. And in temperate climes, especially, there are plants that begin to bloom only when the days lengthen.

Day length could also affect the latitudinal distribution of animals. For example, most vertebrates do not breed just at any time of year, but instead have a well-defined breeding season. The timing of the reproductive cycle, while partly the result of a physiological rhythm within the animal, is usually keyed to some environmental stimulus, commonly the day length. Many familiar little song-

The long, scaly legs of herons may transport algae, fish eggs, and small aquatic organisms from one pond to another. At left, a mounted specimen of the great blue heron in the Museum of Natural History at Regina, Saskatchewan, Canada. (Canadian Government Travel Bureau) Occasionally a parasite modifies the range of its host. In 1945 a blight killed a million cedar trees on Bermuda. Above are blight-killed cedars on Castle Island, Bermuda. (Bermuda News Bureau)

birds, which proverbially nest in the spring, can be brought into reproductive condition at any time of year, even midwinter, by providing artificial illumination in a way that duplicates the increasing day length of spring. (Poultry farmers have long used artificial illumination in autumn and winter to stimulate unseasonal egg production.) Light as a stimulus to reproduction is probably more important in mammals and birds than in the other vertebrates, which are more strongly influenced by temperature. It is worth noting that when the reproduction of an animal is triggered by light, the organism is responding not to the amount of light received but to the day-to-day change in the amount.

The components of the physical environment are not all climatic; some relate to the soil. Elsewhere we have noted several effects of elevation, and topographic relief, on distribution. Chemical properties of the soil are also important in controlling distribution. Bog plants often require a highly acid soil, and forest plants a fairly acid one. Gardeners may use an especially acid fertilizer around Azaleas, Camellias, hollies, and Magnolias, the cultivated varieties of which are derived from forest-dwelling ancestors. Farmers may add lime to some soils, to reduce acidity and yield a better harvest of certain crops; and the owner of a new home may have trouble in establishing shrubs around the house foundations, where bits of mortar and concrete make the soil highly alkaline rather than acid.

Major constituents of soil are quartz, of which most sand is composed; aluminum silicate, the basis of most clays; calcium carbonate, of which limestone and chalk are composed; and humus, which includes a variety of organic substances. According to the predominating constituent, a soil may be sandy, clayey, calcareous, or organic; and each type is unsuitable for some plant species but ideal for others.

Of course plants take up water from the soil, and such chemicals as nitrogen, sulfur, phosphorus, potassium, calcium, and magnesium. Small quantities of other chemicals, such as iron, zinc, manganese, copper, boron, and molybdenum, may also be necessary for normal plant growth. An individual plant may grow, albeit poorly, where some needed chemical is in short supply; but a species is not likely to thrive in an area unless the soil provides the necessary minerals. A sandy soil is especially apt to be lacking in important chemicals.

Climate and plant cover interact to modify the soil. The process is a slow one; but if three areas have long been covered respectively with steppe, prairie, and deciduous forest, each area will have developed its own type of soil. The spread of plants from one area to the next could be inhibited by areal differences in soil.

Little is known about the effect of soil types on animal distribution. Soil conditions often affect animals indirectly, through the plants. Any animal needs a variety of chemicals, and these—except for atmospheric gases—are received ultimately if not directly from plants. Frequently an animal is restricted to some particular kind of soil, but usually there is a good possibility that the restriction actually reflects the nature of the plant cover. A more direct relation with the soil may characterize some burrowing animals. Occasionally the distribution of some animal group can be linked rather clearly to a soil condition. Thus the snails, which build their shells from lime, are often lacking from acid bogs where no lime is present, and are exceptionally common around limestone outcrops.

The soil is full of microscopic organisms such as bacteria, fungi, and roundworms. These play a highly significant role in modifying the soil and rendering it fit for plant growth.

In a great many plants, such as the pines, oaks, beech, red maple, groundgrowing orchids, Indian pipe, gentians, Azaleas, Rhododendrons, and various heaths, an intimate relationship is established between the plant roots and certain soil fungi. A structure called a mycorrhiza is formed; it is a union, on the microscopic level, of fungus threads and the cells of the host plant. Mycorrhizas (the name means "fungus-root") also develop between fungi and various lower plants. Apparently the fungus makes nitrogen available to its host, and also takes up salts and passes them on. In return the fungus is nourished by carbon compounds from the host. Mycorrhizal fungi require a soil that is acid, sometimes exceedingly so; and the restriction of a plant to acid soil may at times reflect the needs of the fungus. This is thought to be the case with blueberries, Rhododendrons, and Azaleas, for example.

An intimate relationship exists between legumes—members of the bean and pea family—and certain bacteria. These bacteria live in nodules on the legume roots, and they alter the soil in a way that makes nitrogen more readily available to the plants. This type of bacterial activity may add more than 250 pounds of available nitrogen to an acre of ground. Unlike mycorrhizal fungi, these bacteria cannot tolerate a highly acid soil.

The biotic environment obviously affects the distribution of a parasite. Thus a certain tiny crustacean lives only as a parasite on the gills of the brook trout, and so ranges (like its host) from Saskatchewan to Labrador, and southward in the Appalachians; whereas a certain parasitic insect, the crab louse, lives on man and so may be expected just about anywhere. Many parasites must alternate between two hosts. For example, the malaria organism alternates between man and an Anopheles mosquito, existing only where both hosts live in proximity to each other. It might be added that most organisms harbor parasites which spread with their host. For example, a little songbird, flitting about your garden, is actually a "flying zoo." It may well be carrying fungi; amoebas, flagellates, spirochaetes, trypanosomes, and other Protozoa; tongue worms; flukes; roundworms; spiny-headed worms; tapeworms; leeches; bugs, fleas; feather lice; louse flies; ticks; and mites; along with viruses which are on the borderline between living things and lifeless chemicals.

Occasionally a plant or animal is removed from an area by a disease. Best known illustration is the American chestnut, which was overtaken by a blight. Several of the American elms are now threatened by Dutch elm disease. The tsetse fly and a malady it carries may render parts of central Africa unfit for a few native animals; certainly in some areas it prohibits the year-round raising of cattle.

Dietary restrictions obviously limit animal distribution. In an extreme case, the koala or "Australian teddy bear" eats only the leaves of certain Eucalyptus trees, and so cannot range more widely than the trees.

One type of biotic interaction seems potentially important in modifying the distribution of any organism; this is competition between species. Plants compete for growing space, moisture, soil nutrients, the attention of pollinating insects, and the light that triggers photosynthesis. Animals compete for food, shelter, territory, nesting or denning sites and materials, and the occupation of limited environmental situations.

As an illustration of competitive relationships, consider the events involved with the introduction of the English sparrow into the United States. Of European origin, the bird was liberated in America in the spring of 1851, when eight pairs were turned loose in Brooklyn, New York. Other small flocks were turned loose in later years and at other localities. Hardy, extraordinarily fecund, aggressive toward other birds, and tolerant of man's presence, the English sparrow in a few decades became a pest throughout much of the United States.

The native bluebird suffered from the spread of the sparrow. The bluebird, too, was tolerant of man, often living in orchards and farmyards; it nested in tree holes, hollow fence rails, and boxes put up by rural or suburban people. The bluebird migrated southward in the winter; and when it returned north in the spring it found most of the available nesting sites already usurped by the pugnacious, cold-hardy, early-breeding sparrow. In 1895 an unusually severe cold wave killed migrating bluebirds by the thousands all through the central and Gulf states, but the sparrow was not greatly affected.

It seemed as though the bluebird would vanish, its place to be taken by the English sparrow, which was better adapted to the prevailing environmental conditions. But then the environment changed: the automobile replaced the horse. The great flocks of sparrows had been nourished largely by grain spilled from feed sacks or contained in horse droppings. When horse-drawn vehicles became a thing of the past, the sparrows declined; the bluebird, mainly an insect eater, made a fairly good recovery. The sparrow-bluebird story has been somewhat oversimplified but does provide an insight into the workings of a competitive relationship. Such relationships have been of special biogeographic significance in times of marked change in the physical environment; for the changes may have strongly favored one species or group over another, while simultaneously opening corridors of dispersal to loose one biota upon another.

Competitive relationships are so common in nature that we are likely to overlook cooperative ones. Thus a lichen is not a single kind of plant but rather an intimate union of a fungus (supportive and water retentive) with an alga (food-producing). This mutually beneficial relationship has much to do with a lichen's ability to grow where other plants cannot—on a rock face, for example, or in the high Arctic. And to cite another famous "partnership," termites can eat wood only because it is digested for them by a variety of microscopic organisms whose sole "habitat" is the termite digestive tract.

Thus distribution is modified by the interaction of evolutionary, innate, and environmental factors.

What about man as yet another biogeographic factor? The effects of man's activities will be considered in a later chapter. At this point we shall say only that man may change the distributional patterns that have resulted from the interplay of evolutionary, innate, and environmental factors. Nature proposes, man then disposes.

32 *"For all that here on earth we dreadfull hold, / Be but as bugs to*

fearen babes withall, / Comparèd to the creatures in the sea entrall."*

DISTRIBUTION IN THE SEA

It would be intellectually satisfying to partition the sea into something approximating the faunal Realms and Regions. Efforts have been made to do so, beginning with Ludwig Schmarda who in 1853 extended his zoogeographic scheme to both land and sea. However, important aspects of marine distribution are obscured when oceans and continents are embraced within a single conceptual framework, for "land" and "sea" are not equivalent subdivisions of the world. Nearly 71 per cent of the earth's surface is ocean. Land is concentrated in the northern hemisphere with extensions southward; sea is concentrated in the southern hemisphere with extensions northward. Some land masses are separated from each other, but all the seas are connected; a cliché mentions "the Seven Seas" but in broad view there is only the World Ocean.

Most terrestrial organisms live on the surface of the ground. A few burrow, or inhabit underground waters, but never live at any great depth. Flying animals rarely attain any great altitude, and in any event must alight at times; wind-drifting seeds, fern spores, and baby spiders are but life-history stages of organisms that are essentially earthbound. But in contrast, the sea is inhabited from top to bottom, from the sunlit surface to the black depths.

Living things in the sea are exposed to environmental rigors that do not exist on land. Consider an organism that lives in one of the ocean's deep trenches,

---

* EDMUND SPENSER, *Faerie Queene.*

396

say 33,000 feet below the surface. Here the water pressure is 14,700 pounds per square inch—a thousand times greater than air pressure at sea level on land. The organism spends its life in very cold water that bears a burden of dissolved salts and other chemicals. There is no light save the occasional glow of a luminescent animal; the night is endless, and old as the sea itself. There are no green plants to produce food; there is only a "rain" of dead things from the sun-lit waters more than six miles above.

The biota of the sea is quite unlike that of the land. The vertebrates are represented in the sea chiefly by fishes; among the other vertebrate groups only a few mammals and reptiles are thoroughly marine. Invertebrates are abundant and diverse in the sea, which harbors many groups that have few or no representatives on land: lancelets, sea squirts, sea lilies, starfishes, brittle stars, sea cucumbers, acorn worms, mesozoan worms, ribbon worms, sea spiders, horseshoe crabs, sea worms, squids, octopuses, tusk shells, lamp shells, sponges, jellyfishes, corals, sea anemones, comb jellies, and dozens of others known only by their scientific names. On the other hand, invertebrate groups that abound on land or in fresh water may be poorly represented in the sea. Nearly or quite lacking from salt water are the insects, mites, spiders, scorpions, daddy longlegs, millipedes, centipedes, slugs, rotifers, leeches, and tongue worms, among others.

Higher plants have had little success in the salt water; hardly a dozen genera are fully marine. Many of the lower plants are also far better represented on land or in fresh water than in the sea; examples include the fungi, lichens, liver-worts, mosses, and ferns. But the fire algae, brown algae, and red algae—plantlike if not true plants—are largely marine. Some one-celled, plantlike organisms, drifting in the upper waters, carry on the photosynthetic process that yields starches and sugars; they are the chief producers of food in the sea.

In the following discussion of marine biogeography, statements are intended to apply to the entire biota, not just to the higher animals and higher plants.

From a standpoint of physical oceanography, an ocean is a structure with three main parts. From the edge of the land, the water depth increases gradually to about 600 feet; beyond this point the drop-off is more rapid, down to about 10,000 feet; and then there is another drop-off, to the ocean bottom. The three zones are known respectively as the continental shelf, continental slope, and ocean basin. If an ocean is defined as a physiographic feature with its own shelf, slope, and basin, then there are three oceans in the world: the Atlantic, the

Pacific, and the Indian. Other "seas" or "oceans" are but appendages to some one of these.

However, one cannot recognize at the outset three biotic areas corresponding to the three oceans. The continental shelf is a highly distinctive environment, really a world of its own within the sea.

The shelf (which is here taken to stretch from the low tide mark to the upper edge of the continental slope) shares in the vicissitudes of continental history. It is influenced by uplifting or downwarping of a coast, and by the rise or fall of sea level. Materials washed off the land are deposited chiefly on the shelf, although some reach the slope as well. The shelf receives the fresh water that runs off the land, and so the waters over the shelf are less saline than those over the slope and basin. Of course the circulation in the sea tends to distribute the dissolved substances. However, the sea is constantly receiving fresh water all around its periphery, and even a small difference in salinity will make for a difference in biota. This is because life in a salt solution poses problems of physiological adaptation.

Water tends to move in the direction of a stronger salt concentration, and a marine organism must somehow avoid losing its body water to the sea. Many marine invertebrates and some vertebrates get around this problem by maintaining their body fluids at a salinity equaling that of sea water. But an organism must excrete its wastes, and this usually requires a movement of water from the environment into the organism, and then out again with a burden of dissolved wastes. A balance can be struck between the two needs, but the striking of it does not often give the organism much leeway in the degree of salinity that it can tolerate.

A drought on land may greatly reduce the amount of fresh water that pours into the shallows of the sea; and in any event, the normal seasonal fluctuation in precipitation will result in seasonal variation of salinity in the near-shore waters of the sea. Therefore the organisms that cannot tolerate much variation in salinity are excluded from the shelf, which is taken over by the more tolerant groups.

Temperature is another important factor controlling marine distribution. The shelf and its waters are affected by air temperature, to a greater degree than are the continental slope and the ocean basin. Accordingly, the shelf is left to the organisms that can tolerate the variations of water temperature.

Near-shore waters are also likely to be distinctive as regards density, for this

The Mediterranean, virtually tideless, has no intertidal zone. A strong current constantly washes its shores. The seaworn Arbatax cliffs of Sardinia. (Italian State Tourist Office)

characteristic of the water bears a fixed relationship to temperature and salinity. The density of the water has a few biogeographic effects. For example, many tiny organisms of the sea, passive drifters, are protected to some degree by spines; in denser water they can develop larger and heavier spines, and still be light enough to float. But to look at this situation in reverse, a more heavily armed species is limited to denser water. Other organisms have developed special flotation devices, and in denser water these need not be so large or elaborate; but a species with lesser development of flotation devices is restricted to denser water.

Light may penetrate very clear sea water for a depth of 1,000 feet and more; but light sufficient to permit photosynthesis rarely exists much deeper than 500 feet, and very few green plants or plantlike organisms reach this depth. Light may not penetrate to the bottom over all parts of the shelf, for the overlying water often is turbid with dissolved materials from the land; but in general, much of the shelf can support rooted plants, while the plant life of the slope and basin is limited to species that drift in the upper waters.

Light is also necessary for vision, and many animals can see adequately with far less illumination than is required for plant growth. The most keen-eyed of fishes can see at a depth of as much as 1,650 feet. Some minute, shrimplike organisms, which rise to the surface at night and return to the depths in the daytime, can distinguish night from day at a depth of about 2,650 feet. But these are cases of extreme visual acuity, and animals to whom vision is important are concentrated on the shelf. It is worth noting that vision serves other functions than the detection of food, mates, and enemies; in many animals of the sea, the daily cycle of activity and repose, and a seasonal cycle of movement, may be keyed respectively to daily and seasonal changes in illumination.

Pressure increases with depth, and of course the extreme pressures will be attained on the slope and especially the basin, not on the shelf.

Evidently, then, the shelf differs from the rest of the sea in several of the main factors that control marine distribution: the temperature, salinity, density, and pressure of the water; the degree of illumination; and the nature and amount of material washed down from the land. The shelf makes up less than 8 per cent of the sea, but it harbors a large majority of the marine organisms. In a consideration of distribution in the sea, attention should first be given to the shelf biota, as apart from other organisms.

The stretch of water that separates the Indian Ocean from the Pacific is interrupted by a great number of islands, large and small. The islands are often separated by deep channels, but these have been no barrier to the dispersal of shelf organisms, many of which have a drifting or swimming stage in their life-history. Thus oysters seem fixed immovably on rocks and reefs; but a female oyster may produce 60 million eggs, and after fertilization these hatch into larvae that swim and drift for perhaps two or three weeks before settling down. Not only larval motility but also past changes in the relation of land to sea have per-mitted an interchange of shelf organisms between the western Pacific and the eastern Indian Ocean. When the view is restricted to the shelf biota, the salt waters of the world need be divided only into the Atlantic and the Indo-Pacific.

But really wide stretches of deep water are not readily crossed by many of the shelf organisms. Slope and basin constitute a barrier to shelf species. (Ob-viously land blocks the spread of marine organisms; but it may be surprising to learn that most such organisms also find the main part of the ocean to be a barrier of comparable effectiveness.) Thus the shelf biota of the world has four main components: eastern Atlantic (which also includes the Mediterranean), western Atlantic (which also includes the Gulf of Mexico), eastern Indo-Pacific, and western Indo-Pacific.

It might be asked why shelf organisms could not follow shallow water north-ward up the Pacific coast of North America, and then through the Aleutian or Bering Strait region to the Pacific coast of Asia; or why they could not pass from the Atlantic to the Pacific by rounding Cape Horn at the southern tip of South America. However, the shelf organisms, while comparatively tolerant of variation in the water temperature, are not unlimited in this regard. There are species and groups confined to tropical waters, others to subtropical, warm-temperate, cool-temperate, subarctic or arctic waters. Passage between, say, the eastern Pacific and the western is possible only to species or groups that are adapted for life under the temperature conditions that prevail in the Bering Sea. As for passage around southern South America, it will be recalled that the Pacific coast thereof, much more than the Atlantic, is chilled by a current of Antarctic origin. And in southern Africa, the Atlantic coast is chilled by a northwardly flowing current while the Indo-Pacific coast is warmed by a southwardly flowing one; and so there is not much movement of shelf organisms around the Cape of Good Hope.

The effect of temperature on the distribution of shelf organisms can best be understood by examining some coast in detail, say the Atlantic coast of the United

States. A great variety of shelf species, widespread in the West Indies, range no farther north than the Florida Keys or the lower east coast of Florida. Just east of the Keys and paralleling them is the only living coral reef in waters of the continental United States. It is decorated with sea fans, sponges, coon oysters, sea whips, sea feathers, elkhorn and staghorn corals, stinging corals, feather-duster worms, and colorful algae. Great lobsters, spiny but clawless, hide in holes of the reef, while the flattened slipper lobster prowls the bottom. Schools of white grunt mill about, along with French grunt, dog snapper, yellow-tailed snapper, red squirrelfish, spadefish, bluehead wrasse, parrotfish, queen angelfish, sergeant major, butter hamlet, blue tang, horned cowfish, porkfish, filefish, hawkfish, and neon goby. An occasional barracuda passes silently by. Sharks nose about the reef; they may be hammerhead, blacktip, lemon, nurse, bull, or tiger shark. Lying quietly on the bottom are spotted goatfish, scorpion fish, sea cucumbers, basket starfish, and huge sea urchins whose long, black spines so easily break off in the leg of a careless diver. After dark, snake eels and morays slither from their hiding places; red coral shrimp appear, and the banded coral shrimp picks the teeth of the green moray eel and is never gulped down.

Some of the above-mentioned organisms live only about coral reefs, and this restriction alone will prevent their occurrence much farther north along the Atlantic coast; for the reef-building corals live only where the water is clear, rather strongly saline, and never cooler than about 68 degrees Fahrenheit. The restriction of these corals to clear water may in turn reflect the needs of certain one-celled organisms that live within the bodies of the coral polyps. These organisms, which need light for photosynthesis, utilize carbon dioxide given off by the coral animals and supply the latter with oxygen and perhaps glucose.

Other shelf animals and plants, not limited to coral reefs but needing at least moderately warm water, follow the Atlantic coast northward about to the latitude of Cape Hatteras, North Carolina. This is the latitude where the warm Gulf Stream, which had been moving northward near the coast, begins to turn eastward to cross the Atlantic. Above Cape Hatteras the warming influence of the Gulf Stream is far less evident. Nevertheless, some shelf organisms range still farther up the coast, about to Cape Cod, Massachusetts. Here the cold Labrador Current, of Arctic origin, intrudes from the north.

Extreme southern Florida, Cape Hatteras, and Cape Cod are the three major termini in the northward distribution of the more southerly shelf organisms. Conversely, they are the major termini in the southwardly distribution of the more northerly organisms. Some of the more northerly species of the western Atlantic, the ones ranging southward in the New World only to the general vicinity of Cape Cod, live also along the coast of northern Europe and the British Isles; a familiar example is the green clam worm. Or in some cases a strictly American species, ranging no farther south than the New England states, has a close counterpart in northern Europe; thus the American lobster has a European "twin." Such distributions impart a certain biogeographic unity to an entire continental shelf.

We have recognized four main biotic areas as regards the distribution of shelf organisms—eastern and western Atlantic, eastern and western Indo-Pacific —and under present climatic conditions only a small minority of the shelf species could extend their range from one of these areas to another. Yet the seas, like the continents, have known periods when mild temperatures were more widespread than at present, and interchange between two adjacent areas has not always been as difficult as it is today.

The four areas are biotically more distinct now than ever in the past. The farther back in time, the more similar the areas in their biota. It looks as though the four biotas began to differentiate around the beginning of the Cretaceous, and the circumstance calls to mind the suggestion that a single super-continent, Pangaea, broke up in the Jurassic Period which immediately preceded the Cretaceous. If ever the earth had but a single continent, then it would have had but a single continental shelf, and perhaps but one shelf biota.

Among the four areas with a presently distinctive shelf biota, the biotic relationships are not what might be anticipated. The western Atlantic, that is the Atlantic coast of the Americas, has closest relationship not with the eastern Atlantic but with the eastern Indo-Pacific, that is the Pacific coast of the Americas. This state of affairs reflects the former existence of the Panama Portal, a broad seaway that once linked the Caribbean with the Pacific. When the portal was open, many shelf organisms passed through it. When the portal was closed, many a species was split into two populations which diverged as time went by. Today numerous shelf organisms of the Caribbean have counterparts on the other side of the Isthmus of Panama. But the portal was open only to species of tropical waters and was utilized especially by groups with excellent potentialities for dispersal. There are still biotic differences between the Atlantic and the

(Facing page) At top, a coral reef in tropical waters of the Bahamas. The fishes are grunts (left). Sea fans are conspicuous (right). (Bahama News Bureau) The warmth of the Gulf Stream permits coral reefs to reach Bermuda. Brain corals and sergeant-major fishes are conspicuous in this view (below) of a Bermuda reef. (Bermuda News Bureau)

Below, left, gannets nest on a rocky cliff of Newfoundland. (Canadian Government Travel Bureau) Right, the sandy coast of Hilton Head Island, South Carolina. (South Carolina State Development Board)

Pacific coasts of the New World, differences less marked toward the tropics. The Tethys Sea facilitated movement of marine organisms from the eastern Atlantic to the western Pacific. As noted in an earlier chapter, this portal has been closed since the Oligocene Period.

Just above the continental shelf is the intertidal zone, the strip of land between the high tide mark and the low. Organisms of this zone have a problem that does not face the shelf species: at regular intervals they must live out of water. They must endure certain environmental rigors of the sea when the tide is in, and of the land when it is out. The organisms that can do this form a very distinctive assemblage, and in marine biogeography the intertidal biota should be considered apart from the shelf biota. In dispersal abilities and disabilities, the intertidal organisms are about on a par with those of the shelf, and so one can recognize four areas that are distinctive as regards the intertidal biota: eastern and western Atlantic, eastern and western Indo-Pacific.

Because the intertidal animals and plants must live at intervals out of water, they are exposed to vagaries of air temperature to a greater degree than are the shelf species, which live submerged. Also, an ocean current has less effect on temperature in the intertidal zone than over the shelf. Therefore the intertidal organisms are distributed more in accordance with temperature values that exist on land, not those that prevail in the sea. For example, the intertidal species that reach Florida from the tropical West Indies do not range very far northward up the coast of that state; in general they range only about as far northward as do the West Indian land plants. The warm Gulf Stream may flow on northward to Cape Hatteras, but the circumstance is not very important to a tropical intertidal species that has to endure air temperature at every low tide.

The composition of the intertidal biota is influenced by the nature of the coast. As an illustration, limpets are flattened mollusks that attach themselves to rocks so strongly as to defy the mightiest breakers. With thick shell tightened into place, a limpet easily survives exposure to air without danger of desiccation. But a diverse limpet fauna is to be expected only where the coastline is rocky. On the other hand, little intertidal crabs that burrow in sand are lacking where bare rocks extend into the sea.

The coast of eastern Canada, and of eastern United States south to Cape Cod, is predominantly rocky. Between Cape Cod and New York City there is a transition from a rocky coast to one that is predominantly sandy. Southward from

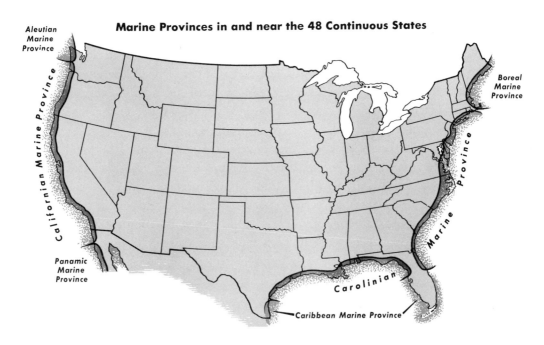

Marine Provinces in and near the 48 Continuous States

New York City, sandy shores become increasingly common and continue along the Atlantic coast to southern Florida. The upper Keys of Florida are of coral, the lower Keys of a limestone that also crops out coastally at Miami. The Gulf coast of the United States is a mixture of mud flats, marshes, and sand. The Pacific coast of the United States is mostly steep and rocky, with sheer cliffs rising out of the sea.

Along a single intertidal stretch there may be several different environments. Conditions in a tide pool are quite unlike those that obtain on the face of a rock, and each habitat has a different biota. In fact, the organisms that live on the landward side of an intertidal rock are not the same ones that live on the seaward side, where dashing waves require special "holdfast" structures. Thus marine distribution often is more a matter of ecology than of geography.

Various parts of the world are fairly distinct as regards both intertidal and shelf biotas. It is therefore possible to recognize Marine Provinces, usually separated from one another by broad zones of transition or intermingling. Six of the Marine Provinces are particularly interesting to students of the North American saltwater biotas.

(1) The Boreal Marine Province stretches along the Atlantic coast of North America from Labrador southward to Maine. (2) The Carolinian Marine Province reaches from the south shore of Cape Cod to northeastern Florida, and from northwestern Florida into Texas. It is interrupted by the southward prolongation of the Florida peninsula. The southern tip of Florida falls in (3) the Caribbean Marine Province, a tropical subdivision that reaches as far as Bermuda and Brazil.

On the Pacific coast of North America (4) the Panamic Marine Province, another tropical subdivision, extends from northern Ecuador northward into southern Baja California and the Golfo de California. Some of its components straggle farther north. (5) The Californian Marine Province stretches from northern Baja California into Washington waters. Still farther north, (6) the Aleutian Marine Province embraces the Pacific coast of Canada and southern Alaska.

Other Marine Provinces exist. Largest of them is the Indo-Pacific Marine Province, which reaches from the east coast of Africa eastward through Indonesia and northern Australia to Hawaii and Sala-y-Gomez. Three very interesting Marine Provinces are the Japanese, the South African, and the Mediterranean; but they will not be treated here.

Turning now to the slope and basin, a distinction must be made at the outset between a biota that is associated with the bottom and a biota that is largely independent of the bottom, drifting or swimming free in the upper waters.

In the upper waters live a great variety of organisms, both plant and animal. They are mostly small and often one-celled, usually passive drifters but sometimes with weak powers of locomotion. The assemblage is called plankton, a term coined from a Greek word that means "wandering." Especially important plankton are certain one-celled plantlike organisms, the diatoms. The individual diatom usually forms a protective shell out of silica, which it extracts from the sea water. When diatoms die, their bodies fall to the bottom, where the shells may form a thick deposit. (The well-known Fuller's earth is but a diatomaceous deposit in some land area that was once covered by the sea.) Diatoms are important because they are the major producers of food in the sea. Provided with chlorophyll, they produce sugars and starches from carbon dioxide and water, and also have a food reserve of oil. Thus they stand at the base of many a food chain.

Of course food chains exist also on land; but at times they are particularly clear cut in the sea. One simple marine food chain may be described. The Labrador Current, like cold currents generally, bears a rich load of nitrates and phosphates, picked up from the ocean bottom. Any gardener knows that plants thrive from an application of nitrates and phosphates; and the plantlike diatoms are no exception. In the reaches of the Labrador Current they multiply; there may be 400 million diatoms in a cubic yard of sea water. Tiny shrimp, and even tinier crustaceans called copepods, glut themselves on the diatoms. Shrimp and copepods in turn are fed upon by young herring, which latter fall prey to codfish. (One more link could be added to this chain, for man operates cod fisheries.)

All over the upper waters of the sea, diatoms and other tiny green organisms produce food which supports animals. The latitudinal and longitudinal distribution of the plankton assemblage is limited chiefly by temperature, and various of its components die when they drift into water that is too warm or too cool. The planktonic producers are limited in their vertical distribution by the penetration of light into the water. While bottom-dwelling plants in exceptionally clear water may live at a depth exceeding 500 feet, the drifters seldom thrive much below 300. Thus the herbivores and carnivores, supported directly or indirectly by the planktonic producers, are also concentrated in the upper 300 feet of water. The assemblage is largely restricted to waters over the slope and basin because the shelf waters frequently offer unsuitable conditions of temperature, salinity, density, and turbidity.

A curious situation exists in the Atlantic, with reference to the distribution of plankton. A great, circular "river" is formed in the sea by the Gulf Stream, the West Wind Drift, the Canaries Current, and the North Equatorial Current. At the center of this circle, between the Bahamas and the Azores, is the famed Sargasso Sea. It is simply an expanse of nearly stagnant water, in which float masses of an alga called sargasso weed. The masses, brought in by eddies off the main current, are not enormous as legend would have it; most clumps of sargasso weed you could scoop up in a bucket. A small number of animals, from snails and shrimp to crabs and little fishes, dwell in the clumps and are often camouflaged to resemble the weed. Yet the Sargasso Sea is remarkable not for living things but for a comparative lack of them. Nutrients needed for plant growth are brought to the surface from the bottom chiefly by cold currents, through a process called upwelling. In the sun-warmed Sargasso Sea there are no moving masses of cold water, no cold currents that move either horizontally or vertically; thus there is a scarcity of planktonic producers, and so a scarcity of herbivorous and carnivorous consumers. In a way the Sargasso Sea is a marine counterpart of a terrestrial desert.

We have discussed the shelf, the intertidal zone, and the sunlit upper waters. A fourth major environment is the deep, dark water below the planktonic zone. We may call this environment the mid-depths. The invention of sonar focused attention upon the mid-depths. Sonar signals are broadcast through the water, and they bounce back from the bottom or from things they encounter before reaching the bottom. In all the oceans, sonar echoes revealed the existence by day of a "false bottom" at a depth of several hundred feet, where the real bottom was known to be several thousand feet below. At night this "false bottom," as scanned by sonar, would move upward to the surface. Investigation finally showed what was actually moving: vast clouds of tiny fishes, whose swim bladders reflect the sonar impulses. Food being scarce below the plankton level, these fishes move up at night, and retire again to the depths at daybreak. Some crustaceans move up and down in like fashion, and so do the giant squids, largest of invertebrates.

Also important below the plankton zone are large, active, torpedo-shaped fishes, such as mackerel, oceanic bonito, albacore, bluefin tuna, wahoo, Spanish mackerel, striped marlin, swordfish, cod, and various sharks. These fishes cruise about in an almost constant search for the thinly scattered food items. (A tuna, for example, swims at a continuous rate of about nine miles per hour, and may cover a million miles in its lifetime.) Most fishes of the mid-depths will also raid the surface aggregations, and may at times come into shallow water over the shelf.

The farther down in the black depths of the sea, the scarcer become the living things. Yet a few species dwell at astonishingly great depths. The so-called abyssal zone begins around 3,000 feet and extends down to about 20,000. Fishes predominate in the abyss. They are mostly carnivores of grotesque shape. Often they are equipped with an enormous mouth, horrendous teeth, and a distensible gullet into which they can pack a victim larger than themselves. Some of the deep-sea fishes have luminescent organs, which serve to lure prey and perhaps to enable the sexes to find each other at breeding time.

In the abyss the basis of the food chains is a rain of dead organisms from the surface and the mid-depths. This material goes to support small invertebrates and fishes, which in turn are devoured by the more actively predaceous species. Very few abyssal fishes are more than a few inches long; food is too scarce in the abyss to permit the existence of large organisms. The abyssal environment is much the same from ocean to ocean, and has remained relatively unchanged for countless ages. Perhaps for this reason, a large proportion of the abyssal groups have reached all the oceans.

Sonar and related devices have revealed that the sea bottom is not smooth, but topographically more rugged than the land. Deep trenches have been discovered in the ocean floor. Those in the Atlantic do not much exceed 27,000 feet in depth, but several in the Pacific exceed 30,000; and the deepest of all, the Mariana Trench, reaches 35,800. The trenches do not occupy the middle of the ocean basins; they are found near land, sometimes off the shores of steep coastal volcanic ranges, but more often on the ocean side of archipelagos that in turn lie not far from a continental mainland.

The discovery of the trenches led to the recognition of an environmental zone below the abyssal. It was dubbed the hadal, a term suggesting Hades and derived from Greek root words meaning "without sight." Man has reached the bottom even of the Mariana Trench, in a device called a bathyscaphe, and has further explored the trenches by underwater television and deep-sea camera. The hadal zone has proven to be inhabited at least by sea cucumbers, bivalve mollusks, arrowworms, proboscis worms, sea anemones, and fishes. But biological exploration of the deep sea has just begun, and it would be premature to say whether or not the hadal biota is fairly homogeneous from trench to trench.

The greatest mountain ranges of the world arise from the ocean floor; they form a continuous system with a total length of 40,000 miles, extending into all the oceans. A few of their peaks rise above water but most are submerged. The biogeographic importance of these ranges is unknown; their biological exploration is awaited with great interest.

Thus the main biogeographic divisions of the sea are really the major ecological ones: the intertidal zone, the shelf, the sunlit upper waters of the open sea, the dark mid-depth waters of the open sea, the abyssal zone, and the hadal zone.

33 *"Well may we affirm, that every part of the world is habitable!*

Whether lakes or brine or . . . warm mineral springs . . . even the surface of

perpetual snow—all support organic beings."* *"They know the tundra of*

Siberian coasts, / And tropic marshes by the Indian seas; / They know the

clouds and night and starry hosts / From Crux to Pleiades."**

WATERS: BRACKISH, FRESH, SUBTERRANEAN, AND HOT

From a standpoint of biogeography it is hard to say where the land ends and the sea begins. The high tide mark has been taken to separate the marine environment from the terrestrial, but this boundary is one of convenience and is not honored by every organism. Distribution where land and sea meet is best understood by a close look at some particular locality, and for this purpose we may select a spot on the Atlantic coast of peninsular Florida, where the marine biota and the terrestrial biota are both rich.

At this locality—perhaps it is near Cape Kennedy—many of the common plants, such as the pines, saw palmettos, cabbage palms, live oaks, Magnolias, and hickories, do not grow where their roots are likely to be flooded with salt or

---

* CHARLES DARWIN ** PAI TA-SHUN, *Wild Geese.*

brackish water, or where their leaves will be exposed to windblown salt spray. But other plants, such as the sea oxeye, the Christmas berry, and the cottonseed bush, withstand considerable salinity, and grow almost down to high tide mark. The oxeye and Christmas berry in fact must require some salinity, for they do not grow very far inland; but the cottonseed bush is nearly as common at inland localities as along the coast. A narrow-leaved cattail sprouts in ponds and ditches where the water is slightly brackish; a gigantic leather fern tolerates even higher salinity and grows almost down to high tide mark.

At our locality there are broad flats, sometimes very dry or occasionally flooded with the rainwater that runs off the higher ground, but often inundated with a brackish mixture of runoff and tidal water. On these flats grow black rush, glasswort, saltwort, and several grasses, all confined to this environment between the land and the sea. White mangrove forms stands of small trees a short distance above high tide mark. Black mangrove grows farther out, on the average; its stands are flooded, at least in part, by an occasional tide that is exceptionally high. Red mangrove, chiefly intertidal, generally grows still farther out, within reach of most high tides; stiltlike roots hold the red mangrove trees well above the water. A short distance offshore, manatee grass and turtle grass—both higher plants although not actually grasses—form submarine meadows that are exposed only by the lowest tides. Where the water is clear, they may grow at such a depth that low tide will never expose them.

Huge land crabs scuttle about even beneath the pines and palmettos, but must go back to the sea to breed. Fiddler crabs, chiefly intertidal, occasionally venture into the saltmarshes or even farther inland. Diamondback rattlesnakes ramble into the saltmarsh from nearby higher areas, to feed upon cotton rats; both the snake and its prey are common farther inland as well as along the coast. A harmless watersnake lives in mangrove swamp and saltmarsh. A small turtle, the diamondback terrapin, lives in the shallow sea water but nests on sandy strands; its eggs are safe from marine predators but not from cotton rats, raccoons, and stray dogs. Barnacles and even oysters may grow on the terrapin's shell and are carried out of their normal medium when the reptile comes ashore to nest. Great sea turtles may live far out in the ocean, yet they too are bound to the land by the habit of nesting on sandy beaches.

An interesting situation exists as regards the fishes that live where fresh water meets salt. At our locality, perhaps fifty or sixty species of fishes might be found in the shallow sea water not far below the low tide mark, and nearly as many species in thoroughly fresh water a mile inland; but only two or three species will be found to inhabit the brackish coastal marshes, estuaries, and other aquatic situations neither fully fresh nor fully saline. This circumstance comes about because the intermediate situations are subject to great variation in salinity, at times being inundated with sea water and at other times with rain-water. Only a small minority of fishes are able to withstand this fluctuation in salinity. However, the few that can do so are likely to be extremely common, for they have all to themselves this environment and its resources. At our locality, as in many other parts of the world, the fishes of the brackish water mostly belong to the family Cyprinodontidae: the killifishes, topminnows, pupfishes, and their allies.

One might in fact say that, in general and throughout much of the world, only a comparatively few species of animals or plants live in the brackish waters, or in areas that lie above high tide mark but not beyond occasional exposure to sea salt. The species that are present are often individually numerous. From the standpoint of evolution, the zone of brackish water has served as a barrier, one that few taxonomic groups have transgressed. Fishes, mollusks, and crustaceans abound in both fresh and salt water, the last two groups also having many terrestrial members; but seldom can we point to a genus, or even a family, that is well represented in the sea and also in fresh water or on land.

As mentioned in an earlier chapter, the earth offers three major environments: land, salt water, and fresh water. Each of these certainly imposes a different set of demands upon living things; but, at least as regards the main patterns of animal distribution, the land and the fresh water are readily treated together. In general, freshwater animals cross sea barriers no more freely than do land animals (and often less freely). The faunal Realms, Regions, and Subregions are based upon the distribution of freshwater vertebrates as well as terrestrial ones.

The freshwater plants (which include emergents as well as rooted aquatics and free-floaters) are distributed somewhat in accordance with the outlines of the floral Kingdoms, Subkingdoms, and Provinces. However, a remarkably high percentage of freshwater plants are of extremely wide distribution. Hornwort, Elatine, marsh pennywort, duckweed, floating heart, Ludwigia, parrot's-feather, naiad, water lily, pondweed, beak rush, widgeon grass, eelgrass—each belongs to a genus represented on almost all continents. Wolffia, which includes the smallest of all seed-bearing plants, is similarly distributed.

As a matter of fact, many individual species of aquatic plants range over several continents. Some familiar examples are the common cattail, the coontail, sawgrass, reed canary grass, several bulrushes, and horned pondweed. The common reed is the most widespread of all the higher plant species, being abundant in practically all lands both temperate and tropical.

It is not certain why freshwater plants should often range so widely, but a few suggestions may be offered. The freshwater habitat is discontinuous; that is to say, bodies of fresh water are separated from one another by expanses of land. It is reasonable to expect some very effective dispersal mechanism among plants that cannot grow on land, and that perforce live at many small and widely scattered localities.

It is also suspected, and in some cases demonstrated, that waterfowl play an important role in distributing the seeds of freshwater plants. Mud caked on the scaly legs of a heron may contain seeds of many different water plants. In the case of the floating duckweeds, several of which are nearly worldwide, the entire plant body adheres to the feet of a duck or goose, which may then carry the plant for hundreds of miles.

As noted in an earlier chapter, seeds are rarely carried across wide expanses of sea in the intestines of a bird. However, many waterfowl fly for vast distances over the land, stopping en route to feed in lakes and streams, and no doubt voiding some plant seeds at many stops. It is easy to see how over the millennia the seeds of freshwater plants could be spread widely in duck droppings.

But it does not suffice to describe distribution in fresh water solely in terms of the faunal and floral subdivisions; there are many kinds of freshwater habitats, each with a characteristic biota. In the United States, for example, one might list the clear, swift, cold, rocky brooks of the uplands; the slower creeks of the lowlands; the larger rivers; the Mississippi River, in a category of its own by virtue of its large size; and intermittent streams, which sometimes go dry or at least break up into isolated stretches of water. Then there are large lakes such as the Great Lakes; the more common small lakes; ponds; bogs, which are well on the way to becoming land rather than water; and various intermittently flooded basins. There are also rainwater pools and puddles; lagoons and oxbows along rivers; springs, where ground water wells up to the surface; and a variety of other minor types such as hot springs, underground waters, and cave pools.

It is particularly useful in biogeography to distinguish between the flowing water and the still, for the two offer very different conditions of existence. In a

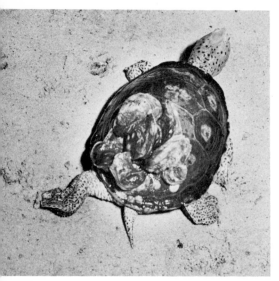

Sea birds forage on land and sea, nest on the former. Terns predominate on this beach (above) in South Carolina. (South Carolina State Development Board) At left, the diamondback terrapin lives in the sea but nests above the high tide mark. This specimen, from Cedar Keys, Florida, harbors a growth of oysters. (Ross Allen's)

body of water with little or no current, there develops a planktonic biota which, as in the ocean, includes some diatoms along with other tiny green organisms that carry on photosynthesis. The planktonic producers are highly important in a body of still water, not only forming the basis of many food chains but also oxygenating the water during the daylight hours. By way of contrast, in a flowing stream the only producers are the rooted or otherwise fixed plants; there is no planktonic assemblage.

Although the planktonic producers give off oxygen into the water each day, the oxygen content may sometimes become very low in a small pond. This is because bottom debris, and especially dead plant material, takes up oxygen as it decays. Small, mucky ponds are inhabited by organisms that require very little oxygen, or else that can surface and breathe air. Stream-dwelling organisms, adapted for life in water that is constantly oxygenated at least to a fair degree, often could not survive in a pond. The beginning aquarist, wishing to keep some local fishes in a small fishbowl without oxygenating apparatus, soon learns to dip for specimens only in the ponds and ditches, where he may capture the hardy little pygmy sunfishes, or various topminnows, killifishes, and live-bearing killifishes; for the daces, darters, and other stream-dwellers usually die in a fishbowl, if they live to reach it.

Special problems beset stream-dwelling organisms, for they must cope with the current. Plants must be well anchored, and not fragile. Animals must sleep (or rest if of species that do not sleep) in spots where they will not be washed downstream; they must forage, elude enemies, and reproduce in a fluid medium that is constantly urging them in one direction only. Fishes that swim in the current are characterized by a shape that offers minimum resistance to the flowing water; they are "streamlined" indeed. And these fishes spend most of their life heading into the current, as they must do if streamlining is to be of advantage.

Fishes are not the only vertebrates to show anatomical modifications that are correlated with the type of water. Larval salamanders can be categorized as either "pond type" or "stream type," and a river-dwelling turtle often has a shell that is low and smooth as compared with that of a pond-dwelling relative. Nor is streamlining the only anatomical modification associated with life in swift water. Most fishes have a swim bladder, a small, balloon-like structure that provides buoyancy (among other things); but the darters, attractive little fishes of North America, have lost this organ. Darters live mostly in swift brooks, where buoyancy is a drawback. In short, life in a stream requires morphological and behavioral adaptations unlike those needed for life in still water. The biota of a pond is not interchangeable with the biota of a nearby creek.

Distribution in fresh water is also affected by the acidity or alkalinity of the water. Swamps and bogs are usually very acid, and so too may be a stream that arises from them. In the lowlands of the eastern United States some rivers are popularly called "black water" streams, so dark are they with dissolved plant acids. On the other hand, a stream that arises from springs in limestone country may be alkaline, and so too may be a nearby pond if it does not accumulate much plant debris. Some organisms live only in the alkaline waters, and others only in the acid; but the reasons for such restrictions are not always clear.

Alkaline streams are more apt to be entered by marine organisms than are acid ones. As noted elsewhere, fresh water will invade the tissues of a saltwater organism; but this reaction is not so pronounced if the fresh water contains dissolved limestone. Thus in the big springs and spring runs of central Florida, which well up through limestone, one may see needlefish, sole, mojarra, stingrays, mullet, blue crab, and marine shrimp, living along with typical freshwater species such as largemouth bass, bluegill, shellcracker, and chain pickerel.

The size of a stream also has some bearing on the fauna it can support. Very large freshwater animals usually are confined to the big rivers, and perhaps the lakes along such streams. Examples are the freshwater dolphins of the family Platanistidae, variously inhabiting the Amazon, Orinoco, Rio de la Plata, Ganges, Indus, and Brahmaputra rivers, and a lake on the Yangtze. A really big river system offers a great variety of aquatic habitats from mouth to headwaters, and therefore may support an unusually rich and diverse assemblage of freshwater animals. Thus in the United States the richest and most diverse fish fauna is that of the Mississippi River drainage. Some parts of the Amazon basin harbor four species of crocodilians, separated by differences of habitat and diet; nowhere else in the world can so many crocodilians inhabit a single geographic region.

The still-water biota and the river biota are differently affected by certain historical factors. Periods of aridity wreak special havoc upon pond and bogdwellers, which die out except where rainfall continues to be abundant; for small bodies of water rapidly vanish unless replenished by the rains. In the United States, the still-water organisms are concentrated in the coastal lowlands of the

Southeast, which escaped serious desiccation during the Cenozoic. The stream-dwellers are also concentrated in this region, but less markedly so. Unlike organisms that inhabit ponds and bogs, the species of large streams are not necessarily dependent upon local precipitation for the perpetuation of their habitat. As an illustration, eastern Egypt is a desert with exceedingly few animals or plants that live in lakes, ponds, or bogs; but it has a fairly rich stream biota in the Nile, which arises 2,500 miles away in the central lake region of Africa.

It may be asked how freshwater animals disperse from one body of water to another. In the first place, many such animals actually are capable of movement over land. This is true of the alligator, freshwater turtles, aquatic snakes, newts, amphiumid salamanders, crayfishes, snails, and leeches, to name but a few. On a rainy night, especially, some water-dwellers can move overland through wet grass or across soggy ground; even one fish, the common eel, has been known to leave the water under these conditions.

But how do thoroughly aquatic animals disperse? There are several known ways. On one occasion in Marion County, Florida, a great deal of sand was dug from a flatwoods area, for road fill. A shallow basin was left behind, and it became filled with rainwater. Although there were no other ponds nearby, the basin soon teemed with mosquito fish and young bluegills, as well as a few red-finned pike and pond sirens. None of these move overland. Investigation revealed many crayfish burrows opening into the pond bottom. Excavation at the pond and elsewhere showed the flatwoods to be riddled with crayfish burrows, a veritable subway system that must permit aquatic organisms to spread widely in flatwoods. It is also thought that fish eggs and tiny aquatic organisms, perhaps enmeshed in filaments of algae, might cling to the legs of herons or other birds, and so be carried from pond to pond. (But the mosquito fish, the species most often to appear suddenly in newly flooded basins, is not transported in any such way, for it is a live-bearer.)

Another method of dispersal was strikingly revealed on the night of September 8, 1950, when a hurricane struck the Cedar Keys area on the Gulf coast of Florida. Torrential rains soon flooded the coastal lowlands. Between U.S. Highway 19 and the Gulf, a distance of twenty miles, a continuous sheet of water rushed from the higher ground to the sea, scouring out the ditches and pond bottoms. Two feet or more of rushing water covered the Cedar Keys road in most spots. Garfish, black bass, bowfins, warmouth, catfishes, pickerel, sirens, amphiumid salamanders, newts, freshwater turtles, and alligators, all routed from their biotopes, swam across the road. Organisms from creeks, ponds, bogs, and swamps were temporarily intermingled; and when the floodwaters subsided, many an animal must have had to settle down far from the spot where it had been living.

The spread of river-dwelling organisms is often facilitated by changes in the drainage patterns of the land. A detailed map will reveal that the headwaters of one stream may closely interdigitate with those of another. Headwater streams tend to lengthen in a headward direction; and often one stream, perhaps containing more water, or working into softer ground, or descending a steeper slope, may undercut the divide that separates it from a neighbor, thus capturing and diverting some of the water that belonged to the latter. And the "pirate" stream receives from the "beheaded" stream not only water but also some living organisms. Stream piracy has been of common occurrence, and helps to explain why many a river species, although unable to cross land or salt water, yet lives today in two or more stream drainages.

It must also be remembered that bodies of water now separated may once have been connected. For example, Lake Winnipeg and Lake of the Woods, in Canada, are but remnants of an enormous body of fresh water—larger than all the present Great Lakes together—that existed in the Pleistocene at a time when drainage toward Hudson Bay was blocked by glacial ice. And from previous remarks on the drying of the central and western United States in the latter Cenozoic, one might guess that the lakes and streams of that region were formerly interconnected to a greater extent than they are today. Thus historical factors must be given considerable weight when effort is made to explain the distribution of freshwater organisms.

One freshwater habitat is of minor significance but is so interesting that it must receive mention. Here and there in both the Old World and the New, a few animals live in subterranean waters. The plethodontid salamanders, catfishes, amblyopsid fishes, crayfishes, small crustaceans called isopods and ostracods, freshwater shrimp, and flatworms are among the groups to have invaded the subterranean waters of the United States. They have done so only in areas where limestone lies at or near the surface of the ground. Limestone deposits are often riddled with channels, which may be full of ground water or may appear above the water table as caves. In limestone country many of the ponds and lakes continue to receive rainwater, yet have no outlet; their drainage is mostly subterranean, into the aforesaid channels. Plant material, reaching the channels from lakes and lime sinks, forms the basis of the subterranean food chain. Only in limestone country could freshwater organisms find both food and channels far below the surface of the ground.

Among subterranean animals is a plethodontid salamander called *Haideo-triton*. Blind, white, and with elongated legs, it was first pumped up from a 200-foot well near Albany, Georgia, but was later found also in caves near Marianna, Florida. A similar-looking plethodontid was first discovered in water from an artesian well, 188 feet deep, at San Marcos, Texas. A well near San Antonio, Texas, yielded a blind, white catfish called *Trogloglanis*, remarkable for its toothless, paper-thin jaws. Several wells near San Antonio have produced another subterranean catfish, named *Satan*; it too is blind and white, but has strong jaws and teeth. There are quite a few blind, white crayfishes known from various underground waters of the Gulf Coastal Plain, the Ozark region, the Tennessee Valley country, and elsewhere. The most grotesque of them is *Troglocambarus*, which lives clinging upside down to the ceilings of submerged caverns in a small area of north-central Florida. A limestone region, extending from Tennessee and Kentucky to southern Indiana and southern Illinois, is noted for blind, white fishes of the family Amblyopsidae, as well as for subterranean crayfishes and smaller crustaceans. Certain of these latter live on the crayfishes, not as parasites but as uninvited nibblers at the crayfishes' food.

Caves, and the subterranean waters to which they often lead, may provide a refuge for some organisms during periods of aridity; a refuge not only for aquatic species but also for nonaquatic ones that require a cool, moist environment— ferns, frogs, salamanders, and camel crickets, for example.

Here and there in the world are hot springs. In some of them the water is barely tepid, but in others it is superheated far beneath the ground, and may burst forth as geysers of steam and boiling water. The most extensive area of hot springs is Yellowstone National Park in Wyoming, but there are other large areas in New Zealand, Algeria, Iceland, and Europe, while lesser ones are scattered over several continents and islands. Hot springs are of biogeographic interest in revealing the amazing adaptability of life. The water of hot springs is usually impregnated with various chemicals, often calcium carbonate, silica, sodium sulfate, sodium chloride, and gypsum. The water is low in oxygen but

A *small lake: Blue Lake in the Cascades of Oregon. (Oregon State Highway Department)*

A "room" in Carlsbad Caverns, New Mexico. (New Mexico State Tourist Bureau)

high in hydrogen sulfide and sometimes sulfur dioxide; sulfuric acid forms in the water, often to strength that will char wood. In some pools, suspended mineral matter may accumulate in such quantity that the water becomes pasty, forming a mud geyser or "paint pot." Yet in the unearthly environment of the hot springs, life exists.

Various algae are most successful in withstanding a high water temperature. In hot, calcareous water some of the green algae live at 146° F., and nongreen ones at 160°. In silica-charged water, some green algae withstand 171°, and nongreen 194°. This last temperature is nearly that at which water boils in Yellowstone National Park (altitude 6,000 to 7,000 feet). Animals cannot tolerate as high a temperature as can some of the lower plants. The most heat resistant of animals are some of the microscopic protozoa; several types of them have been found living at temperatures between 122 and 130°.

Aquatic beetles have been found at 115°, midge larvae at 124. The larvae of soldier flies, horseflies, brine flies, water bugs, caddis flies, dragon flies, and Mayflies have been found in hot springs, but not at temperatures tolerated by the aforesaid beetles and midge larvae. Several kinds of small crustaceans inhabit the streams that arise from hot springs, but do not live at impressively high temperatures. Some pond snails withstand 113°, a water mite about 115.

In general, the invertebrates that inhabit hot springs belong to widespread genera that have also invaded brackish water, alkali pools, and other rigorous environments.

Vertebrates do not exist in truly hot water. At Yellowstone, tadpoles of the spotted frog (a *Rana*) were found at 106°, a record for frogs. Ludwig Schmarda reported turtles in water of 111° at Tunis, but one suspects the reptiles did not long remain immersed. Legend has it that the fishes of the hot springs live in water so hot that it cooks them when they die; but this is not the case. At Yellowstone the fishes live in water that has cooled somewhat; few tolerate more than 86°, and exposure to 100° probably would be fatal to any of them in a short while.

34 *"Nature's polluted / There's man in every secret corner of her /*

Doing damned wicked deeds."* "Mind is ever the ruler of the universe."**

"The path of civilization is paved with tin cans."*** **FROM**

APE-MAN TO SPACE-MAN

Now we come to man. As our attention has been focused on plants and animals, let us see what he has done to their distribution; then we can investigate the spread of man himself.

It was previously suggested that the range of an organism reflects evolutionary, innate, and environmental factors. Man cannot alter the evolutionary factors, but often neutralizes their results by introducing species far beyond their natural range. As an illustration, man has taken the nutria, an aquatic rodent, from South America to many parts of the United States; the European hare to Ontario; the European rabbit to Pennsylvania and Washington; the wild boar from Germany to North Carolina; the fallow deer from Europe to Nebraska; the kangaroo rat from the southwestern United States to Ohio. He has taken the European fallow deer and roe deer, the Indian black duck and nilghai, the Sardinian mouflon, and the Asiatic serow to Texas; the Mexican armadillo, the Central American jaguarondi, the Asiatic sambar deer, the Indian axis deer, the Indian rhesus monkey, and the South American squirrel monkey to Florida. He has stocked the Greenland musk-ox and Siberian reindeer in Alaska; the Barbary sheep in New Mexico; the South American chincilla in California.

Many game birds and sport fishes, along with game and fur-bearing mam-

* THOMAS BEDDOES, nineteenth-century poet.   ** PLATO   *** ELBERT HUBBARD

mals, have been carried far outside their natural ranges. The worldwide trade in small, colorful fishes has led to the establishment of various species outside their native lands. On one occasion some live-bearing killifishes, belonging to a Central American species, were received by a California aquarium in a shipment from the Philippines; investigation revealed that they had been liberated near Manila by a fish fancier who got his stock originally from Egypt! The mosquito fish or Gambusia, native to eastern and central United States, has been planted in ponds around the world as a mosquito control measure.

The bullfrog of eastern United States, valued as food, has been carried to western North America, Puerto Rico, Cuba, and several parts of the Old World; the giant toad of the New World mainland tropics has been carried to Florida, the West Indies, Hawaii, the Philippines, Tinian, Guam, and other cane-growing regions, in the hope that it would control insects in the sugarcane fields. A variety of small frogs and lizards, mainly of West Indian origin, have become well established in southern Florida; some probably arrived as stowaways, but others escaped from, or were freed by, pet dealers and fanciers.

On the French Riviera, horned "toads" (lizards) from the southwestern United States dash about under similarly imported cactus. Enormous flocks of budgerigars, or Australian paroquets, are to be seen along the west coast of peninsular Florida. Starlings and English sparrows, both originally imported from Europe (and by the same man!), have spread throughout much of North America. The American gray squirrel is familiar—sometimes too familiar—in parts of the British Isles, and the American muskrat is now distributed more or less continuously all the way across Eurasia from France to Japan.

Bacteria, fungi, green algae, flagellates, rhizopods, sporozoans, turbellarians, flukes, tapeworms, nematodes, snails, slugs, earthworms, ticks, spiders, mites, crabs, crayfishes, bristletails, roaches, true bugs, book lice, sucking lice, biting lice, beetles, flies, mosquitos, ants, bees, fleas, butterflies, moths, ectoproctans, and a great variety of plants—all these are also known to have been spread, at times, by man.

The list of introductions would be far longer, were it not sometimes difficult or impossible for an introduced species to gain a lasting foothold in a new land. (More than 250 species of foreign game birds have been stocked in the United States, with the hope that they would become established for the hunter's benefit; but how many now exist in your state?) The species that do thrive often become intolerably abundant, doing enormous damage and perhaps necessitating expensive control measures. Examples are the red deer in New Zealand; the European brown rat and house mouse in the United States; the Mexican boll weevil in the southern United States; the Mediterranean fruit fly in Florida; the Japanese beetle in the eastern United States; the European corn borer in North America; the South American fire ant in the southern United States; and the German roach or Croton bug in the United States. Nor should we forget the spread of the European white cabbage butterfly in North America; the European gypsy moth in the northeastern United States; the European codling moth in North America; the African giant snail in Hawaii.

And as for noxious plants being of foreign origin, one has only to mention crabgrass, "Canada" thistle, bull thistle, chickweed, dandelion, English plantain, shepherd's purse, deadly nightshade, white pigweed, yellow sorrel, yarrow, wild carrot, live-forever, dog fennel, moneywort. What a lot of weeds; and all of them of Old World—generally European—origin. It has been estimated that 70 per cent of the world's "pests," plant and animal, earned their reputation only after they had been introduced into new lands. As often as not, a pest was introduced intentionally, with the idea that it would prove useful or attractive. The European rabbit was stocked in Australia to provide meat and hides, and in fifty years its numbers had increased to two billion; it devoured native and crop plants, drank the waterholes dry, and noticeably altered the landscape over vast tracts. It was finally controlled by an introduced virus disease, but a disease-resistant strain of rabbits may now be emerging. The Asiatic mongoose was brought to Jamaica to kill the (harmless) snakes and the canefield rats, but turned instead to waterfowl, ground-nesting birds, and poultry; native and crop plants succumbed to insects, which multiplied as the birds declined. Such episodes provide a costly demonstration of the interrelationships that exist within a plant and animal community.

Man also exterminates plants and animals. The United States biota has suffered egregiously in this regard. In Europe, for all its lengthy habitation by primitive and civilized man, only one vertebrate species was actually exterminated, the aurochs (and its stock was bred into a surviving strain of domestic cattle). But European man, reaching North America and tempted to recklessness by nature's largesse in the new land, rapidly exterminated mammals, birds, and fishes. The sea mink, Steller's sea cow, Carolina paroquet, Labrador duck, heath

Man's activities change the face of the earth. Below, former grassland in southern Saskatchewan has been converted to farmland. (Canadian Government Travel Bureau)

(Facing page) A small minority of animal and plant species are valued and conserved by man. At top, deer at a feeding station in Sam Houston State Park, Louisiana. Longleaf pine (below), replanted in Louisiana, can eventually be cut for timber. (Louisiana State Tourist Commission)

hen, great auk, Eskimo curlew, passenger pigeon, emerald trout, regal silver trout, and San Gorgonio trout have all vanished within historic times. The West Indian wolf seal, Guadalupe fur seal, red wolf, sea otter, black-footed ferret, bison, trumpeter swan, ivory-billed woodpecker, California condor, whooping crane, and Gila trout were brought almost to the point of extinction, although a few of these made partial recovery when later given protection. The American flamingo, roseate spoonbill, snail kite, and Zenaida dove were nearly or quite extirpated from the United States, although persisting elsewhere. Drastically reduced were the respective ranges of the moose, wapiti, white-tailed deer, mule deer, pronghorn, mountain sheep, grizzly bear, gray wolf, and wolverine. Four Hawaiian birds—a duck, a gallinule, a coot, and a stilt—are endangered, and the Hawaiian goose todays exists mainly in captivity.

Man also changes the innate characteristics of organisms. This can happen unintentionally, as in the case of the disease-resistant rabbits. Widespread use of the pesticide "DDT" continually favors the more resistant flies over the less resistant; the result is the development and spread of a fly that can withstand larger doses of the chemical. In like fashion some disease-producing organisms have developed strains that are highly resistant to certain antibiotics.

For thousands of years man has intentionally directed and accelerated the evolution of numerous animals and plants, turning them into the cultivated or domestic types quite unlike their forebears. Most garden flowers have been bred selectively, for characteristics lacking in the wild prototypes. Plants altered by man will seldom spread unaided. Often in a wood or field one encounters an old, abandoned homestead, and how pitiful is the former garden, where now only a few blossoms keep up the struggle against the encroaching natural vegetation. But in contrast, domestic animals often run wild. Feral swine are a serious pest in parts of the southern United States. Cattle, horses, goats, dogs, and cats have all done well—sometimes too well—on their own. Especially on islands are introduced domestic animals likely to exterminate the wild birds and lizards.

Man has changed the distribution of certain organisms through some very ingenious methods of tampering with innate characteristics. A few years ago in Florida, males of the screwworm fly were raised in vast numbers, and sterilized by radiation from a cobalt reactor. Strewn over the countryside from airplanes, the sterile captive-raised males mated with most of the wild females, whose reproductive activity thus came to naught. The screwworm rapidly vanished from large areas where it had previously been an important pest in the livestock industry.

But man affects distribution chiefly by modifying the environment in a large number of ways. It has been suggested that the atmosphere now holds heat better because its carbon dioxide content has been significantly increased through the burning of coal, wood, and petroleum products. It has also been suggested that combustion of leaded gasoline adds lead iodide particles to the atmosphere in sufficient quantity to bring about condensation of water vapor into rain.

Of increasing importance is the fallout of radioactive materials, produced and blown into the upper air by the continual testing of nuclear exposives. Thus in far-northern Europe, radioactive fallout is taken up and concentrated by lichens, which in turn are fed upon by reindeer. The deer are eaten by the Lapps, those hardy little people of the tundra. Some of the Lapps have become radioactive, beyond a "safety level" as conceived by the United States Federal Radiation Council. Obviously the makeup of a natural community could be altered even by a very light fallout, depending on the way plants concentrate the radioactive materials, and the tolerance of the animals that receive these substances via the food chain. Forests, and the small plants that grow beneath the forest trees, are especially susceptible to radiation damage; weed fields and insects are particularly resistant.

Research is being made into ways of controlling the climate. Wetlands are drained and deserts are irrigated. Huge basins are flooded behind hydroelectric dams. Vast tracts are deforested or burned; processes of erosion are accelerated as the natural cover is removed. Plant and animal communities are altered and perhaps obliterated, replaced by cultivated fields and cities. Grasslands are given over to the raising of cattle, sheep, or other domestic herbivores. Streams, once limpid, are muddy with eroded soil and urban pollution. Coastlines are altered by dredging, filling, and the construction of dikes and canals; and the ocean receives the burden of the rivers, whatever its nature. Scattered tracts of land, theoretically set aside as parks and sanctuaries, are succumbing rapidly to commercial interests.

Man's modification of the environment is obvious to anyone who travels in the United States, or who stays home and reads newspapers; the topic need not be elaborated here. One point might be made: It is popularly thought that less urbanized or less industrialized countries, especially those in the forested tropics, are still largely in an unspoiled condition; but such is not the case. It is surprising to learn that in parts of Central America, South America, and the Old World tropics, erosion and deforestation have progressed to a degree rivaling anything to be seen in the United States. Also, a nation emerging from colonial status sometimes celebrates by rescinding all game laws, with resultant decimation of many animal species.

So much for man as a factor influencing the distribution of other organisms. Now what about the spread of man himself?

Biogeographic principles are inadequate to explain man's dispersal. The following brief remarks, on nature and man in a little-known part of Mexico, reveal how the movements of *Homo sapiens* cannot be studied within the same conceptual framework as the movements of other organisms:

For years it was rumored that a giant land turtle inhabited central Coahuila, in the Chihuahuan Desert of northern Mexico. In 1959 the reptile was brought to scientific attention; it proved to be a gopher tortoise, more related to the Florida species than to the other species of southwestern North America. This relationship was interesting because a Coahuilan box turtle also seemed to have a near ally in eastern North America. Central Coahuila next yielded a soft-shell turtle with certain resemblances to the Florida species. A cricket frog and two water snakes also helped to give the Coahuilan fauna an "eastern" aspect. Re-markably, a band of Seminole Indians live in this same part of Coahuila. Seminoles are the Indians who still inhabit southern Florida, and it is astonishing to find some of them in northern Mexico, especially in the exact area where the fauna includes several animals of southeastern affinity.

If asked to account for the presence of the animals in the Chihuahuan Desert, the zoogeographer would probably open with remarks on mid-Cenozoic times, when an equable climate permitted various animal groups to spread widely. He would then comment on the cooling trends of the latter Cenozoic, when many groups fell back to the south; and on the Pliocene drying, which brought desert and grassland to many parts of southwestern North America. He would emphasize, as we did in an earlier chapter, that many groups vanished from the southwest although surviving in the better-watered southeast. But here and there in the southwest, local conditions partially counteracted the trend toward desiccation. (Numerous springs arise along the eastern flank of the Sierra del Carmen in central Coahuila, and the latter area forms an "island" of grassland and chaparral in the midst of a desert.) Finally, the zoogeographer would draw attention to the climatic shifts of the Pleistocene, some of which permitted the extension of fairly humid corridors into northern Mexico, corridors along which some eastern animals were able to reach localities that have since become oases.

But what would a student of American Indian history have to say about the presence of Seminoles in Coahuila? His account would run like this: Certain Indians, living mostly in Georgia and Alabama, resented domination by the militaristic Creeks, and so moved into Florida where they acquired the name "Seminole." In Florida the Seminoles gave shelter to runaway slaves, were caught up by the Spanish-British-American rivalries in the New World, and occupied lands which were coveted by the white man. Thus the Seminoles became embroiled in a series of wars, during which a resistance leader named Wildcat came to the fore. A dominant personality, Wildcat inspired the Seminoles in battle for years, but was eventually deported to a reservation in what later became Oklahoma. Here, disappointed in his effort to be elected tribal chief, and finding little support for his plan to form a coalition of tribes, he rallied a band of Seminoles, Kickapoos, and former slaves, and struck out for Mexico where he offered his military services to the Mexican government in return for land, seed corn, and agricultural implements. The offer was accepted; the Kickapoos soon absconded with all the livestock, but the Seminoles and their Negro allies remained and conducted successful campaigns against revolutionists, smugglers, Apaches, Tonkawas, Kiowas, and Comanches. But lest these hot-blooded Seminoles reopen a dispute with the United States, the Mexican government withdrew them from the border country and settled them in central Coahuila, at a spot where corn agriculture was possible.

Note the words we must use in order to tell the Indians' story: domination and the resentment thereof, militarism, slavery, international rivalries, covetousness, war, leadership, dominant personalities, disappointment, abscondence, and government. This is not the vocabulary of biogeography. The study of man's dispersal is a separate science, anthropogeography. And of course man is further unique in having spread through artificial means, far beyond the limits that would have been set by his innate, physical characteristics; and in having negated many effects of the natural environment through the use of fire, clothing, shelter, weaponry, agriculture, machinery, and medicine.

As a matter of fact, the use of tools and fire, and probably language, ante-dates *Homo sapiens*, the species to which all living men belong. It is instructive to review the fossil record, to determine when primates first began to spread by virtue of cultural attainments rather than by mere physical capabilities.

The primates constitute an order of mammals, with about two hundred living species: man and the great apes; the lesser apes or gibbons; the Old World monkeys, including baboons; the New World monkeys; the lemurs; and a few other primitive, somewhat lemur-like types. Primates probably evolved from insectivores in the Cretaceous. This is not demonstrated by the fossil record; but in the next Period, the Paleocene, primates of a lowly sort are widespread, and are still very much like insectivores in several details of their anatomy. Indeed, there persist today, in the rainforests of southeastern Asia, some little mammals that are as much insectivore as primate. Called tree shrews, they are superficially ratlike in appearance, with sharp claws, a long snout, and a moist, dark nose with bristly whiskers. You would not suspect their relationship to higher primates (unless taking a clue from tree-shrew disposition, described as "wrathful, glut-tonous, and libidinous"). During a part of the Paleocene, tropical forests ex-tended southward to southern Africa, and northward at least to Germany and France. Before the Period was out, early primates had passed between the Old World and the New, ranging at least from Europe to North America.

In the Eocene, lemur-like primates diversified in both Eurasia and North America; but few of them survived beyond that Period. From Burma comes the remains of an animal more monkey than lemur; and we might guess that lemurs fell back as monkeys diversified. (Madagascar, the only land with surviving lemurs, was never reached by monkeys. The tree shrews, and the few other primitive primates that live today on the Asian or African mainland, are noc-turnal and so in little competition with monkeys, which are mostly day prowlers.) It is suspected that the New World monkeys, very different from the Old World ones, evolved independently from some lemur-like stock of the New World.

By the Oligocene, apes had appeared in the Old World, and they began to diversify there. Only one line of descent concerns us, a group of apes called the dryopithecines. From late Oligocene through Miocene times these apes radiated widely, leaving fossil remains in such diverse places as Kenya and Egypt; Turkey and the Russian Caucasus; Spain, France, Italy, and Germany; even Denmark and the northeastern corner of India near the China border. This spread is impressive, but cannot be credited with any assurance to unusual mental attainments; for Miocene climate on the whole was mild, forests were widespread, sea and mountain barriers less pronounced than they are now. A dryopithecine ape known as *Proconsul*, from the Miocene of Africa, was probably ancestral to the chimpanzees and the gorilla that survive today. Another descendant of the dryopithecines was the Pliocene ape *Oreopithecus*; its skeleton, found in an Italian coal bed, was described in some newspapers as a "fossil man." Although manlike in some characters, *Oreopithecus* was an evolutionary dead end, and left no modern descendants. But one dryopithecine ape, *Ramapithecus*, which ranged at least from southern Africa to India in late Miocene times, is regarded as being on a line that leads to man.

Our attention now focuses on southern and eastern Africa. Not that Asia was necessarily out of the primate picture; but the paleontologist must take his fossils where he finds them. A dead animal usually decays; it will fossilize only if sheltered, or covered up soon after death. In southeastern Africa sandstorms often buried dry bones; around lakes and waterholes the remains of dead animals might sink into muck; occasional floods deposited sediments in thick layers. Caves preserved the remains of anything that died in them, or that were dragged into them by carnivores. Thus the fossil beds of Africa, rich in material that throws light on man's ancestry, have been closely scrutinized by anthropologists, geologists, petrologists, palynologists, paleontologists, and pedologists, as well as by geochemists who are concerned with dating.

The most important African fossils are the australopithecines, or man apes. They were about four to five feet tall and rather chimplike, but they walked erect and had a larger brain than any modern ape. They fall into two general types. One of these, apparently an evolutionary dead end, was larger and more robust, with big, heavy teeth which suggest a vegetarian diet. (This is the type whose beetle-browed skull adorns a recent Tanzania postage stamp.) The other type, smaller and with teeth more like ours, may have been more of a carnivore, and on the line of human ancestry.

The oldest australopithecine remains are from near Lake Rudolph in Kenya. A modern dating technique, developed from atomic energy studies, reveals their age to be about 2,500,000 years. At Olduvai Gorge in Tanzania, australopithecines of varying age have been excavated, along with their manufactures. The deepest, oldest level at Olduvai, about 1,800,000 years old, has yielded a circle of stones, unknown in function. The stones could have formed the base of a shelter made from branches, but in any event they were set in place by an intelligent being. The deeper levels also yield definite tools, rude hammerstones; some of these tools were made of stone that did not crop out locally, but only at localities miles away.

Nearly as old as the stone ring are the remains of an australopithecine, a man-ape of the smaller type. He was probably the early tool-maker. (But the more robust, big-toothed man-ape also left his remains in a stratum of nearly comparable age.)

At other localities in southern Africa, cave debris indicates that the man-apes killed and ate a large variety of animals, saving heavy bones, dagger-like horns, and sharp splinters of bone for use as tools and weapons. Much charcoal in the caves suggests that the man-apes used fire, whether or not they could make it.

Thus with the advent of the australopithecines we have already left the realm of biogeography and entered that of anthropogeography, to deal with species whose existence is not entirely at the mercy of nature's vagaries. Fragmentary remains suggest that the australopithecines reached Asia, but more paleontological work is needed before we can say just how far they ranged or what barriers they crossed.

Apes and lower primates are generally provided with heavy canine teeth and the muscles to use them effectively. The dentition is especially formidable in the males, who defend their families or bands. But in the australopithecines the teeth, muzzle, and associated musculature are much reduced, suggesting a long dependence on weapons, presumably of wood, bone, and horn. Anthropogeography may antedate even the man-apes.

Although moving steadily away from biogeography, we cannot just stop our narrative with the australopithecines. Next to be considered is a species called *Homo erectus.* His story is a remarkable one. Not long after Darwin had extended evolutionary theory to include man, a teen-ager named Eugène Dubois reasoned that something more primitive than man, but more advanced than ape, should have left its fossil bones in Indonesia. Young Dubois got a post with the Dutch army in Indonesia, where, amazingly, he found what he was looking for. His find was dubbed *Pithecanthropus erectus,* the "upright ape-man." We have since placed it in our own genus, *Homo,* and uncovered additional remains in Africa, Asia, and Europe.

About 500,000 years ago in Europe, this near-man, *Homo erectus*, lived right up to the edge of the great ice sheet during a peak of glaciation, enabled to do so through his knowledge of fire, shelter, and clothing. He made hand axes of flint, chipped to a definite pattern; and as centuries went by he slowly changed and improved the pattern. In other words, he had the ability to pass a learned skill from generation to generation; the power of speech could be inferred from this.

Homo erectus was a mighty hunter. A cave he once occupied, near Peking in China, yielded the bones of sixty species of animals he had eaten, among them rhino and mammoth elephant. The cave was littered with the charcoal of his cooking fires, and with flint brought from localities far away. *Homo erectus* was also found at Olduvai, in a deposit a little more than 500,000 years old. An older deposit there, nearly 1,000,000 years old, yielded a skull similar to that of the near-man, but more primitive.

The evolution of the higher primates was not a straight line but a much-branched one, with offshoots that flourished for a time and then vanished. Here we need not try to follow the tangled growth. What is important is the trend from dryopithecine to australopithecine to manlike types, with an intelligent use of tools, weapons, and fire at least as far back as the australopithecines. By the time of *Homo erectus*, cultural attainments were permitting prehumans to spread far beyond the limits that would have been set by their innate, physical capabilities. From Olduvai to Peking is 6,000 miles as the crow flies, but the near-man had to walk it; he survived in deserts and grasslands and tropical rain-forests and the tundra that bordered the glacial ice. And there is evidence that he hunted by fire drives; the blazes he set must often have swept for miles, altering the plant and animal communities.

Homo erectus persisted until about 300,000 years ago. Two skulls, one from England and the other from Germany, are somewhat intermediate between this species and modern man, *Homo sapiens*. The skulls date from between 200,000 and 300,000 years ago. But the style of flint work that the near-man finally produced, the so-called Acheulean flint industry, was passed on to our own species, who did not improve on it for another 200,000 years.

The earliest known *Homo sapiens* remains are those of the Neanderthal variety, named for Germany's Neander Valley where the first skeleton was found. Neanderthal man has often been portrayed by nonscientific writers as a

Bushmen (above) were once widespread in Africa but today survive only in the far south of the continent. (South African Tourist Corporation)

(Facing page) American Indians, of Eurasian origin, had spread over most of the New World by 10,000 years ago. They were able to invade tundra, coniferous and hardwood forests, desert, grassland, tropical forest, swampy lowlands, and seacoasts. Pictured is a Chippewa woman in Wisconsin (upper figure) and a Seminole woman in Florida. (Upper photo, Wisconsin Conversation Department; lower, Ross Allen's)

hulking, malefic brute, but he was far from being that. He used fire, made shelters, and experimented with rude stone construction. He eventually developed a new style of flint work, the Mousterian flint industry, which included the world's first spear points of chipped stone. He took care of his injured and aged, and developed some sort of cult involving the bear. He evidently believed in a hereafter, for he buried his dead with considerable care, sometimes aligned in an east-west position and accompanied by tools and other offerings.

At the peak of the last glaciation, Neanderthal man was distributed from western Europe eastward into Siberia, and southward into southern Africa. But other races of man had been developing in other parts of the Old World. By about 35,000 years ago the Mongoloid, Negroid, and Caucasoid types—the three that predominate today—are foreshadowed in the fossil record; and as these spread, the Neanderthal race vanished.

The development and spread of man's racial types was more complicated than the foregoing comments might suggest. For example, the Bushmen, who constitute a distinct racial type, were once widespread in Africa, but were pushed into the inhospitable deserts of the south by the spread of the more successful Negroids. The Australian aborigines reached their present location about 12,000 years ago, coming out of southern Asia; the Veddas of Ceylon may represent stay-at-home descendants of the same stock. The dwarfish little Negritos of the Southwest Pacific must once have been widespread, but today they have been pushed back into the mountains and jungles from the Philippines to New Guinea. The races of man moved around so determinedly, and interbred so frequently, that it is hard to trace all the meanderings. Whence came the Ainu, the hairy, light-skinned people who preceded the Mongoloids in Japan and elsewhere? What were the Lapps, racially, before they began to mix with Caucasoids in northern Scandinavia? The Papuans of New Guinea, the Melanesians of the South Pacific, the Pygmies of Africa—in what isolation did they originate their physical characteristics, and where have they wandered?

It would be fascinating to follow man's spread, but another book would be needed. We shall pursue the topic only to say a little about the peopling of the New World. In the main, the American Indians are Mongoloids of an early type. There are tantalizing hints of man in the New World 20,000 years ago, perhaps even 30,000; but conclusive proof is lacking. About 12,000 years man was widespread in North America east of the Rockies. The giant beasts of the Pleistocene were still alive, and these early Americans—Paleo-Indians, they are called—preyed extensively upon the mammoth and the mastodon, as well as camels, horses, and ground sloths. At this time the Paleo-Indians made a distinctive kind of spear point, a beautifully chipped blade with a concave base and a channel or "flute" on each face. Called the Clovis point (after a New Mexican town where it was first noted), its distribution helps us trace the spread of its makers. A different sort of flint blade, the bayonet-shaped Sandia Cave point, may be older than the Clovis, but is far less widespread or abundant.

It can scarcely be doubted that America was peopled chiefly from Asia by way of Bering Strait. (Marine geologists think it unlikely that Bering Bridge was above water in late Pleistocene times.) But Clovis points are distributed from the Atlantic westward to the Rockies; neither they nor Sandia Cave points can be traced across into Siberia. Sandia Cave points have close counterparts in the Solutrean flint industry of western Europe. Perhaps, then, there was some movement of people to the New World across the North Atlantic at an early time. Some New World flint industries a bit younger than the Clovis, say around 8,000 years old, are more readily traced between Alaska and Siberia. One New World language, that of the Eskimo and Aleut, is represented also in Siberia; the ancestral Eskimos spread along the shores of the Arctic Ocean from northern Asia all the way to Greenland.

Ancestral Indians had reached southern South America by about 9,000 years ago, if not sooner. As the New World became more heavily populated through natural increase, movement across Bering Strait virtually ceased. American archeology is chiefly concerned with the development and spread of ideas within the New World; but there are a few important exceptions.

A little less than 5,000 years ago, a strange new culture appeared on the Pacific coast of Ecuador, without local antecedents. It was essentially a transplantation of the Middle Jomon culture of Japan, and originated probably in southern Kyushu, the southernmost island of Japan. Whether explorers or storm-tossed fishermen, the Jomon Japanese probably followed the great sweep of current that passes from eastern Asia to northwestern South America. One of their most important introductions to the New World was the art of pottery making. The biogeographer wonders what domestic plants and animals the Japanese brought from the Old World. The coconut? The banana? The Lagenaria bottle-gourd, of demonstrable antiquity on both sides of the Pacific? Some kind of cotton? The rumpless bantam chicken that lays blue eggs?

There was an apparently authentic find of a Greco-Roman coin at an ancient

Indian site in Mexico. The murals of Italian Pompeii, buried by volcanic ash nearly 1,900 years ago, include a painting of a pineapple which someone surely brought from the New World tropics. Cultural similarities hint at prehistoric American connections with China, India, and the Mediterranean area. However, we shall note here only the transoceanic movements that have been demonstrated by extensive and well-documented archeological excavations. In this connection the Vikings must be mentioned.

In 1966, around Columbus Day, some people were highly upset to learn that the Vikings had discovered the New World long before Columbus. The Norsemen began pushing into North America more than five hundred years before the Italian navigator made his famous voyage under Spanish auspices. One Norse settlement in southwestern Greenland was begun in A.D. 982. Norse-Eskimo intermarriage eventually became extensive, and Greenland Eskimos came to tip their knives, adzes, and harpoons with forged iron of Norse manufacture.

Around A.D. 1300 there were 4,000 Norsemen in Greenland, and archeologists have excavated their big settlement at Godthäb, across the Davis Strait from Labrador. Items excavated included some that were obtained by the Norsemen farther west in North America: lumps of anthracite coal and spear points of quartzite, neither occurring naturally on Greenland although present in Newfoundland and Labrador.

Just recently, the remains of an old Viking settlement have been explored at L'Anse au Meadow on the northern tip of Newfoundland. Excavators have found a longhouse, a probable steam bath, the slag of iron smelters, bone and ivory chessmen, and a spinning weight. This Norse village in Canada dates from the tenth century.

Soon after Columbus's voyage, there began a movement of Europeans and Negro slaves into the New World. We shall not discuss this great migration, beyond a comment that the physical environment exerted some brief and minor control over the distribution of the immigrants. In the southeastern United States, much movement toward the interior was up the rivers by boat, and was barred by the Fall Line, the old Cretaceous shoreline where the first falls and rapids are encountered. And so on many a river a community was started at the head of navigation. It is no coincidence that the Cretaceous shoreline is still marked by the cities of Tuscaloosa, Columbus, Macon, Augusta, Columbia, Richmond, and Washington.

The Appalachian uplift did not long prove a barrier to westward migration overland in North America; but the European immigrants, essentially a forest people, were baffled by the plains. For a time they preferred to reach wooded California by the long sea route around South America, rather than venture into the dry grasslands and deserts; but soon they developed techniques that permitted crossing of the arid lands.

Today, of course, man is venturing even into the Antarctic, the depths of the sea, and outer space. A pioneer astronaut, who also entered into deep-sea exploration, said that the ocean depths were far more hostile to man than outer space.

Some man-made devices now repose on the moon and on Venus; others have sent back data on the Venusian, Martian, or lunar environment. Man himself will probably set foot on both the moon and Mars before the present century is out.

Will he find life on other worlds? It would be pointless to guess, when we are so close to knowing for sure. At this time we should do no more than summarize investigations into the subject. Biochemists have shown that complex organic molecules, the chemicals of life, are formed more readily than was previously realized. Biogeographers have shown that certain living things of earth normally exist in very harsh environments; some have survived in laboratory duplications of lunar or Martian conditions.

Meteorites made of carbonaceous material have struck the earth. Chemical examination has revealed them to contain complex organic substances of a kind we usually attribute to the presence of living things. Microscopic examination of these meteorites has revealed structures somewhat resembling algal and protozoan cells; two of these structures have even been given generic and specific names. (But some workers believe that the contents of meteorites, both the chemicals and the apparent fossils, are of nonliving origin.)

Alfred Russel Wallace, who was not earthbound in his speculations, concluded that the Martian surface should be much like the lunar, pocked with craters of varying sizes. In 1965 the space probe Mariner IV proved this to be the case. Photographs sent back by the probe revealed no trace of life; but then our weather satellites, encircling earth, rarely pick up any sign of terrestrial life, not even vegetation.

Polar ice caps (or frost caps) on Mars vary seasonally. When they begin to melt back in the Martian summer, dark areas spread over the planet, as though dormant vegetation were awakening; and a dark border follows the retreating edge of the ice cap. Spectroscopic analysis indicates the dark areas to be possibly organic, but not chlorophyll-bearing vegetation as we know it.

The Mariner II probe revealed Venus to be very hot beneath its cloud cover, but a relatively cool spot was also found, perhaps a mountain peak.

The density, pressure, and temperature of Jupiter's atmosphere, as estimated without direct probing, would permit the formation of organic chemicals.

There is one life-form that almost surely will be encountered on other worlds: earth-evolved bacteria that rode man's space-probing vehicles. There are bacteria that need no oxygen whatsoever; that can resist extreme cold; that can live in an astonishing variety of substances, even rocket fuel. The chances of bacterial survival and proliferation are regarded as no better than fair on the surface of the moon, but excellent on Mars.

What about intelligent life on other worlds? There is no hint of it. In 1877 the Italian astronomer Giovanni Schiaparelli thought he saw a network of lines on the surface of Mars. He called the lines "canali," meaning "channels." Unfortunately, the word was mistranslated as "canals," with a misleading implication of artificial construction. The evidence of channels, even as natural features, has not been demonstrated to the satisfaction of all astronomers. Surprisingly, the Mariner IV probe has not settled the controversy as to whether any linear features exist on Mars.

Is it necessary to say anything here about "flying saucers"? Only this, perhaps: Witches and warlocks, goblins and fairies, snowmen and saucer-men, unicorns and werewolves—these inhabit murky corridors of the human mind, where they occupy habitats left open by an absence of facts, and where they are to be studied by the psychologist rather than the biogeographer. Studies on them help us to understand Neanderthal man, capering about a bear skull and choosing to reject the conclusion that it is merely the cranium of a dead animal.

But in spite of false starts into demonology and alchemy, necromancy and prophecy, numerology and astrology, the pursuit of enlightenment has gone on. The mainstream of this pursuit will carry man, along with a few of his favored plants and animals, and his bacterial hangers-on, to other worlds.

"When I see the list of books which I read and abstracted, I am surprised at my industry." **THE LITERATURE OF BIOGEOGRAPHY**

When a scientist makes a discovery, he describes it in print; thus other workers have chance to evaluate his contention. Any science could therefore be thought of as an edifice whose building blocks are books and technical papers. Since biogeography overlaps several major fields of study—zoology, botany, geology, paleontology, climatology, ecology, and more—its literature is enormous, and can only be sampled here. The following list of references is intended primarily for the general reader, who might wish to look further into biogeography or to pursue certain topics at greater length. Secondarily it may serve the purpose of the advanced student, who seeks to learn the authority on which various statements are made.

A. Biogeography of the Land

It should be borne in mind that biogeography, like most sciences, has advanced rapidly in the last few years. Fresh biogeographic insights are provided especially by continuing work in paleontology, geology, taxonomy, and biological collecting.

1. Bartholomew, J. G., W. E. Clarke, and P. H. Grimshawe. *Bartholomew's Physical Atlas.* Vol. V. *Atlas of Zoogeography.* Edinburgh, John Bartholomew and Co., 1911.

* CHARLES DARWIN, on his study of domestic animals and plants.

445

2. Cain, Stanley A. *Foundations of Plant Geography*. New York, Harper and Brothers, 1944.

3. Croizat, Léon. *Manual of Phytogeography*. The Hague, Junk, 1952.

4. —————— *Panbiogeography*. 3 vols. Caracas, L. Croizat, 1958.

5. Dansereau, Pierre. *Biogeography: An Ecological Perspective*. New York, Ronald Press, 1957. [Concerned chiefly with plants and the ecological approach to distribution.]

6. Darlington, Philip J., Jr. *Zoogeography: The Geographical Distribution of Animals*. New York, John Wiley and Sons, Inc., 1957. [Concerned with vertebrates, and exemplifying the historical approach to distributional studies.]

7. De Beaufort, L. F. *Zoogeography of the Land and Inland Waters*. London, Sidgwick and Jackson Ltd., 1951.

8. George, Wilma. *Animal Geography*. New York, Dover Publications, Inc., 1962.

9. Gleason, Henry A., and Arthur Cronquist. *The Natural Geography of Plants*. New York, Columbia University Press, 1964. [A readable account treating North America in special detail.]

10. Good, Ronald. *The Geography of the Flowering Plants*. New York, John Wiley and Sons, Inc., 1964. [Exemplifying the taxonomic approach to plant distribution.]

11. Hesse, Richard, W. C. Allee, and K. P. Schmidt. *Ecological Animal Geography*. New York, John Wiley and Sons, Inc., 1951. [The ecological approach to vertebrate and invertebrate distribution.]

12. Neill, Wilfred T. *Biogeography: The Distribution of Animals and Plants*. Boston, D. C. Heath and Co., 1964. [A very brief summary in the Biological Sciences Curriculum Study series, intended for the guidance of biology teachers.]

13. Polunin, Nicholas. *Introduction to Plant Geography*. New York, McGraw-Hill Book Co., Inc., 1960.

14. Romer, Alfred S. *Vertebrate Paleontology*. Chicago, University of Chicago Press, 1966. [Summarizes the fossil record of vertebrates.]

15. Schimper, A. F. W. *Plant Geography upon a Physiological Basis*. New York, Stechert-Hafner Service Agency, Inc., 1964.

16. Schmidt, Karl P. "Faunal Realms, Regions, and Provinces," *Quarterly Review of Biology*, Vol. 29, No. 4 (1954), pp. 322–31. [Brief summary, with maps, of faunal subdivisions in modern view.]

17. Wulff, E. V. *An Introduction to Historical Plant Geography*. Waltham, Mass., Chronica Botanica Co., 1943.

B. Guides to the Vertebrates

Handbooks, manuals, and guides often treat life-histories as well as distribution. This is all to the good since the two are interrelated in a variety of ways.

1. Austin, Oliver L., Jr. *Birds of the World*. New York, Golden Press, 1961.
2. Bishop, Sherman C. *Handbook of Salamanders*. Ithaca, N.Y., Comstock Publishing Co., 1943. [Covers the species of the United States, Canada, and Baja California.]
3. Blair, W. F., A. P. Blair, Pierce Brodkorb, F. R. Cagle, and G. A. Moore. *Vertebrates of the United States*. New York, McGraw-Hill Book Co., Inc., 1957.
4. Burt, William H. *A Field Guide to the Mammals*. Boston, Houghton Mifflin Co., 1952. [Covers North America, north of Mexico.]
5. Carr, Archie. *Handbook of Turtles*. Ithaca, N.Y., Cornell University Press, 1952. [Covers the United States, Canada, and Baja California.]
6. Cochran, Doris. *Living Amphibians of the World*. Garden City, N.Y., Doubleday and Co., Inc., 1961.
7. Conant, Roger. *A Field Guide to the Reptiles and Amphibians of Eastern North America*. Boston, Houghton Mifflin Co., 1958.
8. Editors of *Life*. *A Guide to the Natural World and Index to the "Life" Nature Library*. New York, *Time*, Inc., 1965. [Covers major taxonomic groups of both plants and animals.]
9. Gilliard, E. Thomas. *Living Birds of the World*. Garden City, N.Y., Doubleday and Co., Inc., 1958.
10. Goin, C. J., and O. B. Goin. *Introduction to Herpetology*. San Francisco, W. H. Freeman and Co., 1962.
11. Hall, E. R., and K. R. Nelson. *The Mammals of North America*. 2 vols. New York, Ronald Press, 1959.
12. Herald, Earl. *Living Fishes of the World*. Garden City, N.Y., Doubleday and Co., Inc., 1961.
13. Hickman, Cleveland P. *Integrated Principles of Zoology*. St. Louis, Mo., C. V. Mosby Co., 1961.
14. Hylander, Clarence J. *Fish and Their Ways*. New York, The Macmillan Co., 1964.
15. Lanham, Url N. *The Fishes*. New York, Columbia University Press, 1962.
16. Mertens, Robert. *The World of Amphibians and Reptiles*. New York, McGraw-Hill Book Co., Inc., 1960.
17. Norman, John R. *A History of Fishes*. New York, Hill and Wang, 1947.
18. Oliver, James A. *The Natural History of North American Amphibians and Reptiles*. Princeton, N.J., D. Van Nostrand Co., Inc., 1955.
19. Perlmutter, Alfred. *Guide to Marine Fishes*. New York, New York University, 1961.
20. Peterson, Roger T. *A Field Guide to the Birds*. Boston, Houghton Mifflin Co., 1947. [Covers land and water birds of eastern United States.]
21. —— *A Field Guide to Western Birds*. Boston, Houghton Mifflin Co., 1961. [Covers land and water birds of western United States.]

22. Pettingill, Olin S., Jr. *A Laboratory and Field Manual of Ornithology*. Minneapolis, Minn., Burgess Publishing Co., 1956. [Includes considerable information on distribution.]
23. Sanderson, Ivan T. *Living Mammals of the World*. Garden City, N.Y., Doubleday and Co., Inc., 1956.
24. Schultz, L. P., and E. M. Stern. *The Ways of Fishes*. New York, D. Van Nostrand Co., Inc., 1948.
25. Smith, Hobart M. *Handbook of Lizards*. Ithaca, N.Y., Comstock Publishing Co., 1946. [Covers the United States and Canada.]
26. Stebbins, Robert C. *Amphibians and Reptiles of Western North America*. New York, McGraw-Hill Book Co., Inc., 1954. [Covers the area north of Mexico.]
27. Sterba, Gunther. *Freshwater Fishes of the World*. New York, Viking Press, 1962.
28. Walker, Ernest P. (and 6 others). *Mammals of the World*. 3 vols. Baltimore, Johns Hopkins Press, 1964.
29. Wright, A. H., and A. A. Wright. *Handbook of Frogs and Toads of the United States and Canada*. Ithaca, N.Y., Comstock Publishing Co., Inc., 1949.
30. ———, and ———. *Handbook of Snakes of the United States and Canada*. 2 vols. Ithaca, N.Y., Cornell University Press, 1957. [A third volume, the bibliography, was published by the authors at Ithaca in 1962.]
31. Young, J. Z. *The Life of Vertebrates*. New York, Oxford University Press, 1962.

C. Guides to the Seed-Bearing Plants

In addition to the following references, also see B 8 (above).

1. Bailey, Liberty H. *Manual of Cultivated Plants*. New York, Macmillan Co., 1949.
2. Benson, Lyman, and R. A. Darrow. *The Trees and Shrubs of the Southwestern Deserts*. Albuquerque, N.M., University of New Mexico Press, 1954.
3. Blackburn, Benjamin. *Trees and Shrubs in Eastern North America*. New York, Oxford University Press, 1952.
4. Blake, S. F., and A. C. Atwood. *Geographical Guide to the Floras of the World*. New York, Hafner Publishing Co., Inc., 1942. [Essentially a bibliography of regional floral studies.]
5. Brown, William H. *The Plant Kingdom*. Waltham, Mass., Blaisdell Publishing Co., Inc., 1935.
6. Fernald, Merritt L. *Gray's Manual of Botany*. Eighth (Centennial) Edition. New York, American Book Co., 1950. [Covers ferns and seed-bearing plants of the central and northeastern United States and adjacent Canada.]
7. Gilkey, Helen M. *Handbook of Northwest Flowering Plants*. Portland, Ore., Binfords and Mort, 1961.
8. Gleason, Henry A. *The New Britton and Brown Illustrated Flora of the North-*

eastern United States and Adjacent Canada. 3 vols. New York, Hafner Publishing Co., Inc., 1963.

9. ———, and Arthur Cronquist. *Manual of Vascular Plants of Northeastern United States and Adjacent Canada.* Princeton, N.J., D. Van Nostrand Co., 1963.
10. Hitchcock, A. S. *Manual of the Grasses of the United States.* U.S. Department of Agriculture, Miscellaneous Publications, No. 200 (1951).
11. Hutchinson, John. *Families of Flowering Plants.* Vol. 1. *Dicotyledones.* New York, Oxford University Press, 1964.
12. Hylander, Clarence J. *The World of Plant Life.* New York, The Macmillan Co., 1956. [Covers lower plants as well as seed-bearers.]
13. Jepson, Willis L. *Manual of the Flowering Plants of California.* Berkeley, Cal., University of California Press, 1960.
14. Rickett, Harold W. *Wild Flowers of the United States.* Vol. 1. *The Northeastern States.* New York, McGraw-Hill Book Co., Inc., 1966.
15. Rydberg, P. A. *Flora of the Rocky Mountains and Adjacent Plains.* New York, Hafner Publishing Co., Inc., 1954.
16. Sargent, Charles S. *Manual of the Trees of North America.* Cambridge, Mass., Riverside Press, 1933.
17. Seymour, E. L. D. (ed.). *The Garden Encyclopedia.* New York, W. H. Wise and Co., 1936.
18. Shreve, Forrest, and I. L. Wiggins. *Vegetation and Flora of the Sonoran Desert.* 2 vols. Stanford, Cal., Stanford University Press, 1964.
19. Small, John K. *Manual of the Southeastern Flora.* Chapel Hill, N.C., University of North Carolina Press, 1953.
20. Went, Frits W., and the Editors of *Life. The Plants.* New York, *Time,* Inc., 1963.

D. *The Earth and the Natural Landscape*

Also see A 10 and A 11.

1. Atwood, W. W. *The Physiographic Provinces of North America.* Waltham, Mass., Blaisdell Publishing Co., 1940.
2. Beiser, Arthur, and the Editors of *Life. The Earth.* New York, *Time,* Inc., 1962.
3. Editors of *Life,* and Lincoln Barnett. *The World We Live In.* New York, *Time,* Inc., 1955.
4. Farb, Peter. *Face of North America: The Natural History of a Continent.* New York, Harper and Row, 1963.
5. Fenneman, Nevin M. *Physiography of Western United States.* New York, McGraw-Hill Book Co., Inc., 1931.

6. Fenneman, Nevin M. *Physiography of Eastern United States.* New York, McGraw-Hill Book Co., Inc., 1948.
7. Finch, V. C., and G. T. Trewartha. *Physical Elements of Geography.* New York, McGraw-Hill Book Co., Inc., 1949.
8. Josephy, Alvin M., Jr. (ed.). *The American Heritage Book of Natural Wonders.* New York, American Heritage Publishing Co., Inc., 1963. [Interesting portrayals of the American landscape before it was much altered.]
9. Mather, Kirtley F. *The Earth beneath Us: The Fascinating Story of Geology.* New York, Random House, 1964.
10. Monkhouse, F. J., and A. V. Hardy. *The North American Landscape.* New York, Cambridge University Press, 1965.
11. Pangborn, Mark W., Jr. *Earth for the Layman.* Washington, D.C., American Geological Institute, 1957. Supplement, 1960.
12. Shimer, John A. *This Sculptured Earth: The Landscape of America.* New York, Columbia University Press, 1959.
13. Thornbury, William D. *Regional Geomorphology of the United States.* New York, John Wiley and Sons, Inc., 1965.

E. In the Field and the Laboratory

The following references pertain especially to the collecting, preserving, naming, and studying of plants and animals, as well as the preparation and use of maps. In this last connection, also see D 7.

1. Anderson, Rudolph M. *Methods of Collecting and Preserving Vertebrate Animals.* National Museum of Canada, Bulletin No. 69 (1948).
2. Camp, W. H., H. W. Rickett, and C. A. Weatherby (eds.). "International Rules of Botanical Nomenclature," *Brittonia,* Vol. 6, No. 1 (1947), pp. 1–120.
3. Conant, Roger. "The Queen Snake, *Natrix septemvittata,* in the Interior Highlands of Arkansas and Missouri, with Comments upon Similar Disjunct Distributions," *Proceedings of the Academy of Natural Sciences of Philadelphia,* Vol. 112, No. 2 (1960), pp. 25–40.
4. Dorf, Erling. "Climatic Changes of the Past and Present," *Contributions from the Museum of Paleontology, University of Michigan,* Vol. 13, No. 8 (1959), pp. 181–210.
5. Frey, David G. "Pollen Succession in the Sediments of Singletary Lake, North Carolina," *Ecology,* Vol. 32, No. 3 (1951), pp. 518–33.
6. Grosvenor, Melville B. (ed.). *National Geographic Atlas of the World.* Washington, D.C., National Geographic Society, 1963.
7. Haag, William G. "The Bering Strait Land Bridge," *Scientific American,* Vol. 206, No. 1 (1962), pp. 112–23.

8. Hibbard, Claude W. "An Interpretation of Pliocene and Pleistocene Climates in North America," *Michigan Academy of Science, Arts, and Letters,* Report for 1959–60, pp. 5–30.

9. Hitchcock, C. B., and F. Debenham (eds.). *Reader's Digest Great World Atlas.* Pleasantsville, N.Y., Reader's Digest Association, 1963.

10. Hopkins, David M. "Cenozoic History of the Bering Land Bridge," *Science,* Vol. 129 (3362) (1959), pp. 1519–28.

11. Jaeger, Edmund C. *A Source-Book of Biological Names and Terms.* Springfield, Ill., Charles C. Thomas, 1944.

12. Lobeck, Armin K. *Things Maps Don't Tell Us: Adventures into Map Interpretation.* New York, The Macmillan Co., 1956.

13. Loveridge, A., and B. Shreve. "The 'New Guinea' Snapping Turtle (*Chelydra serpentina*)," *Copeia,* No. 2 (1947), pp. 120–23.

14. Mayr, Ernst. *Systematics and the Origin of Species.* New York, Columbia University Press, 1942.

15. Neill, Wilfred T. "Historical Biogeography of Present-Day Florida," *Bulletin of the Florida State Museum, Biological Sciences,* Vol. 2, No. 7 (1957), pp. 175–220.

16. ——— "*Hemidactylium scutatum.*" P. 2 *in* W. J. Riemer (ed.), *Catalogue of American Amphibians and Reptiles.* Bethesda, Md., American Society of Ichthyologists and Herpetologists, 1963.

17. Savory, Theodore. *Naming the Living World.* New York, John Wiley and Sons, Inc., 1963.

18. Schenk, E. T., and J. H. McMasters. *Procedure in Taxonomy.* Stanford, Cal., Stanford University Press, 1948.

19. Schuchert, Charles. *Historical Geology of the Antillean-Caribbean Region.* New York, John Wiley and Sons, Inc., 1935.

20. Simpson, George G. "The Principles of Classification and a Classification of Mammals," *Bulletin of the American Museum of Natural History,* Vol. 85 (1945).

21. ——— "Evolution, Interchange, and Resemblance of the North American and Eurasian Cenozoic Mammalian Faunas," *Evolution,* Vol. 1, No. 3 (1947), pp. 218–20.

22. ——— "History of the Fauna of Latin America," *American Scientist* (July 1950), pp. 361–89.

23. Stoll, N. R., and 6 others (eds.). *International Code of Zoological Nomenclature Adopted by the XV International Congress of Zoology.* London, International Commission on Zoological Nomenclature, 1961.

24. "Symposium: Linnaeus and Nomenclatorial Codes," *Systematic Zoology,* Vol. 8, No. 1 (1959), pp. 1–47.

25. Troughton, Ellis. *Furred Animals of Australia.* Sydney, Australia, Angus and Robertson Ltd., 1943.

26. *United States National Museum 1966 Annual Report*. Washington, D.C.
27. Wells, H. G. *The Outline of History*. New York, The Macmillan Co., 1921.
28. Woodring, W. P. "Caribbean Land and Sea through the Ages," *Bulletin of the Geological Society of America*, Vol. 65 (1964), pp. 719–32.

F. *Faunal Subdivisions*

This section pertains to the development of zoogeographic concepts to their modern form. Also see A 1, A 16, and especially A 6 on zoogeography; the atlases E 6 and E 9; E 7, E 10, E 19, E 20, E 21, and E 28 on barriers and bridges between continents; and A 11, D 2, and D 3 on climatic zones of the earth. Certain early works, mentioned in the text, are rare and are not cited. For many early titles, see the respective bibliographies of A 6, A 16, and B 28; also note F 8, F 14, F 17, and F 21, below.

1. Allee, W. C., A. E. Emerson, O. Park, T. Park, and K. P. Schmidt. *Principles of Animal Ecology*. Philadelphia, Pa., W. B. Saunders Co., 1949.
2. Bates, Marston, and the Editors of *Life*. *The Land and Wildlife of South America*. New York, *Time*, Inc., 1964.
3. Bergamini, David, and the Editors of *Life*. *The Land and Wildlife of Australia*. New York, *Time*, Inc., 1964.
4. Bourliere, Francois, and the Editors of *Life*. *The Land and Wildlife of Eurasia*. New York, *Time*, Inc., 1964.
5. Carr, Archie, and the Editors of *Life*. *The Land and Wildlife of Africa*. New York, *Time*, Inc., 1964.
6. Darwin, Charles R. *On the Origin of Species by Means of Natural Selection . . .* London, John Murray, 1859.
7. Dunn, Emmett R. "The Herpetological Fauna of the Americas," *Copeia*, No. 3 (1931), pp. 106–19.
8. Engelmann, Wilhelm. *Bibliotheca Historico-Naturalis, 1700–1846*. New York, Stechert-Hafner Service Agency.
9. Farb, Peter, and the Editors of *Life*. *The Land and Wildlife of North America*. New York, *Time*, Inc., 1964.
10. Freuchen, Peter, and Finn Salomonsen. *The Arctic Year*. New York, G. P. Putnam's Sons, 1958.
11. Gilmore, Raymond M. "Fauna and Ethnozoology of South America." pp. 345–464 in J. H. Steward (ed.), *Handbook of South American Indians*, Vol. 6. Bureau of American Ethnology Bulletin No. 143 (1950).
12. Heilprin, Angelo. *The Geographical and Geological Distribution of Animals*. New York, Appleton Co., 1887.

13. Leopold, A. Starker, and the Editors of *Life. The Desert.* New York, *Time*, Inc., 1962.
14. Meisel, Max. A *Bibliography of American Natural History.* 3 Vols. Brooklyn, N.Y., Premier Publishing Co., 1924–1929.
15. Ripley, S. Dillon, and the Editors of *Life. The Land and Wildlife of Tropical Asia.* New York, *Time*, Inc., 1964.
16. Schmidt, Karl P. "On the Zoogeography of the Holarctic Region," *Copeia*, No. 3 (1946), pp. 144–52.
17. Smith, R. C. *Guide to the Literature of the Zoological Sciences.* Minneapolis, Minn., Burgess Publishing Co., 1958.
18. "Symposium: The Darwin-Linnaeus Year," *University of Kansas Science Bulletin*, Supplement to Vol. 42 (1962).
19. Umbgrove, J. H. F. *Structural History of the East Indies.* New York, Cambridge University Press, 1949.
20. Wallace, Alfred R. *The Geographic Distribution of Animals.* 2 vols. London, The Macmillan Co., 1876.
21. Wood, C. A. *An Introduction to the Literature of Vertebrate Zoology.* London, Oxford University Press, 1931.

G. *Floral Subdivisions*

This section relates especially to the development of phytogeographic concepts, and the differences between the higher plants and the higher animals as regards their age and their capabilities for dispersal. Also see A 10, A 14, A 17, C 20, and F 1. Rare, early works are not included; most are listed in the respective bibliographies of A 5 and A 10.

1. Colbert, Edwin H. *Dinosaurs: Their Discovery and Their World.* New York, E. P. Dutton and Co., Inc., 1961.
2. Dorf, Erling. *Upper Cretaceous Floras of the Rocky Mountain Region. II: Flora of the Lance Formation at Its Type Locality, Niobrara County, Wyoming.* Carnegie Institution of Washington, Publication No. 508 (1942).
3. Elton, Charles S. *The Ecology of Invasions by Animals and Plants.* New York, John Wiley and Sons, Inc., 1958.
4. Estes, Richard. *Fossil Vertebrates from the Late Cretaceous Lance Formation [,] Eastern Wyoming.* University of California Publications in Geological Sciences, Vol. 49 (1964).
5. Farb, Peter, and the Editors of *Life. Ecology.* New York, *Time*, Inc., 1963.
6. Howell, A. Brazier. "Agencies Which Govern the Distribution of Life," *American Naturalist*, Vol. 56, pp. 428–38 (1922).

7. Seward, A. C. *Plant Life through the Ages.* Cambridge, The University Press, 1941.

H. Australia, South-Temperate Africa, and the Far-Southern Lands

The following references pertain to Chapter 10. Also see the phytogeographies, especially A 10; and A 1, A 6, A 11, B 1, B 6, B 9, B 23, B 27, B 28, C 20, E 25, F 2, F 3, F 5, and F 11.

1. Adamson, R. S., and T. M. Salter (eds.). *The Flora of the Cape Peninsula.* Cape Town, Juta and Co. Ltd., 1950.
2. Allan, H. H. "A Consideration of the 'Biological Spectra' of New Zealand," *Ecology,* Vol. 25, No. 2 (1937), pp. 116–52.
3. Brown, R. N. R. "Antarctic and Sub-Antarctic Plant Life and Some of Its Problems." Pp. 343–52 in Problems of Polar Research, American Geographical Society, Special Publication No. 7 (1928).
4. Bryant, Herwil M. "Biology at East Base, Palmer Peninsula, Antarctica," *Proceedings of the American Philosophical Society,* Vol. 89 (1945), pp. 256–69.
5. Camp, W. H. "Distribution Patterns in Modern Plants and the Problems of Ancient Dispersals," *Ecological Monographs,* Vol. 17 (1947), pp. 123–26, 159–83.
6. Darlington, Philip J., Jr. *Biogeography of the Southern End of the World.* Cambridge, Mass., Harvard University Press, 1965.
7. Davis, D. H. S. (ed.). *Ecological Studies in Southern Africa.* The Hague, Dr. W. Junk, 1964.
8. Editors of *Life,* and Lincoln Barnett. *The Wonders of Life on Earth.* New York, Time, Inc., 1960.
9. Keast, A. (ed.). *Biogeography and Ecology in Australia.* The Hague, Dr. W. Junk, 1959.
10. Ley, Willy, and the Editors of *Life. The Poles.* New York, Time, Inc., 1962.
11. Mayr, Ernst. "Wallace's Line in the Light of Recent Zoogeographic Studies," *Quarterly Review of Biology,* Vol. 19, No. 1 (1944), pp. 1–14.
12. Osborn, Fairfield. *The Pacific World.* New York, W. W. Norton and Co., Inc., 1944.
13. Sauer, Carl O. "Geography of South America." Pp. 319–44 in J. H. Steward (ed.), *Handbook of South American Indians,* Vol. 6. Bureau of American Ethnology, Bulletin No. 143 (1950).
14. Simpson, Frank A. (ed.). *The Antarctic Today.* San Francisco, A. H. and A. W. Reed, 1952.
15. "Symposium: The Antarctic," *Scientific American* (Sept., 1962).

16. Thieret, John W. "Kerguelen's Cabbage," *Chicago Natural History Museum Bulletin*, Vol. 32, No. 4 (1961), pp. 4–5, 8.
17. Van Oye, P., and J. van Mieghem (eds.). *Biogeography and Ecology in Antarctica*. The Hague, Dr. W. Junk, 1966.

I. In and Near Tropical Africa

The references below relate to Chapter 11. Also see A 10 and other phytogeographies, as well as A 6, A 11, B 1, B 6, B 9, B 10, B 12, B 16, B 23, C 20, F 1, F 5, H 7, and H 8.

1. Bews, J. W. "The South-east African Flora: Its Origin, Migrations, and Evolutionary Tendencies," *Annals of Botany*, 36 (1922).
2. Boughey, A. S. *The Origin of the African Flora*. New York, Oxford University Press, 1957.
3. Brown, Leslie. *Africa: A Natural History*. New York, Random House, Inc., 1965.
4. Carter, T. Donald. "Stalking Central Africa's Wildlife," *National Geographic* (August 1956), pp. 264–86.
5. Darling, F. Fraser. *Wildlife in an African Territory*. New York, Oxford University Press, 1960.
6. ———— "Wildlife Husbandry in Africa," *Scientific American* (November 1960), pp. 123–30, 133–34.
7. Edwards, D. C., and A. V. Bogden. *Important Grassland Plants of Kenya*. London, Sir Isaac Pitman and Sons, 1951.
8. Grzimek, Bernhard. "The Last Great Herds of Africa," *Natural History* (January 1961), pp. 8–21.
9. Hoppe, E. O. "Mt. Kenya Plants," *Nature Magazine* (December 1953), pp. 529–32.
10. Hutchinson, J. B. *A Botanist in South Africa*. London, P. R. Gawthorn Ltd., 1946.
11. Junod, V. I., and I. N. Resnick. *Handbook of Africa*. New York, New York University, 1963.
12. Keay, R. W. J. (ed.). *Vegetation Map of Africa, South of the Tropic of Cancer*. New York, Oxford University Press, 1959.
13. Keynes, Quentin. "St. Helena: The Forgotten Island," *National Geographic* (August 1950), pp. 265–80.
14. Kimble, G. H. T. *Tropical Africa*. 2 vols. Garden City, N.Y., Doubleday and Co., Inc., 1960.
15. McDonald, William A. "The Deadly Glossina," *Natural History* (October 1958), pp. 427–31.

16. Millar, Lynn. "The Seychelles," *Natural History* (October 1966), pp. 48–51.
17. Sibree, James. *A Naturalist in Madagascar.* London, Seeley, Service and Co., 1915.
18. Thompson, B. W. *The Climate of Africa.* New York, Oxford University Press, 1965.
19. White, F. *Forest Flora of Northern Rhodesia.* New York, Oxford University Press, 1962.
20. Zahl, Paul A. "Mountains of the Moon," *National Geographic,* (March 1962), pp. 412–34.

J. In and Near Tropical Asia

References for Chapter 12. Also see A 10 and other phytogeographies, and A 11, A 16, B 1, B 9, B 12, B 16, B 27, B 28, C 4, F 15 (especially), F 19, and H 12.

1. Aubert de la Rüe, E., F. Bourliere, and J. P. Harroy. *The Tropics.* New York, Alfred A. Knopf, 1957.
2. Brown, W. H. *Vegetation of Philippine Mountains.* Manila, Philippine Bureau of Printing, 1919.
3. Carter, T. D., J. E. Hill, and G. H. H. Tate. *Mammals of the Pacific World.* New York, The Macmillan Co., 1945.
4. Cressey, George B. *Asia's Lands and Peoples.* New York, McGraw-Hill Book Co., Inc., 1951.
5. Dobby, E. H. G. *Southeast Asia.* New York, John Wiley and Sons, Inc., 1951.
6. Farb, Peter, and the Editors of *Life. The Forest.* New York, *Time,* Inc., 1961.
7. Gee, E. P. *The Wild Life of India.* New York, William Collins' Sons and Co. Ltd., 1964.
8. Holttum, R. E. *Plant Life in Malaya.* New York, Longmans, Green and Co., 1954.
9. Kendrew, W. G. *The Climates of the Continents.* Oxford, Clarendon Press, 1961.
10. Merrill, Elmer D. *Plant Life of the Pacific World.* New York, The Macmillan Co., 1945.
11. Myers, George S. "Fresh-water Fishes and East Indian Zoogeography," *Stanford Ichthyological Bulletin,* Vol. 4, No. 1 (1951), pp. 19–27.
12. Richards, Paul W. *The Tropical Rain Forest.* Cambridge, Cambridge University Press, 1957.
13. Stamp, L. Dudley. *Asia.* New York, E. P. Dutton and Co., Inc., 1962.
14. Stebbing, E. P. *The Forests of India.* 3 vols. London, John Lane, 1922–1926.
15. Van Bemmelen, R. W. *The Geology of Indonesia.* Vol. 1 A. The Hague, Martinus Nijhoff, 1949.

16. Wadia, D. N. *Geology of India*. New York, The Macmillan Co., 1953.
17. Wallace, Alfred R. *The Malay Archipelago*. New York, Dover Publications, Inc., 1962.

K. In and Near Polynesia

References for Chapter 13. Also see A 10 and the other phytogeographies; likewise see A 6, J 3, and J 10.

1. Bryan, William A. *Natural History of Hawaii*. Honolulu, Hawaiian Gazette Co., 1915.
2. Editors of Sunset Books. *Hawaii*. Menlo Park, Cal., Lane Book Co., 1964.
3. Gulick, A. "Biological Peculiarities of Oceanic Islands," *Quarterly Review of Biology*, Vol. 7 (1932), pp. 405–27.
4. Henshaw, H. W. *Birds of the Hawaiian Islands*. Honolulu, Thos. G. Thum, 1902.
5. Hillebrand, William. *Flora of the Hawaiian Islands*. New York, Hafner Publishing Co., Inc., 1966. [Reprint of 1888 work.]
6. Keast, Allen. *Australia and the Pacific Islands*. New York, Random House, Inc., 1966.
7. Loveridge, Arthur. *Reptiles of the Pacific World*. New York, The Macmillan Co., 1945.
8. Marshall, J. T., Jr. "Vertebrate Ecology of Arno Atoll, Marshall Islands," Atoll Research Bulletin No. 3 (1951).
9. Mayr, Ernst. "The Zoogeographic Position of the Hawaiian Islands," *Condor*, Vol. 45, No. 2 (1943), pp. 45–48.
10. ——— *Birds of the Southwest Pacific*. New York, The Macmillan Co., 1945.
11. Neal, M. C. *In Gardens of Hawaii*. Honolulu, Bishop Museum Press, 1965.
12. Roughley, T. C. "*Bounty* Descendants Live on Remote Norfolk Island," National Geographic (October 1960), pp. 558–84.
13. Schmidt, Karl P. "Essay on the Zoogeography of the Pacific Islands." Appendix (pp. 275–92) in S. N. Shurcliff, *Jungle Islands*, New York, G. P. Putnam's Sons, 1930.
14. Setchell, W. A. "Pacific Insular Floras and Pacific Paleogeography," *American Naturalist*, Vol. 69 (1935), pp. 289–310.
15. Stone, Benjamin. "Archipelagic Refuge," *Natural History* (November 1963), pp. 33–39. [On endemic plants of Hawaii.]
16. Taylor, William R. *Plants of Bikini and Other Northern Marshall Islands*. Ann Arbor, Mich., University of Michigan Press, 1950.
17. Tinker, Spencer W. *Animals of Hawaii*. Honolulu, Tongg Publishing Co., 1941.

18. Woodford, C. M. "The Gilbert Islands," *Geographical Journal*, Vol. 6 (1895), pp. 325–50.
19. Zimmerman, Elwood. *Insects of Hawaii*. Vol. 1. Honolulu, University of Hawaii Press, 1948. [Includes remarks on other animal groups, and plants.]

L. In and Near the New World Tropics

Two chapters, 14 and 15, may be treated together. In addition see the phytogeographies; also A 6, A 16, B 1, B 6, B 9, B 12, B 16, B 23, B 27, B 28, E 19, E 22, E 28, F 2 (especially), F 7, F 11, H 13, J 1, J 9, and J 12.

1. Allen, Glover M. "Mammals of the West Indies," *Bulletin of the Museum of Comparative Zoology, Harvard College*, Vol. 54, No. 6 (1911), pp. 175–263.
2. Barbour, Thomas. "Contributions to the Zoogeography of the West Indies, with Especial Reference to Amphibians and Reptiles," *Memoirs of the Museum of Comparative Zoology, Harvard College*, Vol. 44, No. 2 (1914), pp. 209–359.
3. Bates, Henry W. *The Naturalist on the Amazon*. New York, E. P. Dutton and Co., Inc., 1910.
4. Bates, Marston. *The Forest and the Sea*. New Work, Random House, Inc., 1960.
5. Beebe, William. *Galapagos: World's End*. New York, G. P. Putnam's Sons, 1924.
6. ———— *High Jungle*. New York, Duell, Sloan, and Pearce, 1949.
7. ———— *Edge of the Jungle*. New York, Meredith Press, 1950.
8. Belt, Thomas. *The Naturalist in Nicaragua*. New York, E. P. Dutton and Co., 1911.
9. Britton, N. L. *Flora of Bermuda*. New York, Hafner Publishing Co., 1965. [Reprint of 1918 work.]
10. Carlson, Fred A. *Geography of Latin America*. New York, Prentice-Hall, Inc., 1951.
11. Darlington, Philip J., Jr. "The Origin of the Fauna of the Greater Antilles, with Discussion of Dispersal of Animals over Water and through the Air," *Quarterly Review of Biology*, Vol. 13, No. 3 (1938), pp. 274–300.
12. Davis, David E. "Annual Cycle of Plants, Mosquitos, Birds, and Mammals in Two Brazilian Forests," *Ecological Monographs*, Vol. 15, No. 4 (1945), pp. 243–95.
13. Davis, Luther C., Jr. "The Amazon's Rate of Flow," *Natural History* (June–July 1964), pp. 14–19.
14. Dawson, E. Yale. "Ecological Paradox of Coastal Peru," *Natural History* (October 1963), pp. 32–37.
15. Grisebach, A. H. R. *Flora of the British West Indian Islands*. New York, Stechert-Hafner Service Agency, Inc., 1942.

16. Gudger, Eugene W. "The Giant Fresh-Water Fishes of South America," *Scientific Monthly* (December 1943), pp. 500–13.

17. Guenther, Konrad. *Naturalist in Brazil.* Boston, Houghton Mifflin Co., 1931.

18. Leopold, A. Starker. *Wildlife of Mexico.* Berkeley, Cal., University of California Press, 1959.

19. Lutz, Bertha. "Wild Life in Brazil," *Natural History* (November–December 1932), pp. 539–50.

20. Matthiessen, Peter. *Cloud Forest: A Chronicle of the South American Wilderness.* New York, Viking Press, Inc., 1961.

21. Myers, George S. "The Amazon and Its Fishes," *The Aquarium Journal* (March, April, May, July–August 1947; February, March 1949).

22. Neill, Wilfred T. "The Vipers of Queimada Grande," *Nature Magazine* (April 1957), pp. 188–190, 220.

23. Schmidt, Karl P. "A Naturalist's Glimpse of the Andes," *Scientific Monthly* (May 1945), pp. 335–46.

24. Skutch, Alexander F. *A Naturalist in Central America.* New York, Devin-Adair Co., 1966.

25. Standley, Paul C. "Trees and Shrubs of Mexico," *Contributions of the United States National Herbarium*, Vol. 23 [in 5 parts] (1920–1926).

26. ———— "Flora of Costa Rica," *Field Museum of Natural History, Botanical Series*, Vol. 18 [in 4 parts] (1937–1938).

27. Steward, Julian H. (ed.). *Handbook of South American Indians.* 6 vols. *Bureau of American Ethnology Bulletin* No. 143 (1946–1950). [Vol. 6, especially, is of biogeographic interest, but other volumes include photographs and descriptions of the environment.]

28. Tate, G. H. H. "Life Zones at Mount Roraima," *Ecology*, Vol. 13, No. 13 (1932), pp. 235–57.

29. West, R. C., and J. P. Augelli. *Middle America: Its Lands and Peoples.* New York, Prentice-Hall, Inc., 1966.

M. *The Northern Edge of the New World Tropics*

These references pertain especially to southern Baja California and southern Florida. Some relate also to the Mexican border country of the United States, where there is a tropical element in the biota. Phytogeographies and zoogeogaphies devote relatively little space to the small areas at the periphery of the tropics in the United States and Baja California. See B 3, B 4, B 20, C 2, C 12, C 18, C 19, D 4, D 6, E 15 (especially), E 19, E 28, F 9, L 1, L 11, L 15, L 18, and L 25.

1. Beard, Daniel B. "Wildlife of Everglades National Park," *National Geographic*, January 1949, pp. 83–116.

2. Blair, W. Frank. "Biotic Provinces of Texas," *Texas Journal of Science*, Vol. 2, No. 1 (1950), pp. 93–117.

3. Blake, Emmett R. *Birds of Mexico*. Chicago, University of Chicago Press, 1953.

4. Bond, James. *Field Guide to Birds of the West Indies*. New York, Macmillan Co., 1947.

5. Cahalane, Victor C. "Everglades—Yesterday, Today and Tomorrow," *National Geographic* (December 1947), pp. 513–17.

6. Carr, A. F., Jr. "A Contribution to the Herpetology of Florida," *University of Florida Publication, Biological Science Series*, Vol. 3, No. 1 (1940), pp. 1–118. [Includes summary of major plant communities in Florida.]

7. Davis, John H., Jr. "The Natural Features of Southern Florida," *Florida State Geological Survey, Geological Bulletin* No. 25 (1943).

8. Dickson, J. D., III, R. O. Woodbury, and T. R. Alexander. "Check List of Flora of Big Pine Key, Florida and Surrounding Keys," *Quarterly Journal of the Florida Academy of Sciences*, Vol. 16, No. 3 (1953), pp. 181–97.

9. Harper, Roland M. "A Preliminary List of the Endemic Flowering Plants of Florida. Part 2. List of Species," *Quarterly Journal of the Florida Academy of Sciences*, Vol. 11, Nos. 2–3 (1949), pp. 39–57.

10. Hock, R. J. "Southwestern Exotic Felids," *American Midland Naturalist*, Vol. 53, No. 2 (1955), pp. 324–28.

11. Hylander, Clarence J. *Wildlife Community: From the Tundra to the Tropics in North America*. Boston, Houghton Mifflin Co., 1965.

12. Nelson, Edward W. *Lower California and Its Natural Resources*. Riverside, Cal., Manessier Publishing Co., 1921.

13. Safford, W. E. "Natural History of Paradise Key and the Nearby Everglades of Florida," *Annual Report of the Smithsonian Institution* (1917), pp. 377–434.

14. Small, John K. "Green Deserts and Dead Gardens," *Journal of the New York Botanical Garden*, Vol. 24 (1923), pp. 193–247.

15. ———— "The Land Where Spring Meets Autumn," *Journal of the New York Botanical Garden*, Vol. 25 (1924), pp. 53–94.

16. "Symposium: The Biogeography of Baja California and Adjacent Seas. Part III. Terrestrial and Fresh-Water Biotas," *Systematic Zoology*, Vol. 9, Nos. 1–4 (1960), pp. 148–232.

N. Arctic and Subarctic Lands

Chapter 17. Also see A 6, A 10 (especially), A 15, E 7, E 21, F 10 (especially), H 10, and M 11.

1. Anderson, Jacob P. *Flora of Alaska and Adjacent Parts of Canada.* Ames, Iowa, Iowa State University Press, 1959.
2. Banfield, A. W. F. "Role of Ice in the Distribution of Mammals," *Journal of Mammalogy,* Vol. 35, No. 1 (1954), pp. 104–107.
3. Dunbar, M. J. "Arctic and Subarctic Marine Ecology: Immediate Problems," *Arctic,* Vol. 6, No. 2 (1953), pp. 75–90.
4. Hansen, Henry P. (ed.). *Arctic Biology.* Corvallis, Ore., Oregon State College, 1957.
5. Kimble, G. H. T., and D. Good. *Geography of the Northlands.* New York, John Wiley and Sons, Inc., 1955.
6. Leopold, A. S., and F. F. Darling. *Wildlife in Alaska, an Ecological Reconnaissance.* New York, Ronald Press Co., 1953.
7. Longstaff, T. G. "An Ecological Reconnaissance in West Greenland," *Journal of Animal Ecology,* Vol. 1, No. 2 (1932), pp. 119–42.
8. Milne, J. L., M. Milne, and the Editors of *Life. The Mountains.* New York, *Time,* Inc., 1962.
9. Polunin, Nicholas. *The Circumpolar Arctic Flora.* New York, Oxford University Press, 1959.
10. Porsild, A. E. *Illustrated Flora of the Canadian Arctic Archipelagos.* Ottawa, National Museum of Canada, 1957.
11. Wiggins, Ira L., and J. H. Thomas. *Flora of the Alaskan Arctic Slopes.* Toronto, University of Toronto, 1961.
12. Wynne-Edwards, V. C. *Freshwater Vertebrates of the Arctic and Subarctic.* Fisheries Research Board of Canada, Bulletin No. 94 (1952).

O. Eurasia, Between the Arctic and the Tropics

Four chapters, 18 through 21, are considered together here. Also see A 10, B 1, B 6, B 9, B 12, B 16, B 23, B 27, B 28, C 12, E 7, E 21, F 4 (especially), F 13, J 3, J 4, J 6, J 9, J 13, N 5, and N 8.

1. Berg, Leo S. *Natural Regions of the U.S.S.R.* New York, The Macmillan Co., 1950.
2. Biel, Edwin R. Climatology of the Mediterranean Area. *University of Chicago, Institute of Metereology, Miscellaneous Reports* No. 13 (1944).
3. Bisch, Jorgen. *Mongolia, Unknown Land.* New York, E. P. Dutton and Co., Inc., 1963.
4. Brangham. N. A. *A Naturalist's Riviera.* New York, Hillary House Publishers Ltd., 1965.
5. Carruthers, Douglas. *Beyond the Caspian. A Naturalist in Central Asia.* Edinburgh, Oliver and Boyd, 1949.

6. Chaney, Ralph W. "Tertiary Forest and Continental History," *Bulletin of the Geological Society of America*, Vol. 51 (1940), pp. 469–88.

7. ——— "Tertiary Centers and Migration Routes," *Ecological Monographs*, Vol. 17, No. 2 (1947), pp. 139–48.

8. Clark, J. G. D. *Prehistoric Europe, the Economic Basis*. London, Methuen and Co., 1952.

9. Cox, E. H. M. *Plant Hunting in China*. New York, William Collins' Sons and Co. Ltd., 1945.

10. Cressey, George B. *Crossroads: Land and Life in Southwest Asia*. Chicago, J. B. Lippincott Co., 1960.

11. Curry-Lindahl, Kai. *Europe: A Natural History*. New York, Random House, Inc., 1964.

12. Darling, F. Fraser. *Natural History in the Highlands and Islands*. New York, William Collins' Sons and Co. Ltd., 1947.

13. Deevey, E. S. "Biogeography of the Pleistocene. Part I. Europe and North America," *Bulletin of the Geological Society of America*, Vol. 60 (1949), pp. 1315–1416.

14. Fernald, M. L. "Specific Segregations and Identities in Some Floras of Eastern North America and the Old World," *Rhodora*, Vol. 33 (1931), pp. 25–63.

15. Field, Henry W. *Bibliography on Southwestern Asia*. Coral Gables, Fla., University of Miami Press, 1956.

16. Gansser, Augusto. *Geology of the Himalayas*. New York, Interscience Publishers, Inc., 1964.

17. Haden-Guest, S., J. K. Wright, and E. M. Teclaff (eds.). *A World Geography of Forest Resources*. New York, Ronald Press Co., 1956.

18. Hoffman, George W. (ed.). *A Geography of Europe*. New York, Ronald Press Co., 1961.

19. Li, Hui-Lin. "Floristic Relationships between Eastern Asia and Eastern North America," *Transactions of the American Philosophical Society*, Vol. 42, No. 2 (1952), pp. 371–429.

20. ——— *The Garden Flowers of China*. New York, Ronald Press Co., 1959.

21. Lindroth, Carl H. *The Faunal Connections between Europe and North America*. New York, John Wiley and Sons, Inc., 1957.

22. Löve, Askell, and Doris Löve. *Atlantic Biota and Their History*. Long Island City, N.Y., Pergamon Press, 1963.

23. Matthews, James R. *Origin and Distribution of the British Flora*. Camden, N.J., Thomas Nelson and Sons, 1962.

24. Newbigin, Marion I. *Mediterranean Lands*. New York, Appleton-Century-Crofts, 1924.

25. ——— *Plant and Animal Geography*. London, Methuen and Co., 1936.

26. Peering, F. H., and S. M. Walter. *Atlas of British Flora*. Camden, N.J., Thomas Nelson and Sons, 1962.

27. Sauer, Carl O. *Agricultural Origins and Dispersals*. New York, American Geographical Society, 1952.
28. Semple, Helen C. *The Geography of the Mediterranean Region*. New York, Henry Holt and Co., 1931.
29. Suskhin, Peter. "Outlines of the History of the Recent Fauna of Palaearctic Asia," *Proceedings of the National Academy of Sciences*, Vol. 2, No. 6 (1925), pp. 299–302.
30. Turrill, W. B. *The Plant-Life of the Balkan Peninsula*. London, Oxford University Press, 1929.
31. Vevers, G. M. *Animals of the U.S.S.R.* London, William Heinemann, 1948.

P. North America, Between the Arctic and the Tropics

Chapters 22 and 23 may be considered together. All entries in Sections A, B, C, and D relate at least in part to temperate North America; thus the present section is comparatively short. Also see E 3, E 4, E 7, E 8, E 10, E 21, F 7, F 9 (especially), F 13, F 16, J 9, M 2, M 11, N 1, N 5, N 8, O 6, O 7, O 13, O 14, O 17, O 19, O 21, O 22, and O 25.

1. Borchert, John R. "The Climate of the Central North American Grassland," *Annals of the Association of American Geographers*, Vol. 40, No. 1 (1950), pp. 1–39.
2. Carpenter, J. R. "The Grassland Biome," *Ecological Monographs*, Vol. 10, No. 4 (1940), pp. 617–84.
3. Freeman, O. W., and H. W. Martin (eds.). *The Pacific Northwest*. New York, John Wiley and Sons, Inc., 1942.
4. Gleason, Henry A. "The Vegetational History of the Middle West," *Annals of the Association of American Geographers*, Vol. 12 (1932), pp. 39–85.
5. Harper, Roland M. *Forests of Alabama*. University, Ala., Geological Survey of Alabama, 1943.
6. Hubbs, Carl L. (ed.). *Zoogeography*. American Association for the Advancement of Science, Publication No. 51 (1958).
7. Linsdale, Jean M. "Environmental Responses of Vertebrates in the Great Basin," *American Midland Naturalist*, Vol. 19, No. 1 (1938), pp. 1–206.
8. Matthiessen, Peter. *Wildlife in America*. New York, Viking Press, Inc., 1959.
9. Munz, P. A., and D. D. Keck. *The California Flora*. Berkeley, Cal., University of California Press, 1959.
10. Palmer, E. Laurence. *Fieldbook on Natural History*. New York, McGraw-Hill Book Co., Inc., 1949.
11. Peterson, R. T., and J. Fisher. *Wild America*. Boston, Houghton Mifflin Co., 1955.

12. Ridley, H. N. *The Dispersal of Plants throughout the World*. Ashford, Kent, L. Reeve and Co., 1930.

13. Shantz, H. L., and R. Zon. *Atlas of American Agriculture: Natural Vegetation*. Washington, D.C., U.S. Department of Agriculture, 1924.

14. Storer, T. I., and R. L. Usinger. *Sierra Nevada Natural History*. Berkeley, Cal., University of California Press, 1963.

15. Swarth, Harry S. "Faunal Areas of Southern Arizona," *Proceedings of the California Academy of Sciences*, Vol. 18, No. 12 (1929), pp. 267–383.

16. Transeau, E. N. "The Prairie Peninsula," *Ecology*, Vol. 16, No. 4, pp. 423–37 (1935).

17. United States Printing Office. World Weather Records 1941–50 (1959).

18. Weaver, J. E. *North American Prairie*. Lincoln, Nebraska, Johnsen Publishing Co., 1954.

19. —— and F. W. Albertson. *Grasslands of the Great Plains*. Lincoln, Nebr., Johnsen Publishing Co., 1965.

20. Webb, William. "Biogeographic Regions of Texas and Oklahoma," *Ecology*, Vol. 31, No. 3 (1950), pp. 426–33.

Q. Ecology

Four chapters, 25 through 28, deal chiefly with ecological matters, and may be treated together. References A 5, F 1, G 5, and M 11 are especially important in this connection. Also see A 9, A 15, B 18, C 18, D 4, F 2, F 3, F 4, F 5, F 9, F 13, F 15, G 3, H 7, H 9, H 10, H 17, J 1, J 6, J 9, J 12, L 12, L 14, M 2, M 7, N 8, P 1, P 2, P 4, P 7, and P 13.

1. Andrewartha, H. G., and L. C. Birch. *The Distribution and Abundance of Animals*. Chicago, University of Chicago Press, 1954.

2. Bailey, Vernon. "Life Zones and Crop Zones of New Mexico," *North American Fauna*, No. 35 (1913).

3. Beadle, N. C. W. "The Misuse of Climate as an Indicator of Vegetation and Soils," *Ecology*, Vol. 32, No. 2 (1951), pp. 343–45.

4. Braun, E. Lucy. *Deciduous Forests of Eastern North America*. Philadelphia, Blakiston Co., 1950.

5. Cain, Stanley S. *Manual of Vegetation Analysis*. New York, Harper and Brothers, 1957.

6. Clements, Frederic E. "Nature and Structure of the Climax," *Journal of Ecology*, Vol. 24, No. 1 (1936), pp. 253–84.

7. —— and V. E. Shelford. *Bio-ecology*. New York, John Wiley and Sons, Inc., 1939.

8. Daubenmire, Rexford F. *Plants and Environment*. New York, John Wiley and Sons, Inc., 1947.

9. Dice, Lee R. *The Biotic Provinces of North America*. Ann Arbor, Mich., University of Michigan Press, 1943.

10. —— *Natural Communities*. Ann Arbor, Mich., University of Michigan Press, 1952.

11. Egler, Frank E. "A Commentary on American Plant Ecology, Based on the Textbooks of 1947–1949," *Ecology*, Vol. 32, No. 4 (1951), pp. 673–94.

12. Evans, Francis C. "Relative Abundance of the Species and the Pyramid of Numbers," *Ecology*, Vol. 31, No. 4 (1950), pp. 631–32.

13. Fautin, R. W. "Biotic Communities of the Northern Desert Shrub Biome in Western Utah," *Ecological Monographs*, Vol. 16 (1946), pp. 251–310.

14. Geiger, Rudolf. *The Climate near the Ground*. Cambridge, Mass., Harvard University Press, 1950.

15. Goldman, E. A., and R. T. Moore. "The Biotic Provinces of Mexico," *Journal of Mammalogy*, Vol. 26, No. 4 (1945), pp. 347–60.

16. Griggs, Robert F. "The Ecology of Rare Plants," *Bulletin of the Torrey Botanical Club*, Vol. 67 (1940), pp. 575–94.

17. Jaeger, Edmund C. *The North American Deserts*. Stanford, Cal., Stanford University Press, 1957.

18. Jenny, Hans. *Factors of Soil Formation*. New York, McGraw-Hill Book Co., Inc., 1941.

19. Just, Theodore (ed.). "Plant and Animal Communities," *American Midland Naturalist*, Vol. 21, No. 1 (1939), pp. 1–255.

20. Kendeigh, S. C. "History and Evaluation of Various Concepts of Plant and Animal Communities in North America," *Ecology*, Vol. 35, No. 2 (1954), pp. 153–71. [A good review, replete with references.]

21. Marbut, C. F. *Soils of the United States, Atlas of American Agriculture, Part III*. Washington, D.C., U.S. Department of Agriculture, 1935.

22. Merriam, C. Hart. "Results of a Biological Survey of the San Francisco Mountain Region and Desert of the Little Colorado, Arizona," *North American Fauna*, No. 3 (1890), pp. 1–136.

23. —— "Laws of Temperature Control of the Geographic Distribution of Terrestrial Animals and Plants," *National Geographic*, Vol. 6 (1894), pp. 229–38.

24. Odum, E. P., and H. T. Odum. *Fundamentals of Ecology*. Philadelphia, W. B. Saunders Co., 1959.

25. Oosting, Henry J. *The Study of Plant Communities*. San Francisco, Cal., W. H. Freeman and Co., 1956.

26. Phillips, J. F. V. "Succession, Development, the Climax, and the Complex Organism: An Analysis of Concepts," *Journal of Ecology*, Vol. 22 (1934), pp. 554–71; Vol. 23 (1935), pp. 210–46, 488–508.

27. Raunkiaer, C. *The Life Forms of Plants and Statistical Plant Geography*. Oxford, Clarendon Press, 1934.

28. Shelford, Victor E. "The Relative Merits of the Life Zone and Biome Concepts," *Wilson Bulletin*, Vol. 57 (1945), pp. 248–52.

29. ——— *Ecology of North America*. Urbana, Ill., University of Illinois Press, 1963.

30. Smith, Hobart M. "An Evaluation of the Biotic Province Concept," *Systematic Zoology*, Vol. 9, No. 1 (1960), pp. 41–44.

31. Thornthwaite, C. W. "An Approach toward a Rational Classification of Climate," *Geographical Review*, Vol. 38, No. 1 (1948), pp. 55–94.

32. Udvardy, M. F. D. "Notes on the Ecological Concepts of Habitat, Biotope and Niche," *Ecology*, Vol. 40, No. 4 (1959), pp. 725–28.

33. Weaver, J. E., and F. E. Clements. *Plant Ecology*. New York, McGraw-Hill Book Co., Inc., 1938.

34. Whittaker, R. H. "A Consideration of Climax Theory: The Climax as a Population and Pattern," *Ecological Monographs*, Vol. 23, No. 1 (1953), pp. 41–78.

35. Young, Frank N. "In Defense of the Concepts of Major and Minor Habitats in Approaching Biological Problems," *Quarterly Journal of the Florida Academy of Sciences*, Vol. 21, No. 1 (1958), pp. 92–100.

R. The World of the Past

Chapters 29 and 30 are taken together. References of special importance here include A 14, A 17, E 4, E 8, G 7, and H 8. Also see A 6, E 5, E 10, E 15, E 19, E 21, E 22, E 28, F 7, F 12, G 1, G 2, G 4, H 5, H 6, O 6, O 7, and O 13.

1. Beerbower, J. R. *Search For the Past*. Englewood Cliffs, N. J., Prentice-Hall, Inc., 1960.

2. Brooks, C. E. P. *Climate through the Ages: A Study of the Climatic Factors and Their Variation*. New York, McGraw-Hill Book Co., 1949.

3. Denison, R. H. "A Review of the Habitat of the Earliest Vertebrates," *Fieldiana: Geology*, Vol. 11 (1956), pp. 359–457.

4. Du Toit, A. L. *Our Wandering Continents: An Hypothesis of Continental Drifting*. New York, Hafner Publishing Co., Inc., 1937.

5. Emiliani, Cesare. "Ancient Temperatures," *Scientific American* (February 1958), pp. 54–63.

6. Glaessner, Martin F. *Principles of Micropaleontology*. New York, John Wiley and Sons, Inc., 1947.

7. ——— "Pre-Cambrian Animals," *Scientific American* (March 1961), pp. 72–78.

8. Harland, W. B., and M. J. S. Rudnick. "The Infra-Cambrian Ice Age." *Scientific American* (August 1964), pp. 28–36.

9. Heezen, Bruce C. "The Rift in the Ocean Floor," *Scientific American* (October 1960), pp. 98–110.

10. Hurley, Patrick M. *How Old Is the Earth?* Garden City, N. Y., Doubleday and Co., Inc., 1959.

11. Menard, Henry W. "The East Pacific Rise," *Scientific American* (December 1961), pp. 52–61.

12. Öpik, Ernst J. "Climate and the Changing Sun," *Scientific American* (June 1958), pp. 85–92.

13. Romer, Alfred S. *Fish Origins—Fresh or Salt Water?* London, Pergamon Press Ltd., Papers in Marine Biology and Oceanography, pp. 261–80 (1955).

14. ———— "The Early Evolution of Land Vertebrates," *Proceedings of the American Philosophical Society*, Vol. 100, No. 3 (1956), pp. 157–67.

15. Runcorn, S. K. (ed.). *Continental Drift*. New York, Academic Press, 1962.

16. Schwarzbach, Martin. *Climates of the Past*. New York, D. Van Nostrand Co., Inc., 1963.

17. Wilson, J. Tuzo. "Continental Drift," *Scientific American* (April 1963), pp. 86–100.

18. Wright, H. E., Jr., and D. G. Frey (eds.). *The Quaternary of the United States*. Princeton, N.J., Princeton University Press, 1965.

S. Factors Influencing Distribution

The literature of this topic is diverse, being concerned with the environment as well as the organism's response to it. See especially A 11, A 15, L 11, P 12, Q 8, and Q 21. Also see B 18, B 31, D 3, E 7, E 15, F 1, G 5, G 6, O 6, O 7, Q 1, and Q 23.

1. Baker, H. G. "Self-Compatibility and Establishment after 'Long Distance' Dispersal," *Evolution*, Vol. 9 (1955), pp. 347–49.

2. Beck, Stanley D. *Animal Photoperiodism*. New York, Holt, Rinehart, and Winston, Inc., 1963.

3. Birch, L. C. "The Role of Weather in Determining the Distribution and Abundance of Animals," *Cold Springs Harbor Symposia on Quantitative Biology*, Vol. 22 (1957), pp. 203–18.

4. Blair, W. Frank (ed.). *Vertebrate Speciation*. Austin, Tex., University of Texas, 1961.

5. Bonner, James. "Chemistry in Plant Societies," *Natural History* (November 1959), pp. 508–13.

6. Cowles, R. B., and C. M. Bogert. "A Preliminary Study of the Thermal Requirements of Desert Reptiles," *Bulletin of the American Museum of Natural History*, Vol. 83 (1944), pp. 265–96.

7. Crombie, A. C. "Interspecific Competition," *Journal of Animal Ecology*, Vol. 16 (1947), pp. 44–73.

8. Davis, John H. "Proposals concerning the Concept of Habitat and a Classification of Types," *Ecology*, Vol. 41, No. 3 (1960), pp. 537–41.

9. Du Rietz, G. E. "Factors Controlling the Distribution of Species in Vegetation." Ithaca, N.Y., Proceedings of the International Congress of Plant Sciences, 1929.

10. Etkin, William (ed.). *Social Behavior and Organization among Vertebrates*. Chicago, University of Chicago Press, 1964.

11. Fraenkel, G. S., and D. L. Gunn. *The Orientation of Animals: Kineses, Taxes, and Compass Reactions*. New York, Dover Publications, Inc., 1961.

12. Grinnell, Joseph. "Barriers to Distribution as regards Birds and Mammals," *American Naturalist*, Vol. 48 (1914), pp. 248–54.

13. Guppy, H. B. *Plants, Seeds, and Currents in the West Indies and Azores*. London, William and Norgate, 1917.

14. King, Wayne. "The Occurrence of Rafts for Dispersal of Land Animals into the West Indies," *Quarterly Journal of the Florida Academy of Sciences*, Vol. 25, No. 1 (1962), pp. 45–52.

15. Klopfer, Peter. *Behavioral Aspects of Ecology*. Englewood Cliffs, N.J., Prentice-Hall, Inc., 1962.

16. Lack, David. "The Psychological Factor in Bird Distribution," *British Birds*, Vol. 31 (1937), pp. 130–36.

17. Lyon, Marcus W., Jr. "Vertical Distribution of Animals," *Scientific American*, Vol. 110 (1914), p. 432.

18. Maier, N. R. F., and T. C. Schneirla. *Principles of Animal Psychology*. New York, Dover Publishing Co., 1964.

19. Milne, L. J., and M. Milne. *Paths across the Earth*. Evanston, Ill., Harper and Row, Publishers, 1958.

20. Myers, George S. "Salt-Tolerance of Fresh-Water Fish Groups in Relation to Zoogeographical Problems," *Bijdragen Tot De Dierkunde*, Vol. 28 (1949), pp. 315–22.

21. Neill, Wilfred T. "Viviparity in Snakes: Some Ecological and Zoogeographical Considerations," *American Naturalist*, Vol. 98, No. 898 (1964), pp. 33–55.

22. Roe, A., and G. G. Simpson (eds.). *Behavior and Evolution*. New Haven, Yale University Press, 1958.

23. Wynne-Edwards, V. C. *Animal Dispersion in Relation to Social Behavior*. New York, Hafner Publishing Co., 1962.

T. The Oceans

Chapter 32. Also note A 10, B 5, B 14, B 15, B 17, B 19, B 23, B 24, B 28, B 31, C 5, C 20, D 2, D 3, F 10, G 5, H 6, H 8, H 10, L 4, N 3, N 4, R 9, and R 11.

1. Carson, Rachel. *Edge of the Sea*. Boston, Houghton-Mifflin Co., 1955.
2. —— *The Sea around Us*. New York, Oxford University Press, 1961.
3. Dakin, William J. *Great Barrier Reef*. San Francisco, Tri-Ocean Books, 1963.
4. Ekman, Sven. *Zoogeography of the Sea*. New York, The Macmillan Co., 1953.
5. Engel, Leonard, and the Editors of *Life*. *The Sea*. New York, *Time*, Inc., 1961.
6. Galtsoff, P. S. (ed.). *Gulf of Mexico: Its Origin, Waters, and Marine Life*. United States Fish and Wildlife Service, Fishery Bulletin No. 89 (1954).
7. Hedgpeth, J., and H. S. Ladd (eds.). *Treatise on Marine Ecology and Paleocology*. 2 vols. Geological Society of America, Memoir No. 67 (1957).
8. Idyll, Clarence P. *Abyss: The Deep Sea and the Creatures That Live In It*. New York, Thomas Y. Crowell Co., 1964.
9. Krogh, August. *Osmotic Regulation in Aquatic Animals*. New York, Dover Publications, Inc., 1965.
10. MacGinitie, G. E. "Littoral Marine Communities," *American Midland Naturalist*, Vol. 21, No. 1 (1939), pp. 28–55.
11. Miner, Roy W. *Field Book of Seashore Life*. New York, G. P. Putnam's Sons, 1950.
12. Moore, Hilary B. *Marine Ecology*. New York, John Wiley and Sons, Inc., 1958.
13. Neill, Wilfred T. "The Occurrence of Amphibians and Reptiles in Saltwater Areas, and a Bibliography," *Bulletin of Marine Science of the Gulf and Caribbean*, Vol. 8, No. 1 (1958), pp. 1–97.
14. Piccard, J., and R. S. Dietz. *Seven Miles Down: The Story of the Bathyscaph "Trieste."* New York, G. P. Putnam's Sons, 1961.
15. Ricketts, E. F., and J. Calvin. *Between Pacific Tides*. Stanford, Cal., Stanford University Press, 1962.
16. *Symposium: Essays in Marine Biology*. Edinburgh, Oliver and Boyd, 1953.
17. Wilson, Douglas P. *Life of the Shore and Shallow Sea*. London, Nicholson and Watson Ltd., 1937.

U. Fresh and Brackish Waters

Chapter 33. Also see A 10, A 15, B 3, B 5, B 12, B 14, B 15, B 24, B 31, C 12, D 5, D 6, D 12, F 9, F 10, G 5, G 8, Q 24, Q 29, T 6, T 7, T 10, and T 13; and especially A 6, A 7, and Q 10.

1. Boyce, S. G. "The Salt Spray Community," *Ecological Monographs*, Vol. 24, No. 1 (1954), pp. 29–67.
2. Brues, Charles T. "Animal Life in Hot Springs," *Quarterly Review of Biology*, Vol. 2, No. 2 (1927), pp. 181–203.
3. Chapman, V. J. *Salt Marshes and Salt Deserts of the World*. Brooklyn, N.Y., Interscience Press, 1961.

4. Coker, Robert E. *Streams, Lakes, Ponds*. Chapel Hill, N.C., University of North Carolina Press, 1954.
5. Edmondson, W. T., (ed.). *Ward and Whipple's Fresh-Water Biology*. New York, John Wiley and Sons, Inc., 1959.
6. Leopold, L. B., K. S. Davis, and the Editors of *Life*. *Water*. New York, *Time*, Inc., 1966.
7. Needham, J. G., and P. R. Needham. *Guide to the Study of Freshwater Biology*. San Francisco, Holden-Day, Inc., 1962.
8. Pratt, Henry S. *A Manual of the Common Invertebrate Animals Exclusive of Insects*. Chicago, A. C. McClurg and Co., 1929.
9. Reid, George K. *Ecology of Inland Waters and Estuaries*. New York, Reinhold Publishing Co., 1961.
10. Vandel, A. *Biospeleology, the Biology of Cavernicolous Animals*. Long Island City, N.Y., Pergamon Press, Inc., 1966.

V. Lower Organisms

Although lower organisms are not covered in any detail by the present work, attention is called to some of the literature relating to them. Also see A 4, B 13, C 5, C 12, C 20, E 15, F 10, H 15, K 19, N 3, Q 1, R 8, S 3, T 1, T 2, T 3, T 6, T 8, T 11, T 12, T 15, T 16, T 17, U 2, U 4, U 5, U 6, U 7, and U 10; and especially B 8, T 4, and U 8.

1. Baer, J. G. *Ecology of Animal Parasites*. Urbana, Ill., University of Illinois Press, 1952.
2. Bates, Marston. *The Natural History of Mosquitoes*. New York, The Macmillan Co., 1949.
3. Borradaile, L. A., and F. A. Potts. *The Invertebrata*. Cambridge, Cambridge University Press, 1958.
4. Brock, Thomas D. *Principles of Microbial Ecology*. Englewood Cliffs, N.J., Prentice-Hall Co., Inc., 1966.
5. Buchsbaum, R. M., and L. J. Milne. *The Lower Animals: Living Invertebrates of the World*. Garden City, N.Y., Doubleday and Co., Inc., 1960.
6. Calkins, G. N. *The Biology of the Protozoa*. Philadelphia, Lea and Febiger, 1933.
7. Chandler, A. C. *Introduction to Parasitology*. New York, John Wiley and Sons, Inc., 1949.
8. Curran, C. H. *Insects of the Pacific World*. New York, The Macmillan Co., 1946.
9. Drimmer, Fred (ed.). *The Animal Kingdom*. 3 vols. Garden City, N.Y., Doubleday and Co., Inc., 1954.

10. Dunbar, M. J. (ed.). *Marine Distributions*. Toronto, University of Toronto, 1963.

11. Essig, E. O. *Insects and Mites of Western North America*. New York, The Macmillan Co., 1958.

12. Hardy, A. C. *The Open Sea*. Boston, Houghton Mifflin Co., 1956.

13. Hyman, L. H. *The Invertebrates*, 5 vols. New York, McGraw-Hill Book Co., 1940–1959.

14. Kaston, B. J., and E. Kaston. *How to Know the Spiders*. Dubuque, Iowa, William C. Brown Co., 1953.

15. Krieger, Louis C. *The Mushroom Handbook*. New York, The Macmillan Co., 1936.

16. Klots, A. B. *A Field Guide to the Butterflies*. Boston, Houghton Mifflin, 1951.

17. —— and E. B. Klots. *Living Insects of the World*. Garden City, N.Y., Doubleday and Co., Inc., 1959.

18. MacGinitie, G. E., and N. MacGinitie. *Natural History of Marine Animals*. New York, McGraw-Hill Book Co., Inc., 1949.

19. Morgan, A. H. *Field Book of Ponds and Streams*. New York, G. P. Putnam's Sons, 1930.

20. Mozley, Alan. *An Introduction to Molluscan Ecology*. London, H. K. Lewis and Co. Ltd., 1954.

21. Pennak, Robert W. *Fresh-Water Invertebrates of the United States*. New York, Ronald Press, 1953.

22. Savory, Theodore H. "False Scorpions," *Scientific American* (March 1966), pp. 95–100.

23. Schmitt, Waldo L. *Crustaceans*. Ann Arbor, Mich., University of Michigan Press, 1965.

24. Smith, Gilbert M. *The Fresh-Water Algae of the United States*. New York, McGraw-Hill Book Co., 1933.

25. Stirton, R. A. *Time, Life, and Man*. New York, John Wiley and Sons, Inc., 1959. [Includes account of the fossil record of invertebrates.]

26. Weidel, Wolfhard. *Virus*. Ann Arbor, Mich., University of Michigan Press, 1959. [Viruses are not presently considered to be alive, but they miss by such a small margin that a reference to them is not amiss.]

27. Wherry, E. T. *The Fern Guide*. Garden City, N.Y., Doubleday and Co., 1961.

28. Yonge, C. M. *The Sea Shore*. New York, William Collins' Sons and Co. Ltd., 1949.

W. Man and Distribution

Chapter 34: the primates, the spread of man, and the interaction of man with the natural world. Also see A 6, D 2, D 3, F 2, F 3, F 4, F 5, F 9, F 10,

F 11, F 13, F 15, G 5, I 5, I 6, I 13, J 1, J 4, J 5, J 13, J 14, K 11, L 18, L 27, L 29, M 5, N 5, N 8, O 3, O 8, O 9, O 10, O 15, O 17, O 18, O 20, O 24, O 27, P 4, Q 17, R 1, R 18, T 2, T 14, U 6, V 1, V 7, and V 13.

1. Baker, J. N. L. *History of Geographical Discovery and Exploration*. London, George G. Harrap and Co., 1937.
2. Bates, Marston. *The Prevalence of People*. New York, Charles Scribner's Sons, 1955.
3. Bergamini, David, and the Editors of *Life. The Universe*. New York, *Time*, Inc., 1962.
4. Blake, Peter. *God's Own Junkyard: The Planned Deterioration of America's Landscape*. New York, Holt, Rinehart, and Winston, 1964.
5. Boswell, Victor R. "Our Vegetable Travelers," *National Geographic* (August 1949), pp. 145–217.
6. Camp, W. H. "The World in Your Garden," *National Geographic* (July 1947), pp. 1–65.
7. Carr, Donald E. *Death of the Sweet Waters*. New York, W. W. Norton and Co., Inc., 1966.
8. Clark, Andrew H. *The Invasion of New Zealand by People, Plants and Animals*. New Brunswick, N.J., Rutgers University Press, 1949.
9. Cubbedge, Robert E. *Destroyers of America*. New York, Macfadden Bartell Corp., 1964.
10. Dicken, S. N., and F. R. Pitts. *Introduction to Human Geography*. Waltham, Mass., Blaisdell Publishing Co., 1963.
11. East, W. G., and A. E. Moodie (eds.). *The Changing World*. Yonkers, N.Y., World Publishing Co., 1956.
12. Editors of *Life. The Epic of Man*. New York, *Time*, Inc., 1961.
13. Eimerl, S., and I. De Vore. *The Primates*. New York, *Time*, Inc., 1965.
14. Ekirch, Arthur A. *Man and Nature in America*. New York, Columbia University Press, 1963.
15. Goodall, Jane. "My Life among Wild Chimpanzees," *National Geographic* (August 1963), pp. 272–308.
16. Hale, J. R., and the Editors of *Life. Age of Exploration*. New York, *Time*, Inc., 1966.
17. Harper, John L. (ed.). *Biology of Weeds*. New York, John Wiley and Sons, Inc., 1962.
18. Howell, F. C., and the Editors of *Life. Early Man*. New York, *Time*, Inc., 1965.
19. Hoyt, Joseph B. *Man and the Earth*. Englewood Cliffs, N.J., Prentice-Hall, Inc., 1962.
20. Jacks, G. V., and R. O. Whyte. *Vanishing Lands: A World Survey of Soil Erosion*. New York, Doubleday, Doran and Co., 1939.

21. James, Preston E. *The Geography of Man.* Waltham, Mass., Blaisdell Publishing Co., Inc., 1959.

22. Kroeber, A. L. *Cultural and Natural Areas of Native North America.* Berkeley, University of California Press, 1939.

23. —— (ed.). *Anthropology Today.* Chicago, University of Chicago Press, 1953.

24. Leakey, L. S. B. "Finding the World's Earliest Man," *National Geographic* (September 1960), pp. 420–35.

25. Lindemann, W. "Transplantation of Game in Europe and Asia," *Journal of Wildlife Management,* Vol. 20, No. 1 (1956), pp. 68–70.

26. MacGowan, R. A., and F. I. Ordway, III. *Intelligence in the Universe.* Englewood Cliffs, N.J., Prentice-Hall, Inc., 1966.

27. Magness, J. R. "How Fruit Came to America," *National Geographic* (September 1951), pp. 325–77.

28. Mason, R. J. "The Paleo-Indian Tradition in Eastern North America," *Current Anthropology,* Vol. 3, No. 3 (1962), pp. 227–78.

29. Mulvaney, D. J. "The Prehistory of the Australian Aborigine," *Scientific American* (March 1966), pp. 84–93.

30. Neill, Wilfred T. "The Association of Suwannee Points and Extinct Animals in Florida," *Florida Anthropologist,* Vol. 17, No. 1 (1964), pp. 17–32.

31. Pinney, Roy. *Vanishing Wildlife.* New York, Dodd, Mead and Co., 1963.

32. Smith, Marian W. (ed.). Asia and North America: Transpacific Contacts. *American Antiquity,* Vol. 18, No. 3, Pt. 2 (1953).

33. Sullivan, Walter. *We Are Not Alone.* New York, McGraw-Hill Book Co., 1966. [Summary of investigations into the possibility of extraterrestrial life.]

34. Thomas, W. L., Jr. (ed.). *Man's Role in Changing the Face of the Earth.* Chicago, University of Chicago Press, 1956.

35. Vernadsky, W. I. "The Biosphere and the Noösphere," *American Scientist,* Vol. 33, No. 1 (1945), pp. 1–12.

In conclusion, this bibliography skims the literature. However, a majority of the cited works offer their own lists of references; most of the journals mentioned have included other articles of distributional significance; and most of the publishing houses have issued other titles of biogeographic interest.

INDEX